Genetic Engineering: Concepts, Tools and Techniques

Genetic Engineering: Concepts, Tools and Techniques

Edited by **Rosanna Mann**

SYRAWOOD
PUBLISHING HOUSE

New York

Published by Syrawood Publishing House,
750 Third Avenue, 9th Floor,
New York, NY 10017, USA
www.syrawoodpublishinghouse.com

Genetic Engineering: Concepts, Tools and Techniques
Edited by Rosanna Mann

International Standard Book Number: 978-1-68286-123-3 (Hardback)

Printed in the United States of America.

Contents

Preface

Genetic engineering has become a very important field of study with its growing applications in biological engineering, medical science and other related fields. This book brings forth some of the most innovative concepts and elucidates the unexplored aspects of genetic engineering such as advanced artificial synthesis of genes, gene therapy, genetic cloning and applications of genetic engineering in various fields like agriculture, medical and biomedical science, etc. It will also provide interesting topics for research which readers can take up.

This book is the end result of constructive efforts and intensive research done by experts in this field. The aim of this book is to enlighten the readers with recent information in this area of research. The information provided in this profound book would serve as a valuable reference to students and researchers in this field.

At the end, I would like to thank all the authors for devoting their precious time and providing their valuable contributions to this book. I would also like to express my gratitude to my fellow colleagues who encouraged me throughout the process.

Editor

A rapid and inexpensive one-tube genomic DNA extraction method from *Agrobacterium tumefaciens*

Suresh P. Kamble · Madhukar M. Fawade

Abstract Many methods have been used to isolate genomic DNA, but some of them are time-consuming and costly, especially when extracting a large number of samples. Here we described an easy protocol using two simple solutions for DNA extraction from *A. tumefaciens* cells. Compared with the standard protocol, this protocol allows rapid DNA isolation with comparable yield and purity at negligible cost. Following this protocol, we have demonstrated: (1) gDNA extraction was achieved within 15 min; (2) this method was cost-effective, since it only used calcium chloride and lysozyme; SDS, phenol, chloroform and proteinase K were not necessary; (3) the method gave high yield of gDNA (130 ng/loopful culture) compared with standard protocol that was suitable for restriction analysis; (4) the protocol can be carried out in a single test tube and the cells directly from solid media can be used. Thus, this protocol offers an easy, efficient and economical way to extract genomic DNA from *A. tumefaciens*.

Keywords Genomic DNA extraction · Restriction digestion · Calcium chloride · Lysozyme

Abbreviations

gDNA Genomic DNA
LB Luria Bertani broth
kbp Kilo base pairs
$CaCl_2$ Calcium chloride
TE Tris-EDTA
TAE Tris-acetate EDTA

S. P. Kamble (✉)
Center for Biotechnology, Pravara Institute of Medical Sciences, Loni, Ahmednagar, Maharashtra, India
e-mail: suresh.kamble@pmtpims.org

M. M. Fawade
Department of Biochemistry, Dr. Babasaheb Ambedkar Marathwada University, Aurangabad 411004, Maharashtra, India

Introduction

To study the molecular systematics of any organism, high quality DNA is required. The rapid availability of genomic DNA is necessary for cloning genes, selecting recombinant constructs and for taxonomy (Niemi et al. 2001). The cell wall is the main obstacle for quick and easy lysis of *Agrobacterium* cells, and therefore, it must be disrupted for efficient recovery of genomic DNA (gDNA). Conventional methods for gDNA preparation from *Agrobacterium* utilize either enzymatic degradation followed by lysis of cells with detergent or extraction of gDNA with phenol–chloroform (Charles and Nester 1993). When analyzing a large number of samples, these methods are time-consuming and relatively expensive. For quick genotyping, cells can also be lysed by repeated freeze–thaw cycles in a buffer containing Triton X-100 and SDS, followed by extraction of gDNA with chloroform (Harju et al. 2004; Smith and Cantor 1987). Although this method gives good yield, it requires transfer of the sample to a new eppendorf tube after chloroform extraction, which slows down the protocol and makes it inconvenient for simultaneous handling of large number of samples. The above method gives relatively low yield and the results are poorly reproducible. In addition, a large number of cells are required for the protocol.

As calcium chloride is used commonly in *Agrobacterium* transformation protocol to weaken cell walls (McCormack et al. 1998; Mattanovich et al. 1989), we decided to combine it with lysozyme to develop a quick, efficient and robust method for gDNA extraction from *Agrobacterium*.

Materials and methods

Culture maintenance and growth conditions

Agrobacterium tumefaciens strain C58C1 was grown on LB agar plates (10 g/L tryptone, 5 g/L yeast extract, 10 g/L NaCl, pH 7.2, and 15.0 g/L agar) for 24 h at 28 °C. For gDNA isolation, 48 h culture was used. And for isolation of gDNA from liquid culture, *A. tumefaciens* cells were grown overnight at 28 °C at 200 rpm in LB medium (10 g/L tryptone, 5 g/L yeast extract, 10 g/L NaCl, pH 7.2). 3 ml culture was used.

DNA isolation

Eight to ten single colonies of *A. tumefaciens* were picked up from LB plate, suspended in 100 μL of 200 mM $CaCl_2$ and 1 % lysozyme and incubated at 42 °C for 2–5 min. After incubation, 300 μL of 96 % ethanol was added; the samples were mixed briefly by vortexing; and DNA was collected by centrifugation at 13,200 rpm for 5 min. Precipitated DNA was air dried at room temperature for 10 min and dissolved in 50 μL TE; cell debris was spun down by brief centrifugation at 12,000 rpm for 2 min and supernatant containing purified DNA was directly used for the subsequent experiments or stored at −20 °C.

Quantification

The purity and yield of gDNA were assessed spectrophotometrically by calculating the A_{260}/A_{280} and A_{260}/A_{230} ratios and A_{260} values to determine protein impurities and DNA concentration.

Restriction analysis

To test whether the gDNA prepared using this method could be digested with restriction enzyme, 1–2 μg of gDNA from *A. tumefaciens* was incubated with 5U EcoRI in a final volume of 20 μL for 2 h at 37 °C and applied to 1 % agarose gel electrophoresis.

Results and discussion

In the recommended DNA extraction protocol, *A. tumefaciens* cells were lysed by calcium chloride along with lysozyme without the use of phenol, Triton X-100. Since calcium chloride is used to weaken the cell wall and

lysozyme to break up the cell wall (Ledeboer et al. 1976; Chassy 1976; Chassy and Giuffrida 1980), it could directly loose and disrupt the cell wall or nucleus envelop and gDNA was released from the cells. The released gDNA was directly precipitated using 96 % ethanol, omitting phenol chloroform extraction step. This method gave reproducible yields of high quality DNA (Table 1). We also compared our results with the standard method of gDNA extraction protocol (Slusarenko 1990). The obtained genomic DNA by our method and standard method was run in 0.8 % TAE-agarose gel (Fig. 1).

Next, we optimized the protocol to find out the critical components for effective DNA extraction by $CaCl_2$–lysozyme lysis method. We tested different concentrations of $CaCl_2$ and lysozyme in the lysis solution (data not shown). We also used the different incubation time ranges from 2 to 5 min (data not shown). To summarize, we recommend using 200 mM $CaCl_2$ and 1 % lysozyme in the lysis solution and carrying the lysis at 42 °C for 3 min.

We used the gDNA prepared by our method for restriction digestion. The restriction digestion pattern of gDNA clearly showed that gDNA obtained could be digested by EcoRI (Fig. 2). The size of most digested gDNA fragments ranged from 23.13 to 0.5 kbp, while the size of control DNA (lane no. 4 in Fig. 2 and lane no. 2 in Fig. 1) corresponded to more than 23 kbp (Fig. 1). Hence DNA was completely digested and there was no evidence of the presence of nucleases in the sample.

Table 1 Yield and quality of DNA obtained from *Agrobacterium* using recommended method

Method	Mean DNA yield (ng/ml)	A_{260}/A_{280}	$A_{260/230}$
Standard method	150 ± 0.400	1.72	2.185
Recommended method	130 ± 0.325	1.65	1.988

Fig. 1 Genomic DNA isolated from *Agrobacterium* by standard and recommended method. M-*Hind*III digested Lambda DNA as marker; L1-Genomic DNA isolated by standard method; L2-Genomic DNA isolated by recommended method

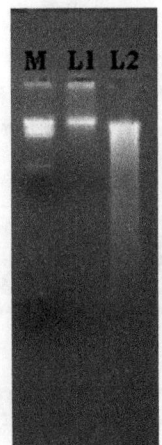

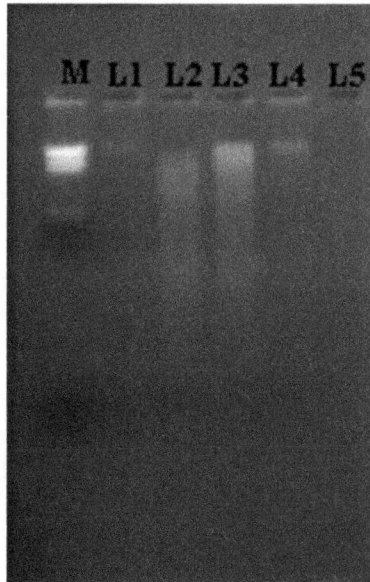

Fig. 2 Agarose gel electrophoresis of *EcoR*I restriction digestion reaction of gDNA by standard and recommended method. M-*Hind*III digested Lambda DNA as marker; L1-uncut DNA from standard method; L2-digested DNA from standard method; L3-digested DNA from recommended method; L4-Uncut DNA obtained by recommended method; and L5-blank

We repeated the restriction digestion experiment over a period of 1–2 months and obtained the same banding pattern which indicated the reproducibility of the results and integrity of the gDNA (Ellsworth et al. 1993).

These restriction digestion results show that no restriction process was inhibited by any components in the DNA preparation. This gDNA extraction method has several advantages. First, the numbers of extraction steps were minimized so the gDNA extraction was achieved within 15 min, while other methods needed at least 5–30 min. Second, the method gave high yield of gDNA compared with standard protocol. Third, this method was cost-effective, since it only uses calcium chloride and lysozyme. SDS, phenol, chloroform and proteinase K were not necessary. Fourth, the protocol can be carried out in a single test tube and the cells directly from solid media cab be used.

Conclusions

We have developed a quick and reliable method for gDNA extraction from *Agrobacterium* that is suitable for restriction digestion. The protocol can be carried out in a single eppendorf tube <15 min and directly from the cells. Finally, it can be used for sequencing, PCR and blotting techniques.

Acknowledgments The authors would like to thank the authorities of Pravara Institute of Medical Sciences (DU), Loni for providing the facilities.

References

Charles TC, Nester EW (1993) A chromosomally encoded two-component sensory transduction system is required for virulence of *Agrobacterium tumefaciens*. J Bacterio 175:6614–6625

Chassy BM (1976) A gentle method for the lysis of oral Streptococci. Biochem Biophys Res Commun 68:603–608

Chassy BM, Giuffrida A (1980) Method for the lysis of gram-positive, Asporogenous bacteria with lysozyme. Appl Environ Microbiol 39(1):153–158

Ellsworth DL, Rittenhouse D, Honeycutt RL (1993) Artificial variation in randomly amplified polymorphic DNA banding patterns. BioTechnique 14:214–218

Harju S, Fedosyuk H, Peterson KR (2004) Rapid isolation of yeast genomic DNA: burst n' grab. BMC Biotechnol 4:8

Ledeboer AM, Krol AJ, Dons JJ, Spier F, Schilperoort RA, Zaenen I, Van Larebeke N, Schell J (1976) On the isolation of TI-plasmid from *Agrobacterium tumefaciens*. Nucleic Acids Res 3(2):449–463

Mattanovich D, Rüker F, Machado AC, Laimer M, Regner F, Steinkuehler H, Himmler G, Katinger H (1989) Efficient transformation of *Agrobacterium spp.* by electroporation. Nucleic Acids Res 17(16):6747

McCormack AC, Elliott MC, Chen DF (1998) A simple method for the production of highly competent cells of *Agrobacterium* for transformation via electroporation. Mol Biotechnol 9(2):155–159

Niemi RM, Heiskanen I, Wallenius K, Lindstrom K (2001) Extraction and purification of DNA in rhizosphere soil samples for PCR-DGGE analysis of bacterial consortia. J Microbiol Methods 45:155–165

Slusarenko AJ (1990) A rapid mini prep for the isolation of total DNA from *Agrobacterium tumefaciens*. Plant Mol Biol Rep 8(4):249–252

Smith CL, Cantor CR (1987) Purification, specific fragmentation, and separation of large DNA molecules. Methods Enzymol 155:449–467

In search of genome annotation consistency: solid gene clusters and how to use them

James J. Davis · Gary J. Olsen · Ross Overbeek · Veronika Vonstein · Fangfang Xia

Abstract Maintaining consistency in genome annotations is important for supporting many computational tasks, particularly metabolic modeling. The SEED project has implemented a process that improves annotation consistencies across microbial genomes for proteins with conserved sequences and genomic context. In this research report, we describe this process and show how this effort has resulted in improvements to microbial genome annotations in the SEED. We also compare SEED annotation consistencies with other commonly used resources such as IMG (the Joint Genome Institute's Integrated Microbial Genomes system), RefSeq (the National Center for Biotechnology Information's Reference Sequence Database), Swiss-Prot (the annotated protein sequence database of the Swiss Institute of Bioinformatics, European Molecular Biology Laboratory and the European Bioinformatics Institute) and TrEMBL (Translated European Molecular Biology Laboratory nucleotide sequence data Library). Our analysis indicates that manual and computational efforts are paying off for the databases where consistency is a major goal.

Keywords Automatic annotation · Protein clusters

Introduction

The primary goal of the SEED Project is to produce accurate annotations for microbial genomes (Overbeek et al. 2005). Maintaining annotation consistency is a second major objective since it facilitates numerous computational tasks, notably the construction of metabolic models. In many contexts, it becomes important to determine, given two assigned functions, whether or not they refer to the same abstract function. To trivially illustrate what we mean by consistency (or lack of it), consider the following list of functions:

1. 50s ribosomal protein l34
2. LSU ribosomal protein L34p
3. Ribosomal protein L34
4. Ribosomal protein L34
5. Ribosomal protein L34 RpmH
6. RpmH
7. rpmH gene product

These are all alternative names of the same function, and they all occur within the public repositories. While heuristic tools can be developed to allow recognition of variants, it is less cumbersome to seek accurate and identical representations of each function. Thus, rather than attempting to computationally determine that these are all equivalent, we have attempted to unify these variants within the SEED Project. For instance, in this case we use only the second annotation from the list above.

J. J. Davis (✉) · G. J. Olsen
Institute for Genomic Biology, MC-195, University
of Illinois at Urbana-Champaign, 1206 W. Gregory Dr.,
Urbana, IL 61801, USA
e-mail: james2@illinois.edu

G. J. Olsen
Department of Microbiology, University of Illinois
at Urbana-Champaign, 601 S. Goodwin Ave.,
Urbana, IL 61801, USA

R. Overbeek · V. Vonstein
Fellowship for Interpretation of Genomes, 15W155 81st St.,
Burr Ridge, IL 60527, USA

R. Overbeek · F. Xia
Mathematics and Computer Science, Argonne National
Laboratory, 9700 S. Cass Ave., Argonne, IL 60439, USA

To be clear, we wish to approach consistency in annotations to support automated construction of metabolic (and more general) models based upon the annotated functions of the genes. We are not concerned with a global standard in nomenclature since sets of terms that accurately and consistently reflect the functions of proteins can be automatically mapped to one another through the associated protein sequences. Secondly, we are not intending to reflect chromosomal location or expression in the function of the protein, but rather the function that it would perform if it was expressed in a cell, again, with a goal toward modeling and metabolic engineering.

Given the goal of representing the activity (or other function) of gene products, the most obvious first step in building and maintaining annotation consistency between genomes is to apply a standard (within the given genome database) nomenclature among proteins with identical primary sequences. In addition, there are many instances where conserved sequence similarity and genomic context offer abundant evidence for annotating a given gene. This report describes simple tools that we have constructed for estimating conserved gene clusters within an operational taxonomic unit (OTU), guiding highly reliable projections of function within the OTU, constructing sets of proteins believed to implement identical functions, and using these sets to estimate the consistency of a set of annotations.

Description of the algorithm

There are many instances where protein-encoding genes with highly conserved amino acid sequence and genomic context can be safely annotated based on the annotation of genomes that have already been sequenced. The following steps describe how we chose our sets of gene clusters with conserved genomic context.

Step 1. The microbial genomes in the SEED database are separated into Operational Taxonomic Units (OTUs). We define an OTU as a set of genomes that are ≥97 % identical in their 16S rDNA genes (e.g. Schloss and Handelsman 2005). At the time of this study (February, 2013), there were 1,386 OTUs represented in the SEED. OTUs containing less than five genomes were omitted from subsequent steps, and this resulted in a total of 100 OTUs, containing 4,117 microbial genomes analyzed in this study.

Step 2. A focus organism representing an OTU is chosen.

Step 3. A set of organisms, moderately related to the focus organism, is chosen. It is necessary to find a set of organisms that are related to the focus organism to determine if the context of each gene is conserved. In this case, closely related strains are avoided because their genomic

context is too strongly conserved, but more distantly related organisms are less conserved and are thus more useful for determining if a given gene has a conserved context. Our set of related organisms is defined as those that are between 50 and 90 % amino acid identity from the focus organism and >90 % identical to one another. Percent identity is determined from a concatenated alignment of aminoacyl-tRNA synthetase proteins (AARS). This alignment includes all of the bacterial and archaeal genomes in the SEED database, and contains all of the AARS proteins except for the asparaginyl-, glutaminyl-, glycyl- and lysyl-tRNA synthetases, which were excluded because they are absent or nonhomologous in many taxa (Woese et al. 2000). From this set of related organisms, a representative set that has less than 90 % protein identity from each other is chosen. It must be noted that there has been extensive horizontal gene transfer among the AARS proteins and that their concatenated alignment does not necessarily provide an accurate phylogeny outside of a given OTU (Woese et al. 2000). We use them in this context because they are among the best-annotated genes in the SEED and their concatenated alignment provides a suitable frame of reference, although almost any highly conserved protein or rRNA alignment with adequate taxonomic representation would suffice.

Step 4. Gene clusters in the genome of the focus organism are chosen for analysis. In order to determine the regions of conserved contiguity, we search for gene sets in which contiguity is maintained in the focus genome and throughout the set of moderately related organisms. This search is performed by taking two genes occurring close to one another in the genome of the focus organism, and determining whether the same pair of genes also occurs in close proximity throughout the genomes of the moderately related set. If there is substantial preservation of contiguity, we treat the two genes in the reference genome as part of a single cluster and these binary connections are used to form larger clusters (using single-linkage clustering). We define substantial preservation of contiguity as follows: for each pair of genes in the reference genome that is separated by less than five intervening genes, we look for bidirectional best hits (BBH) in each of the moderately related genomes. We restrict the usual notion of BBHs (e.g. Overbeek et al. 1999) to genes that have protein products that are reciprocal best hits, similar over 80 % of each protein, and at least 50 % identical over the region of similarity using BLASTP (Altschul et al. 1997). Then in each moderately related genome with a pair of BBH proteins, we look for conserved location of the corresponding gene pair using the same parameters as above (they must have no more than five intervening genes). For a given pair of genes in the focus organism to be considered as a cluster, or as members of a larger cluster, the pair must have a conserved location in 40 % of the genomes of the moderately related set of organisms.

Step 5. Gene clusters are populated. Once we have generated estimates of the gene clusters in the genome of the focus organism, we project these potential clusters (again, very conservatively) to all of the genomes within the same OTU. Here the same parameters from step 4 are used, and we also require that conserved contiguity be detected in at least five genomes or in 20 % of the genomes of the OTU, whichever is larger. We call the set of clustered genes passing all of the above criteria and projecting throughout the OTU a "Solid Cluster". We tabulate these solid clusters in the form of tables in which each row represents a single genome from the OTU, and each column contains one gene in the reference genome and the corresponding BBHs in the other genomes from the OTU. Each column in each of these tables constitutes a "Solid Set" which is believed to be composed of isofunctional homologs.

There are a number of parameters in this approach relating to the definition of "the generation of OTUs", "closeness of gene pairs", "BBHs", and "conserved contiguity". In this report, we do not explore the optimization of each individual parameter. In all cases, we chose relatively conservative values because they are ultimately linked to the automated propagation of gene annotations in the SEED (see below). We certainly acknowledge that loosening these parameters can lead to larger clusters covering more of the genes within the reference genome, but that this may also increase projection errors.

Step 6. The Solid Clusters are retained, and steps 1–4 are repeated for other focal genomes from different OTUs.

Using clusters to evaluate the consistency of annotations

We propose that a manual annotation assigned to an individual protein-encoding gene occurring in a solid set should propagate to all of the protein sequences occurring in the solid set. We have implemented this within the SEED Project in an attempt to project the relatively expensive manual annotations. Thus, a single manual assignment done in a genus in which hundreds of genomes exist (a situation that is rapidly beginning to happen) may induce hundreds of annotation updates.

The existence of a collection of solid sets makes it possible to easily define a number of metrics to measure the consistency of annotations. For a number of annotation efforts, we have chosen to measure two values:

1. Given two genes encoding identical proteins, what is the frequency of identical assigned functions?

2. Given two genes encoding two proteins from the same solid set, what is the frequency of identical assigned functions?

We have computed Solid Clusters for 100 distinct OTUs that were present in the PubSEED. This led to the formation of 73,093 distinct solid sets, with each set believed to contain proteins implementing a common function. Table 1 shows these values for several collections of annotated proteins, which were downloaded in February of 2013 (Lima et al. 2009; Markowitz et al. 2012; O'Donovan et al. 2002; Overbeek et al. 2005; Pruitt et al. 2007). The collections analyzed are IMG (ftp://downloads1.jgi-psf.org/pub/IMG/img_core_v400.tar), RefSeq (ftp://.ncbi.nih.gov/blast/db/FASTA/nr.gz), the SEED (ftp://.theseed.org/misc/annotation/seed.fa), Swiss-Prot (ftp://.uniprot.org/pub/uniprot_sprot.fasta.gz) and TrEMBL (ftp://.uniprot.org/pub/databases/uniprot/current_release/knowledgebase/complete/uniprot_trembl.fasta.gz). In each case, we have tabulated the number of sequences from the publicly distributed collection that have identical protein sequences occurring within solid sets, as well as the two metrics. The data in the table clearly indicate that the efforts expended in Swiss-Prot and the SEED Projects have led to significant advances in annotation consistency.

Overall, the fraction of proteins in each database that are currently represented by solid sets is low, ranging from 0.048 in IMG to 0.206 in Swiss-Prot. This range differs because of the presence of eukaryotic proteins (which are not currently analyzed), the density of genome sequences for a given OTU, and the parameters of the algorithm. The percentage of individual genomes encoding proteins covered by solid sets ranges from 0 to 56 %, with the genome of *Buchnera aphidicola* strain APS having the highest coverage. In general, for OTUs that are rich in genomic data, we observe more proteins encoded by the genome occurring in solid sets. For instance, in *Escherichia coli* K-12 45 % of the proteins encoded by the genome are covered by solid sets. As sequence data continue to accumulate, solid sets will cover a larger fraction of the genomes in more diverse OTUs.

It is important to note that consistency is not the sole goal of most annotation projects. Accuracy of the annotation is clearly more important (Chen et al. 2013). For instance, the eight ribosomal proteins mentioned in the introduction, while inconsistent, could all be viewed as being accurate. Furthermore, they could all be viewed as being consistent in the eyes of an expert annotator. In this report, we have not attempted to assess the absolute accuracy in the databases. Instead we have focused on consistency, primarily to support the automated steps necessary in model building (i.e., that the same string of

Table 1 Measured inconsistencies in annotations

Source of annotations	Fraction of proteins in the database occurring in solid sets	Unique protein sequences occurring in solid sets	Metric 1 (frequency of inconsistency given identical proteins)	Metric 2 (frequency of inconsistency, among members of a solid set)
IMG	0.048	436,872	0.640	0.697
RefSeq	0.022	459,433	0.564	0.625
SEED	0.135	538,181	0.023	0.037
Swiss-Prot	0.206	67,972	0.039	0.039
TrEMBL	0.085	419,239	0.341	0.396

characters in the annotation is assigned to proteins implementing the same abstract function).

The topic of consistency is closely related to the use of a controlled vocabulary. We have chosen to use the SEED functional roles. They have been adopted by the Model SEED metabolic modeling framework which has constructed thousands of metabolic models using the SEED's controlled vocabulary (Henry et al. 2010), and more recently by the US Department of Energy's Kbase project (www.kbase.us). These resources make it possible to automatically reconstruct the metabolic network (or a good approximation of it) from just the list of functional roles associated with the genes in a genome, if (and only if) there exists a consistently used controlled vocabulary and one has a table associating reactions with the functional roles corresponding to the enzymes that catalyze the reactions. The Model SEED and Kbase projects include a precise correspondence between a subset of the SEED functional roles and the reactions these functional roles enable.

Summary

In this report, we have described a simple technology for generating sets of proteins from a single OTU that are believed to implement identical functions. What distinguishes this effort from other well-known projects to construct protein families is that the Solid Clusters are populated very conservatively, leading to sets that only cover proteins encoded by genomes from a single OTU and are of high reliability. Furthermore, since the generation of solid clusters is fully automated, it provides a complementary approach to traditional methods of genome annotation that use hierarchical annotation structures such as SEED Subsystems, GO terms and COGs (Ashburner et al. 2000; Overbeek et al. 2005; Tatusov et al. 2003).

We have made the Solid Clusters, along with the generated sets of proteins available on the PubSEED web site (ftp://ftp.theseed.org/misc/annotation/). We used these sets to evaluate the consistency of existing sets of annotations from a number of sources. We will periodically update the relevant datasets, allowing any group to evaluate their annotations using this metric, and the evaluation of commonly used sources of annotations.

Acknowledgments We wish to thank the other FIG and KBase team members for assistance on this project. We also thank Matthew Benedict for his helpful suggestions. This work was supported by the United States National Institutes of Health, National Institute of Allergy and Infectious Diseases, and Department of Health and Human Services under Grant number HHSN272200900040C; the Office of Science, Office of Biological and Environmental Research, of the United States Department of Energy under contract number DE-AC02-06CH11357, as part of the DOE Systems Biology Knowledgebase and by the University of Illinois Institute for Genomic Biology Fellows Program.

Conflict of interest The authors declare that there are no conflicts of interest.

References

Altschul SF, Madden TL, Schäffer AA, Zhang J, Zhang Z, Miller W, Lipman DJ (1997) Gapped BLAST and PSI-BLAST: a new generation of protein database search programs. Nucleic Acids Res 25:3389–3402

Ashburner M, Ball CA, Blake JA, Botstein D, Butler H, Cherry JM, Davis AP, Dolinski K, Dwight SS, Eppig JT, Harris MA, Hill DP, Issel-Tarver L, Kasarskis A, Lewis S, Matese JC, Richardson JE, Ringwald M, Rubin GM, Sherlock G (2000) Gene ontology: tool for the unification of biology. The Gene Ontology Consortium. Nat Genet 25:25–29

Chen I-MA, Markowitz VM, Chu K, Anderson I, Mavromatis K, Krypides NC, Ivanova NN (2013) Improving microbial genome annotations in an integrated database context. PLoS ONE 8:e54859.

Henry CS, DeJongh M, Best AA, Frybarger PM, Linsay B, Stevens R (2010) High-t throughput generation, optimization and analysis of genome-scale metabolic models. Nat Biotechnol 28:977–982

Lima T, Auchincloss AH, Coudert E, Keller G, Michoud K, Rivore C, Bulliard V, de Castro E, Lachaize C, Baratin D et al (2009) HAMAp: a database of completely sequenced microbial proteome sets and manually curated microbial protein families in UniProtKB/Swiss-Prot. Nucleic Acids Res 37:D471–D478

Markowitz VM, Chen I-MA, Palaniappan K, Chu K, Szeto E, Grechkin Y, Ratner A, Jacob B, Huang J, Williams P et al (2012) IMG: the integrated microbial genomes database and comparative analysis system. Nucleic Acids Res 40:D115–D122

O'Donovan C, Martin MJ, Gattiker A, Gasteiger E, Bairoch A, Apweiler R (2002) High-quality protein knowledge resource: SWISS-PROT and TrEMBL. Brief Bioinform 3:275–284.

Overbeek R, Fonstein M, D'Souza M, Pusch GD, Maltsev N (1999) The use of gene clusters to infer functional coupling. Proc Natl Acad Sci USA 96:2896–2901

Overbeek R, Begley T, Butler RM, Choudhuri JV, Chuang HY, Cohoon M, de Crecy-Lagard V, Diaz N, Disz T et al (2005) The subsystems approach to genome annotation and its use in the project to annotate 1000 genomes. Nucleic Acids Res 33:5691–5702

Pruitt KD, Tatusova T, Maglott DR (2007) NCBI reference sequences (RefSeq): a curated non-redundant sequence database of genomes, transcripts and proteins. Nucleic Acids Res 35:D61–D65

Schloss PD, Handelsman J (2005) Introducing DOTUR, a computer program for defining operational taxonomic units and estimating species richness. Appl Environ Microbiol 71:1501–1506

Tatusov RL, Fedorova ND, Jackson JD, Jacobs AR, Kiryutin B, Koonin EV, Krylov DM, Mazumder R, Mekhedov SL, Nikolskaya AN, Rao BS, Smirnov S, Sverdlov AV, Vasudevan S, Wolf YI, Yin JJ, Natale DA (2003) The COG database: an updated version includes eukaryotes. BMC Bioinform 4:41.

Woese CR, Olsen GJ, Ibba M, Söll D (2000) Aminoacyl-tRNA synthetases, the genetic code, and the evolutionary process. Microbiol Mol Biol Rev 64:202–236

3

Genetic diversity among some canola cultivars as revealed by RAPD, SSR and AFLP analyses

Reda E. A. Moghaieb · Etr H. K. Mohammed · Sawsan S. Youssief

Abstract To assess the genetic diversity among four canola cultivars (namely, Serw-3, Serw-4, Misser L-16 and Semu 249), random amplified polymorphic DNA (RAPD), simple sequence repeat polymorphism (SSR) and amplified fragment length polymorphism (AFLP) analyses were performed. The data indicated that all of the three molecular markers gave different levels of polymorphism. A total of 118, 31 and 338 markers that show 61, 67.7 and 81 % polymorphism percentages were resulted from the RAPD, SSR and AFLP analyses, respectively. Based on the data obtained the three markers can be used to differentiate between the four canola cultivars. The genotype-specific markers were determined, 18 out of the 72 polymorphic RAPD markers generated were found to be genotype-specific (25 %). The highest number of RAPD specific markers was scored for Semu 249 (15 markers), while Serw-4 scored two markers. On the other hand, Serw-3 scored one marker. The cultivar Semu 249 scored the highest number of unique AFLP markers, giving 57 unique markers, followed by Misser L-16 which was characterized by 40 unique AFLP markers, then Serw-3 giving 31 unique markers. While Serw-4 was characterized by the lowest number producing 14 unique positive markers. The dendrogram built on the basis of combined data from RAPD, SSR and AFLP analysis represents the genetic distances among the four canola cultivars. Understanding the genetic variability among the current canola cultivars opens up a possibility for developing a molecular genetic map that will lead to the application of marker-assisted selection tools in genetic improvement of canola.

Keywords Canola · RAPD · SSR · AFLP · Genetic diversity · Molecular markers

Introduction

Canola (*Brassica napus* L.) is considered as the most important source of vegetable oil and protein-rich meal worldwide. It was developed through conventional plant breeding from rapeseed. It ranks the third among the oil crops, following palm oil and soya oil and the fifth among economically important crops, following rice, wheat, maize and cotton (Sovero 1993; Stoutjesdijk et al. 2000). There are increased domestic and export market opportunities for canola oil that can be realized through the development of high-oleic acid canola to replace saturated palm oil in food service applications (Spector 1999; Stoutjesdijk et al. 2000). In addition, high-oleic acid oils are more nutritionally beneficial because oleic acid had cholesterol-lowering properties, whereas saturated fatty acids tend to raise blood cholesterol levels (Stoutjesdijk et al. 2000).

Egypt recently experienced a solid decline in the total oilseed production on account of reduced cotton area, which overwhelmed increases in soybean output (Hassan and Sahfique 2010). This increased demand, and the need for crop diversification, will undoubtedly promote increased acreage of canola in Egypt. According to the Egyptian Ministry of Agriculture and Land Reclamation (2003–07), the seed oil content in the canola cultivar Serw-4 riches 42 %, while the cultivar Serw-3 have 40 % therefore these two cultivars seems to be promising for its

R. E. A. Moghaieb (✉) · E. H. K. Mohammed · S. S. Youssief
Department of Genetics and Genetic Engineering Research Center, Faculty of Agriculture, Cairo University, Giza, Egypt
e-mail: rmoghaieb@cu.edu.eg

E. H. K. Mohammed
Plant Protection Research Institute, Agriculture Research Center, Ministry of Agriculture, Dokki, Giza, Egypt

high oil contents. In Egypt, there are agricultural opportunities to increase canola production by expanding into the new reclaimed regions.

Traditional breeding strategies that have attempted to utilize genetic variation arising from varietal germplasm, induced mutations and somaclonal variations of cell and tissue cultures have met with only limited success (Kebede et al. 2010). Therefore, the methods that evaluate and identify the genotypes more precisely during the growing season, especially at early stages, are preferred by plant breeders (Charcosst and Moreau 2004; Basunanda et al. 2007).

The analysis of genetic variation and relatedness in germplasm are of great value for genetic resources conservation and plant breeding programs to determine the best crosses between different genotypes. Over the years, the methods for assessing genetic diversity have ranged from classical strategies such as morphological analysis to biochemical and molecular techniques (Marijanovic et al. 2009).

In recent years, the identification of *Brassica* cultivars has depended on the application of different DNA markers. As DNA sequences are independent of environmental conditions, identification can be determined at any stage of plant growth (Ahmad et al. 2007; Younessi et al. 2011).

DNA markers reflect directly individual differences at the level of DNA molecules, and cover coding and noncoding regions of the genome (Dandelj et al. 2004). They are not affected by environment, developmental stage, certain tissue and organ, and have high-genomic frequency, high polymorphism and mostly a random genomic distribution (Charcosst et al. 2004; Zeng et al. 2004).

Several molecular techniques have been developed to assess genetic diversity and discriminate between genotypes in different crops. These include restriction fragment length polymorphism (RFLP) (Jaroslava et al. 2002), random amplified polymorphic DNA (RAPD) (Ahmad et al. 2007), amplified fragment length polymorphism (AFLP) (Vos et al. 1995) and microsatellites or simple sequence repeat polymorphism (SSR) (Halton et al. 2002).

The objectives of this investigation were to determine the genetic variability among four canola cultivars (namely Serw-3, Serw-4, Misser L-16 and Semu 249) at the molecular levels using RAPD, SSR and AFLP markers and to use the combined data to construct a phylogenetic tree. The genotype-specific markers were also determined.

Materials and methods

Plant materials

Four canola genotypes, namely Serw-3, Serw-4, Misser L-16 and Semu 249 were kindly provided by Field Crop

Institute, Agricultural Research Center, Ministry of Agriculture, Egypt.

RAPD analysis

DNA extraction

Genomic DNA was isolated from young leaves of greenhouse grown plants using the CTAB method early described in Rogers and Bendich (1985). The quality and quantity of DNA were determined using agarose gel (0.8 %) electrophoresis and spectrophotometer (Table 1).

PCR analysis

PCR reactions were performed in a total volume of 20 µl containing 10 ng DNA, 200 µM dNTPs, 1 µM of 15 arbitrary 10-mer primers (Operon Technology, Inc., Alameda, CA, USA), 0.5 units of Red Hot Taq polymerase (AB gene House, UK) and 10 × Taq polymerase buffer (AB gene House, UK). For DNA amplification Biometra thermal cycler (2720) was programmed as follows: 94 °C for 5 min followed by 35 cycles 94 °C for 1 min, 35 °C for 1 min, 72 °C for 1 min and 72 °C for 7 min.

The amplification products were analyzed by electrophoresis in 1 % agarose in TAE buffer, stained by ethidium bromide and photographed under UV light. The sequence of the tested primers was as follows:

Simple sequence repeat polymorphism analysis PCR reaction mix includes the following: DNA, 10 ng/µl;

Table 1 Names and sequences of RAPD primers used to assess the genetic variability among the four canola culivars

Primers name	Sequence
OPE-A-10	5′-GTGATCGCAG-3′
OPE-B-10	5′-CTGCTGGGAC-3′
OPE-B-17	5′-AGGGAACGAG-3′
OPE-C-02	5′-GTGAGGCGTC-3′
OPE-C-05	5′-GATGACCGCC-3′
OPE-E-04	5′-GTGACATGCC-3′
OPE-G-14	5′-GGATGAGACC-3′
OPE-K-04	5′-CCGCCCAAAC-3′
OPE-K-15	5′-CTCCTGCCAA-3′
OPE-K-10	5′-GTGCAACGTG-3′
OPE-L-04	5′-GACTGCACAC-3′
OPE-M-13	5′-GGTGGTCAAG-3′
OPE-N-13	5′-AGCGTCACTC-3′
OPE-P-09	5′-GTGGTCCGCA-3′
OPE-Q-14	5′-GGACGCTTCA-3′

Table 2 Names and sequences of the SSR loci used to characterize the four canola genotypes

Primer name	Sequence of forward primers	Sequence of reverse primers
RM 206	5'-CCCATGCGTTTAACTATTCT-3'	5'-CGTTCCATCGATCCGTATGG-3'
RM 264	5'-GTTGCGTCCTACTGCTACTTC-3'	5'-GATCCGTGTCGATGATTAGC-3'
RM 561	5'-GAGCTGTTTTGGACTACGGC-3'	5'-GAGTAGCTTTCTCCCACCCC-3'
RM 544	5'-TGTGAGCCTGAGCAATAACG-3'	5'-GAAGCGTGTGATATCGCATG-3'
RM 547	5'-TAGGTTGGCAGACCTTTTCG-3'	5'-GTCAAGATCATTCTCGTAGCG-3'
RM 519	5'-AGAGAGCCCCTAAATTTCCG-3'	5'-AGGTACGCTCACCTGTGGAC-3'
RM 566	5'-ACCCAACTACGATCAGCTCG-3'	5'-CTCCAGGAACACGCTCTTTC-3'

$10 \times$ buffer; 10 mM dNTPs; 50 mM MgCl$_2$; 10 μM each of forward and reverse primers. The PCR profile starts with 95 °C for 5 min followed by 35 cycles of denaturation at 94 °C for 1 min, annealing at 55 °C for 1 min extension at 72 °C for 2 min. A final extension 72 °C for 7 min was included. The PCR products were electrophoresed in a 2 % agarose gels (for SSRs) at 100 V. The gel was then stained in ethidium bromide for 30 min, and then observed on a UV trans-illuminator.

AFLP analysis

Amplified fragment length polymorphism was performed as described by Vos et al. (1995) using the GIBCO BRL system I (Cat. No. 10544) according to the manufacturer's protocol.

Amplified fragment length polymorphism amplification products were separated in a vertical denaturing 6 % polyacrylamide gel in a Sequi-Gen Cell (Bio-Rad Laboratories Inc.) as described by Bassam et al. (1991).

Band scoring and cluster analysis

The RAPD, SSR and AFLP gel images were scanned using the Gel Doc 2000 Bio-Rad system and analyzed with Quantity One Software v 4.0.1 (Bio-Rad Laboratories, Hercules, Co. USA). The bands were sized and then binary coded by 1 or 0 for their presence or absence in each genotype. The systat ver. 7 computer program was used to calculate the pairwise differences matrix and plot the dendrogram among canola cultivars (Yang and Quiros 1993). Cluster analysis was based on similarity matrices obtained with the unweighed pair-group method (UPGMA) using the arithmetic average to estimate the dendrogram (Table 2).

Result and discussion

Recently, the identification of *Brassica* cultivars has depended on the application of different DNA markers. As DNA sequences are independent of environmental conditions, identification can be determined at any stage of plant growth (Ahmad et al. 2007). The RAPD markers are easier and quicker to use and preferred in applications where relationships between closely related breeding lines are of interest. This analysis detects nucleotide sequence polymorphisms in DNA using a single primer of arbitrary nucleotide sequence. DNA profiling with suitable RAPD primers could be used for identification and discrimination between of oilseed rape cultivars (Mailer et al. 1997; Ahmad et al. 2007). In the present work to study the genetic variability among four canola cultivars, RAPD analysis was performed. All the primer tested produced amplification products with variable band number. One hundred eighteen RAPD markers were obtained and out of them 72 were polymorphic (61 %) and can be considered as useful RAPD markers for the four canola cultivars used (Fig. 1; Table 3). The highest number of RAPD bands was recorded for primers OPE-M-13 (11 bands), followed by OPE-Q-14 (10 bands), while the lowest was scored for OPE-C-02 and OPE-G-14 (6 bands).

The genotype-specific RAPD markers for the different canola cultivars used are listed in Table 4; 18 out of the 72 polymorphic RAPD markers generated were found to be genotype-specific (25 %). The highest number of RAPD specific markers was scored for Semu 249 (15 markers), while Serw-4 scored two markers. On the other hand, Serw-3 scored one marker.

Microsatellite or SSR is another PCR-based marker which is preferred by many geneticists and plant breeders because of higher repeatability, co-dominant nature, specificity and having multiple alleles (Cheng et al. 2009). Cruz et al. (2007) employed the SSR markers to characterize the flowering time of spring and winter type *B. napus* L. germplasm. Qu et al. (2012) studied the genetic diversity and performed relationship analysis among the *B. napus* germplasm using SSR markers.

In the present study, seven SSR primer pairs flanking dinucleotide SSR (GA or AG) were used to investigate the level of polymorphism among the four canola

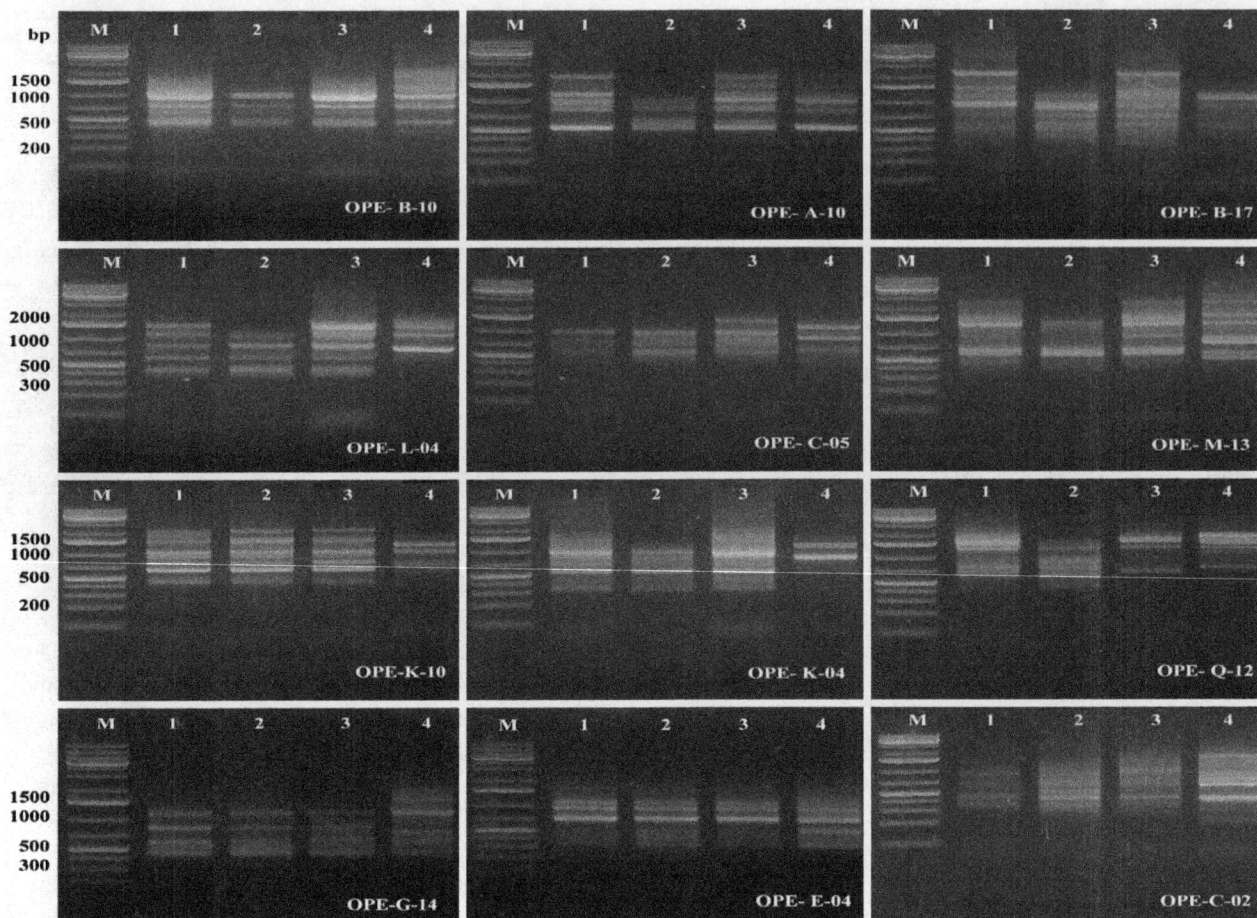

Fig. 1 Genetic polymorphism among canola cultivars as revealed by RAPD analysis. M: 1 kbp plus DNA ladder, 1-4: the canola cultivars Serw-3, Serw-4, Misser L-16 and Semu 249, respectively

Table 3 Total number of scorable bands, polymorphism % and band size of RAPD markers obtained by 15 random primers

Primer	Total scorable bands	Polymorphic bands	Polymorphism (%)	Band size range
OPE-A-10	7	4	57	500–2,000
OPE-B-10	8	2	25	400–2,000
OPE-B-17	9	4	44	250–2,000
OPE-C-02	6	4	62	500–1,300
OPE-C-05	8	3	37	200–1,500
OPE-E-04	7	1	14	300–1,500
OPE-G-14	6	2	33	450–1,500
OPE-K-04	9	7	77	350–2,000
OPE-K-10	7	3	42	450–2,000
OPE-K-15	5	3	60	600–2,000
OPE-L-04	8	5	62	400–1,500
OPE-M-13	11	8	72	500–2,200
OPE-N-13	8	5	62	350–2,000
OPE-P-09	9	5	55	400–1,400
OPE-Q-14	10	8	80	500–2,000
Total	118	72	61	

Table 4 Canola genotypes and their specific RAPD markers

Genotypes	Markers	Total marker
Serw-3	OPE-B-17,400	1
Serw-4	OPE-C-05,500, OPE-Q-14,500	2
Misser L-16	–	–
Semu 249	OPE-B-10,2000,400,OPE-L-04,700,OPE-C-02,650,OPE-M-13,2200,700,OPE-C-05,200,OPE-G-14,1500,700,OPE-K-04,950,OPE-K-10,1400,OPE-Q-14,1300,OPE-N-13,800,OPE-P-09,1400,500	15
Total		18

cultivars. The seven SSR primer sets revealed 31 alleles and 21 out of them were polymorphic (67.7 %). All primers showed different levels of polymorphism except RAM 206, RM 544 and RM 519 which showed no polymorphism among the four canola cultivars (Fig. 2; Table 5). The size of the detected alleles produced from using the SSR primer sets ranged from 75–2000 bp, which reflects a large difference in the number of repeats between the different alleles. These results agreed with

Moghaddam et al. (2009) that could successfully asses- sed the genetic diversity in rapeseed cultivars using RAPD and microsatellite markers.

The AFLP technique has been used for varietal finger- printing, mapping and genetic diversity studies. The major advantage of the AFLP technique over the other techniques is that it generates a larger number of amplified products in a single reaction (Powell et al. 1996). AFLP's on average reveal more polymorphic markers then either RFLP or RAPD's. In *B. napus* Lombard et al. (2000) indicate that two AFLP primer pair combinations were sufficient to distinguish 83 different cultivars from each other.

In the present investigation, the AFLP analysis was performed using three selective primer combinations and generated a total of 338 bands (Fig. 3; Table 6) The number of markers observed per primer combination ranged from 98 to 130. The total accounting marker number was 338 amplified bands, representing 87 % polymorphism and an average number of polymorphic bands per AFLP primer combination ranged from 79 to 91 bands. The highest percentage of polymorphism was obtained with M-CAT/E-AGG (91 %) followed by M-CAC/E-AGG (90 %) then M-CAG/E-ACC (79 %). The M-CAC/E-AGG produced the highest polymorphic

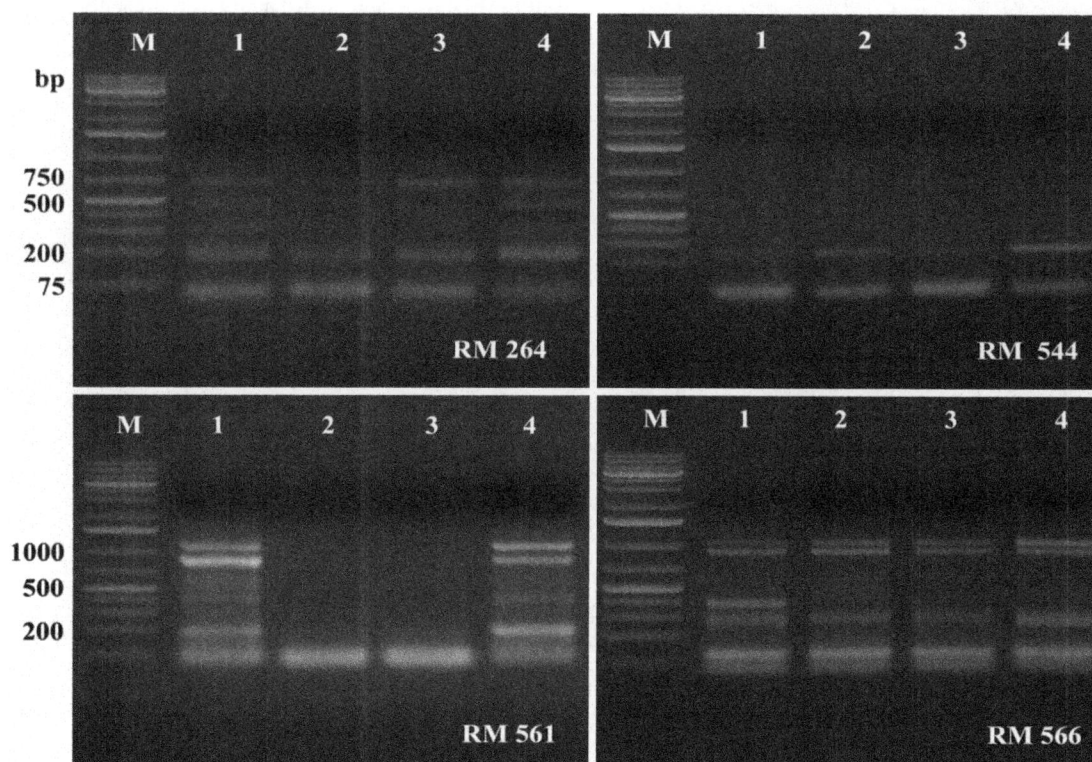

Fig. 2 The genetic variability among canola cultivars as revealed by SSR analysis

Table 5 Total bands, polymorphic bands and polymorphism percentages among canola cultivars as revealed by SSR analysis

Primer name	Total scorable alleles	Polymorphic alleles	Polymorphism (%)
RM 206	2	0	0
RM 264	5	2	40
RM 561	6	5	83
RM 544	2	0	0
RM 547	7	4	57
RM 519	3	0	0
RM 566	6	1	16
Total	31	21	67.7

bands (117), while the lowest number was obtained with M-CAG/E-ACC (78). As shown in Table 6, the size of AFLP fragments generated by the different primer combinations ranged from 471–1044 bp and the polymorphic fragments were distributed across the entire size range (Fig. 3; Table 6).

The unique AFLP markers that characterized the four canola varieties are listed in Table 7. Based on the data obtained, it can be distinguished between the four canola cultivars tested according to the AFLP specific markers. The cultivar Semu 249 exhibited the highest number of unique AFLP markers, giving 57 markers, followed by Misser L-16 which was characterized by 40 unique AFLP markers, then Serw-3 with 31 unique markers. While Serw-4 was characterized by the lowest number producing 14 unique positive markers.

The RAPD; SSR and AFLP based dendrogram group the investigated cultivars into two main clusters. The first cluster included Semu 249, the second cluster is divided into two sub-clusters the first one contains the cultivar Misser L-16 and the second one has the cultivars Serw-3 and Serw-4 (Fig. 4). These results indicate that the cultivars Serw-3 and Serw-4 may possess a high degree of genetic similarity; also their ancestors might be closely related (Fig. 4).

The dendrogram built on the basis of combined data from RAPD, SSR and AFLP analysis represents the genetic distances among the four canola varieties (Fig. 4).

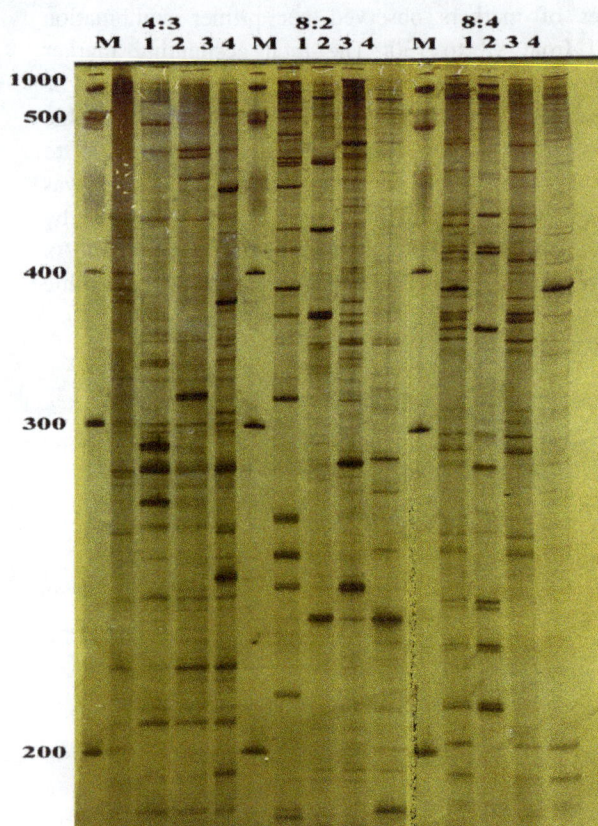

Fig. 3 The genetic diversity among canola cultivars as revealed by AFLP analysis

Conclusion

The results of the present study indicate that DNA markers represent efficient tools for estimating the genetic variability and the genetic relationships among the four canola cultivars. The markers generated are enough to distinguish between the different genotypes used. The genotype-specific molecular markers were determined and these markers can be considered as useful markers for high oil production in canola breeding programs. This opens up a possibility for developing a molecular genetic map that will lead to the

Table 6 Total bands, polymorphic bands and polymorphism percentages among canola cultivars as revealed by AFLP analysis

Primer	Total scorable bands	Polymorphic band	Polymorphism (%)	Band size range	Origin-specific markers
M-CAG/E-ACC	98	78	79	473–979	33
M-CAC/E-AGG	130	117	90	471–983	58
M-CAT/E-AGG	110	101	91	471–1,044	51
Total	338	296	87		142

Table 7 Canola genotype-specific AFLP markers

Genotypes	Markers	Total marker
Serw-3	M-CAG/E-ACC (931, 890, 824, 561, 553, 509)	31
	M-CAC/E-AGG (954, 947, 943, 915, 664, 542, 536, 533, 525, 522, 472)	
	M-CAT/E-AGG (1040, 978, 958, 928, 887, 821, 811, 604, 591, 567, 565, 533, 530, 522)	
Serw-4	M-CAG/E-ACC (888, 592)	14
	M-CAC/E-AGG (933, 869, 823, 810, 681, 653, 611, 471)	
	M-CAT/E-AGG (974, 876, 633, 527)	
Misser L-16	M-CAG/E-ACC (952, 943, 854, 821, 647, 559)	40
	M-CAC/E-AGG (958, 952, 939, 935, 931, 913, 897, 888, 863, 861, 647, 639, 609, 591, 574, 553, 542, 539, 531)	
	M-CAT/E-AGG (951, 947, 932, 865, 862, 794, 735, 587, 580, 567, 534, 520, 503, 501, 471)	
Semu 249	M-CAG/E-ACC (970, 958, 909, 886, 880, 878, 861, 857, 843, 835, 823, 817, 678, 655, 564, 535, 531, 530, 511)	57
	M-CAC/E-AGG (977, 867, 865, 833, 816, 808, 674, 649, 633, 629, 606, 596, 583, 572, 569, 554, 549, 537, 511, 510)	
	M-CAT/E-AGG (869, 814, 788, 778, 763, 750, 741, 732, 717, 711, 689, 683, 666, 662, 661, 621, 546, 494)	
Total		142

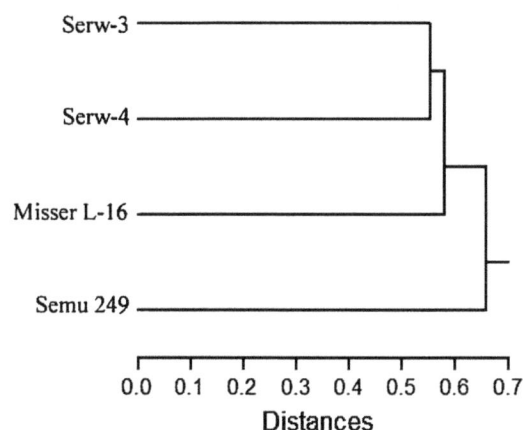

Fig. 4 Clustering of four canola cultivars based on pooled RAPD; SSR and AFLP markers

application of marker-assisted selection tools in genetic improvement of canola.

Acknowledgments The authors express deep sense of gratitude to the Field Crop Institute, Agricultural Research Center, Ministry of Agriculture–Egypt, for providing the canola seed used in this study. Also for the staff member of the Genetic Engineering Research Center, Cairo University for all the support, assistance and constant encouragements to carry out this work.

Conflict of interest The authors declare that they have no conflict of interest.

References

Ahmad N, Munir I, Khan I, Ali W, Muhammad W, Habib R, Khan R, Swati Z (2007) PCR-based genetic diversity of rapeseed germplasm using RAPD markers. Biotechnology 6(3):334–338

Bassam JB, Caetano-Anolles G, Gressho PM (1991) Fast and sensitive silver staining of DNA in polyacrylamide gels. Anal Biochem 196:80–83

Basunanda P, Spiller TH, Hasan M, Gehringer A, Schondelmaier J, Lühs W, Friedt W, Snowdon RJ (2007) Marker-assisted increase of genetic diversity in a double-low seed quality winter oilseed rape genetic background. Plant Breed 126:581–587

Charcosst A, Moreau L (2004) Use of molecular markers for the development of new cultivars and the evaluation of genetic diversity. Euphytica 137:81–94

Cheng XM, Xu JS, Xia S, Gu JX, Yang Y, Fu J, Qian XJ, Zhang SC, Wu JS, Liu KD (2009) Development and genetic mapping of microsatellite markers from genome survey sequences in *Brassica napus*. Theor Appl Genet 118:1121–1131

Cruz VMV, Luhman R, Marek LF, Rife CL, Shoemaker RC, Brummer EC, Gardner CAC (2007) Characterization of flowering time and SSR marker analysis of spring and winter type *Brassica napus* L. germplasm. Euphytica 153:43–57

Dandelj D, Jakse J, Javornik B (2004) Assesment of genetic variability of olive variaties by microsatellite and AFPL markers. Euphitica 136:93–102

Halton TA, Christopher JT, McClure L, Harker N, Henry RJ (2002) Identification and mapping of polymorphic SSR marker from expressed gene sequence of barley and wheat. Mol Breed 9:63–71

Hassan MB, Sahfique FA (2010) Current situation of edible vegetable oils and some propositions to curb the oil gap in Egypt. Nat Sci 8(12):1–7

Jaroslava A, Polakova K, Leisova L (2002) DNA analysis and their application in plant breeding. Czech J Genet Plant Breed 38:29–40

Kebede B, Thiagarajah M, Zimmerli C, Rahman MH (2010) Improvement of open pollinated spring rapeseed (*B. napus*) through introgression of genetic diversity from winter rapeseed. Crop Sci 50:1236–1243

Lombard V, Baril CP, Dubreuil P, Blouet F, Zhang D (2000) Genetic relationships and fingerprinting of rapeseed cultivars by AFLP: consequences for varietal registration. Crop Sci 40:1417–1425

Mailer RJ, Wratten N, Vonarx M (1997) Genetic diversity among Australian canola cultivars determined by randomly amplified polymorphic DNA. Aust J Exp Agric 37:793–800

Marijanovic JA, Kondic SA, Saftic-Pankovic D, Marinkovic R, Hristov N (2009) Phenotypic and molecular evaluation of genetic diversity of rapeseed (Brassica napus L.) genotypes. Afr J Biotechnol 8:4835–4844

Ministry of Agriculture and Land Reclamation, Economic Affairs Sector, Central Administration of Agricultural Economics, Food Balance Sheet (2003–07)

Moghaddam M, Mohammmadi SA, Mohebalipour N, Toorchi M, Aharizad S, Javidfar F (2009) Assessment of genetic diversity in rapeseed cultivars as revealed by RAPD and microsatellite markers. Afr J Biotechnol 8(14):3160–3167

Powell W, Morgante M, Andre C, Hanafy M, Vogel J, Tingey S, Rafalski A (1996) The comparison of RFLP, RAPD, AFLP and SSR (microsatellite) markers for germplasm analysis. Mol Breeding 2:225–238

Qu C, Hasan M, Lu K, Liu L, Liu X, Xie J, Wang M, Lu J, Odat N, Wang R, Chen L, Tang Z, Li J (2012) Genetic diversity and relationship analysis of the Brassica napus germplasm using simple sequence repeat (SSR) markers. Afr J Biotechnol 11(27):6923–6933

Rogers SO, Bendich AJ (1985) Extraction of DNA from milligram amounts of fresh herbarium and mummified plant tissues. Plant Mol Biol 5:69–76

Sovero M (1993) Rapeseed a new oilseed crop for the United States. In: Janick J, Simon JE (eds) New crops. Wiley, New York, pp 302–307

Spector AA (1999) Essentiality of fatty acids. J Am Oil Chem Soc 34:51–53

Stoutjesdijk PA, Hurlestone C, Singh SP, Green AG (2000) High-oleic acid Australian Brassica napus and B. juncea varieties produced by co-suppression of endogenous Delta 12-desaturase. Biochem Soc Trans 28:938–940

Vos P, Hogers R, Bleeker M, Reijans M, van de Lee T, Hornes M, Frijters JP, Peleman J, Kuiper M, Zabeau M (1995) AFLP: a new fingerprinting technique for DNA fingerprinting. Nucleic Acids Res 23:4407–4414

Yang X, Quiros C (1993) Identification and classification of celery cultivars with RAPD markers. Theor Appl Genet 86:205–212

Younessi MH, Nejat AIP, Mohamad T, Abbas M (2011) Phenotypic and molecular analysis of soybean mutant lines through random amplified polymorphic DNA (RAPD) marker and some morphological traits. Afr J Agric Res 6(7):1779–1785

Zeng L, Kwon TR, Liu X, Wilson C, Grieve CM, Gregorio GB (2004) Genetic diversity analyzed by microsatellite markers among rice (Oryza sativa L.) genotypes with different adaptations to saline soils. Plant Sci 166:1275–1285

4

Adenosine deaminase production by an endophytic bacterium (*Lysinibacillus* sp.) from *Avicennia marina*

Kandasamy Kathiresan · Kandasamy Saravanakumar ·
Sunil Kumar Sahu · Muthu Sivasankaran

Abstract The present study was carried out with the following objectives: (1) to isolate the endophytic bacilli strains from the leaves of mangrove plant *Avicennia marina*, (2) to screen the potential strains for the production of adenosine deaminase, (3) to statistically optimize the factors that influence the enzyme activity in the potent strain, and (4) to identify the potent strain using 16S rRNA sequence and construct its phylogenetic tree. The bacterial strains isolated from the fresh leaves of a mangrove *A. marina* were assessed for adenosine deaminase activity by plating method. Optimization of reaction process was carried out using response surface methodology of central composite design. The potent strain was identified based on 16S rRNA sequencing and phylogeny. Of five endophytic strains, EMLK1 showed a significant deaminase activity over other four strains. The conditions for maximum activity of the isolated adenosine deaminase are described. The potent strain EMLK1 was identified as *Lysinibacillus* sp. (JQ710723) being the first report as a mangrove endophyte. Mangrove-derived endophytic *bacillus* strain *Lysinibacillus* sp. EMLK1 is proved to be a promising source for the production of adenosine deaminase and this enzyme deserves further studies for purification and its application in disease diagnosis.

K. Kathiresan · K. Saravanakumar (✉) · S. K. Sahu ·
M. Sivasankaran
Faculty of Marine Sciences, Centre of Advanced Study
in Marine Biology, Annamalai University,
Parangipettai 608 502, Tamil Nadu, India
e-mail: saravana732@gmail.com

K. Kathiresan
e-mail: kathirsum@rediffmail.com

Keywords Mangroves · Endophytic bacteria ·
Adenosine deaminase and 16S rRNA

Introduction

Endophytic bacteria are important source for developing the novel drugs for effective treatment of diseases in humans, plants and animals (Strobel et al. 2004). The mangroves do have colonized endophytic bacteria, but the potential of the bacteria for medicinal enzymes are largely unexplored (Gayathri et al. 2010). Adenosine deaminase-ADA (EC 3.5.4.4) is a zinc-metallo enzyme involved in purine metabolism (Alrokayan 2002, 2007). This enzyme is widely distributed in different species (Bachrach 2004; Pospisilova 2007) but not reported from mangrove endophytes. The genes encoding these deaminases are essential in bacteria and yeast (Losey et al. 2006). This enzyme in particular is of special interest for its role in cellular growth regulation and differentiation (Hershfield and Mitchell 1995). Two types of the adenosine deaminase occur in human as ADA1 and ADA2; Lymphocytes and macrophages cells are known to have the ADA1 (Hirschhorn and Ratech 1980; Srinivasa Rao et al. 2010). Clinical and in vitro studies strongly suggest a mutual relationship between the absence of adenosine enzymatic activity and the immunodeficiency disease (Booth et al. 2006), which is characterized by severe defects in cellular and humoral immunity (Jasmin et al. 2012). The production of adenosine deaminase in the marine environment in particular mangrove biotope is almost non-existent. Hence, the present study was attempted to isolate endophytic bacilli from the mangrove plant *A. marina* for exploring the production of adenosine deaminase and to identify the potent strain by using 16S rRNA sequencing.

Materials and methods

Sample collection and surface sterilization of leaves

Healthy mangrove leaves of *A. marina* (Forsk.) Vierh were collected from Vellar estuary, Parangipettai, Tamil Nadu, India. All the samples were collected in sterile plastic bags and transported aseptically to the laboratory. The leaves were washed in running tap water, followed by 70 % ethanol for 2 min, 2 % sodium hypochlorite containing 0.1 % Tween 20 for 10 s and finally rinsed in distilled water for 2 min.

Isolation of endophytic *Bacillus* sp.

The washed leaves were crushed in a mortar and pestle. About 1 mL of crushed sample was serially diluted up to 10^{-5} dilutions using 12.5 mM potassium phosphate buffer (pH 7.1). For the isolation of endophytic bacillus, 0.1 mL of aliquot from 10^{-2} to 10^{-5} dilutions was inoculated by spread plate method on MRS (de man rogosa sharps) agar medium using sterile L-rod in Petri plate. The plates were incubated at 28 °C for 120 h.

Purification and selection of endophytic bacillus

Morphologically different bacterial colonies were selected and streaked on nutrient agar plates and incubated at 28 °C for 48 h. Five morphologically different colonies were purified by continuous cluttering method and the purified bacterial strains were named as EMLK1, EMLK2, EMLK3, EMLK4 and EMLK5. All the selected isolates were subcultured in nutrient agar slants and preserved in refrigerator at 4 °C.

Confirmation of adenosine deaminase activity in *Bacillus* strains

Production of the enzyme was confirmed by the change in colour from yellow to pale red by spectrophotometer which is due to the release of ammonia by increase in the pH level on modified Czapek Dox agar medium (Eaton et al. 1998; Shanmugam 2011) inoculated with the five *Bacillus* strains separately.

DNA extraction

The modified and standardized method of Sambrook and Russell (2001) was used to extract DNA from broth culture of EMLK1 strain. Briefly, 2 mL of broth culture of the sample in log phase was taken in Eppendorf tube. The tube was spun at 12,000 rpm for 10 min. The supernatant was discarded and the pellet dissolved completely in 500 μL of lysis buffer solution followed by addition of 10 % SDS and 5 μL proteinase K and incubated for 1 h at 60 °C. 250 μL of 5 M NaCl was added and chilled on ice for 10 min. Centrifugation was carried out at 8,000 rpm for 15 min and the supernatant was transferred to 1.5 mL Eppendorf tube. DNA was precipitated by adding chilled absolute ethanol and inverted gently several times. Then it was incubated at −20 °C for 3–4 h followed by centrifugation at 13,000 rpm for 15 min. Supernatant was discarded and the pellet was washed with 70 % ethanol. Pellet was air dried at room temperature and resuspended in 50 μL of TE buffer (pH 8.0) for further use.

16s rRNA amplification and sequencing

16s rRNA amplification reaction was performed using 16S primers 27F and 1492R in a 0.2 mL optical-grade PCR tube (Tarsons, India). 50 ng of DNA extract was added to a final volume of 50 μL of PCR reaction mixture containing 1.5 mM $MgCl_2$, 1× Reaction buffer (without $MgCl_2$) (Fermentas), 200 μM of each dNTPs (Fermentas), 100 pM of each primer and 1.5 U Taq DNA polymerase (Fermentas). PCR was performed in an automated thermal cycler (Lark Research Model L125+, India) with an initial denaturation at 95 °C for 5 min followed by 30 cycles of 95 °C for 30 s (denaturation), 52 °C for 45 s (annealing), 72 °C for 90 s (extension) and 72 °C for 10 min (final extension). PCR product was run on 1 % agarose in TAE buffer (40 mM Tris, 20 mM Acetic acid, 1 mM EDTA [pH 8.0]) to confirm that the right product (1,500 bp) was formed. The PCR product was purified using the QIAGEN PCR purification kit for sequencing. DNA sequencing was carried out using 3730 Genetic analyser, Applied biosystems, USA (Ramachandra Innovis, Chennai, India).

Phylogenetic analyses

Sequence similarity search for the obtained sequence was performed against the non-redundant database maintained by the National Center for Biotechnology Information using the BLAST algorithm (http://www.ncbi.nlm.nih.gov). Then the 16S rRNA sequence of the isolate was aligned with the sequences of selected reference taxa from NCBI and Ribosomal Database Project II (http://rdp.cme.msu.edu) using the ClustalW implemented in the MEGA5 software (Tamura et al. 2011) and the alignment was inspected and adjusted manually where necessary. The aligned sequences were incorporated to construct phylogenetic tree using maximum-likelihood method (Saitou and Nei 1987). All characters were equally weighted and unordered. Initial tree(s) for the heuristic search were obtained automatically by applying Neighbour-Join and BioNJ algorithms to a matrix of pairwise distances estimated using the maximum composite likelihood approach, and then selecting the topology with superior log likelihood value. Alignment gaps were treated as missing data. MP analysis was conducted using a heuristic search. *Bacillus subtilis* (JN609214) was taken as outgroup taxon. The robustness of trees in the maximum-likelihood (ML) analyses was evaluated by 1,000 bootstrap replications.

Fermentation and optimization of reaction condition for deaminase activity

The optimization of the reaction conditions for deaminase activity was analyzed in fermentation medium containing folic acid (0.5 g), dextrose solution (4.5 g) and mineral solution (133 %) which constituted of NH_4NO_3 (0.213 g), $MgSO_4$ (0.027 g), $ZnSO_4$ (0.001 g), $CaCl_2$ (0.007 g), $MnSO_4$ (0.003 g), $CuSO_4$ (0.001 g), $FeSO_4$ (0.013 g), $NaSO_4$ (0.027 g) and NaH_2PO_4 (0.027 g). The response surface methodology was used to statistically optimize the important reaction factors: pH in the range of 4.0–8.0, reaction time (0–30 min), concentrations of adenosine (10–50 μL), and concentrations of microbial cell free culture filtrate (10–50 μL).

Statistical optimization

Deaminase activity was optimized using a standard response surface methodology design also known as central composite design (CCD). The range and the levels of the variables (high and low) considered in the present work are given in Table 1. Adenosine deaminase activity (U L^{-1}) (Y) was taken as the response of the design experiments. The quadratic equation model for predicting the optimal point is expressed according to Eq. 1.

$$Y = \beta_0 + \beta_1 X_1 + \beta_2 X_2 + \beta_3 X_3 + \beta_4 X_4 + \beta_{11} X_1^2$$
$$+ \beta_{22} X_2^2 + \beta_{33} X_3^2 + \beta_{44} X_4^2 + \beta_{12} X_1 X_2 + \beta_{13} X_1 X_3$$
$$+ \beta_{14} X_1 X_4 + \beta_{23} X_2 X_3 + \beta_{24} X_2 X_4 + \beta_{34} X_3 X_4 \quad (1)$$

where X_1 is pH (°C), X_2 is reaction time (min), X_3 is adenosine concentration (μL), X_4 is microbial cell free culture filtrate (μL).

Table 1 Experimental range and levels of independent process variables

Factor	Range and coded value				
	−2	−1	0	1	2
pH	4	5	6	7	8
Reaction time (min)	0	10	20	30	40
Adenosine concentration (μL)	10	20	30	40	50
Microbial cell free culture filtrate concentration (10–50 μL)	10	20	30	40	50

Four factors were studied and their low and high levels of actual values. Thirty-two experiments were conducted in duplicate. Design Expert Version 8.0.6 (Stat Ease, USA) was used for data analysis. The optimum values of the selected factors were obtained by fitting the regression equation and by analyzing the contour and surface plots. The multiple coefficient of determination was used and variability among dependent variables was explained. R^2 and the model equation were used to predict the optimum value and subsequently the interaction between the factors within the specified range was elucidated (Elibol and Ozer 2002). The adenosine deaminase activity was calculated by following the method of the (Giusti and Galanti 1984). Their mean values and 5 % standard errors were calculated.

Results

The surface sterilized leave samples of *A. marina* were subjected for the isolation of endophytic *Bacillus* sp. and tested for the adenosine deaminase enzyme activity. Among the selected five strains only the *Bacillus* strain EMLK1 showed the significant enzyme activity as revealed by zone of inhibition. 16S rRNA sequencing result revealed that the potential strain was *Lysinibacillus* sp. (Genbank accession number: JQ710723). The evolutionary history was inferred by using the Maximum Likelihood method based on the Tamura–Nei model (1993). The tree with the highest log likelihood (−2,548.3764) is shown in Fig. 1. The percentage of trees in which the associated taxa clustered together is shown next to the branches. The tree is drawn to scale, with branch lengths measured in the number of substitutions per site. The analysis involved 10 nucleotide sequences. All positions containing gaps and missing data were eliminated. There were a total of 1,058 positions in the final dataset. Evolutionary analyses were conducted in MEGA5 (Tamura et al. 2007).

Recently, literature data compiled from hundreds of species descriptions has suggested that strains sharing less than 98.8 % sequence similarity belong to different genospecies. Therefore, additional DNA–DNA hybridization study is required to differentiate the identified *Lysinibacillus* strain up to species level more accurately.

The experimental model fitness for deaminase activity was tested by the quadratic model along with the contour error plot and normal probability plot and the ADA activity for each cycle performed as per the experimental design along with experimental response and predicted response. The response surface methodology based on the estimates of the parameters indicated an experimental relationship between the response and input variables expressed by the following quadratic model Eq. (2).

Fig. 1 Molecular phylogenetic analysis of *Lysinibacillus* sp. by maximum likelihood method

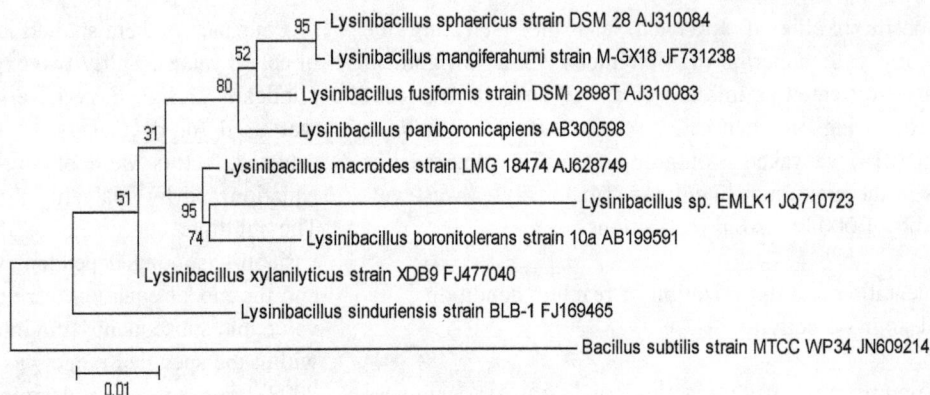

$$\text{ADA activity} = 8.50 + 0.95A + 2.23B + 1.08C$$
$$+ 0.74D - 0.32AB + 0.33AC + 0.58AD$$
$$+ 0.76BC - 0.026BD + 1.18CD \qquad (2)$$

where A, B, C and D are independent variables. Significance of each coefficient presented in Eq. (2) was determined by the student's t test and p-values. The results of the quadratic model for the enzyme activity are in the form of analysis of variance (ANOVA). The value of predicted R^2 and adjusted R^2 is close to 0.92 revealing a high correlation between the observed values and the predicted values. The means that regression model provides an excellent explanation of the relationship between the independent variables (factors) and the response (enzyme activity). The lack-of-fit term was non-significant as it was desired. The non-significant value (0.76) of lack-of-fit observed was more than probability of 0.05 which revealed that the quadratic model was valid for the present study. The statistical optimization revealed the optimized conditions of reaction: pH 6.59, 31.89 min of reaction time, 47.84 μL of adenosine concentration and 40.27 μL of microbial cell free culture filtrate for the better production of the adenosine deaminase enzyme activity.

Discussion

Nowadays, the study of adenine/adenosine deaminases is receiving much attention due to diseases that are induced in the absence or excess of the enzyme activity. The production of the enzyme has been reported in several microbes: *Escherichia coli* (Sugadev et al. 2006), *Enterococcus faecalis* (Pospisilova et al. 2006) *Saccharomyces cerevisiae* and *Schizosaccharomyces pombe* (Lee et al. 2009). Although bacterial bacilli are reportedly producing adenosine deaminase (Jun et al. 1991), the present study reported for the first time the enzyme production by endophytic bacillus, *Lysinibacillus* sp. derived from mangroves. This enzyme activity could be due to the presence of pterin synthesizing biochemical compounds in *Lysinibacillus* sp.

The statistically optimized conditions for enzyme activity in the present study are in accordance with the earlier reports of Sakai and Jun (1978) who reported that optimum pH for the reaction was 5.0–6.0, and the optimum temperature was 55 °C for production of extracellular adenosine deaminase by the endophytic *Streptomyces* sp. The enzyme in *Lysinibacillus* sp. was found to be unstable, but protected from inactivation by ethyl alcohol and this result is in agreement with the results of Sakai and Jun (1978).

Acknowledgments The authors are thankful to the authorities of Annamalai University, Tamil Nadu, India.

Conflict of interest The authors declare that they have no conflict of interest.

References

Alrokayan SAH (2002) Purification and characterization of adenosine deaminase from camel skeletal muscle. Int J Biochem Cell Biol 34:1608–1618

Alrokayan SAH (2007) Role of adenosine deaminase and purine nucleoside phosphorylase in severe combined immunodeficiency disease: a biochemical and molecular study. Biosci Biotechnol Res Asia 4(1):55–58

Bachrach U (2004) Polyamines and cancer: minireview article. Amino Acids 26:307–309

Booth C, Hershfield M, Notarangelo L, Buckley R, Hoenig M, Mahlaoui N et al (2006) Management options for adenosine deaminase deficiency: proceedings of the EBMT satellite workshop. Clin Immunol 123:139–147

Eaton AD, Clesceri LS, Greenberg AE (1998) Standard methods for the examination of water and waste water, 20th edn. American Public Health Association, Washington

Elibol M, Ozer D (2002) Response surface analysis of lipase production by freely suspended *Rhizopus arrhizus*. Process Biochem 38:367–372

Gayathri S, Saravanan D, Radhakrishnan M, Balagurunathan R, Kathiresan K (2010) Bio prospecting potential of fast growing endophytic bacteria from leaves of mangroves and salt marsh plants species. Indian J Biotechnol 9:397–402

Giusti G, Galanti B (1984) Colorimetric method. In: Bergmeyer HU (ed) Methods of enzyme analysis, 3rd edn. Chemie, Weinheim, pp 315–323

Hershfield M, Mitchell B (1995) Immunodeficiency diseases caused by adenosine deaminase deficiency and purine nucleoside phosphorylase deficiency. In: Scriver C et al (eds) The metabolic and molecular basis of inherited disease. McGraw-Hill, New York, pp 1725–1768

Hirschhorn R, Ratech H (1980) Isonzymes of adenosine deaminase. In: Rattazzi MC, Scandalios JG, Whitt GS (eds) Isoenzymes: current topics in biological and medical research, vol 4. Alan R Liss, Inc, New York, pp 131–157

Jasmin HJ, Kaushik VB, Anand BV, Vaidehi RP, Sankalp MS, Dipmala P et al (2012) Value of adenosine deaminase level for the differential diagnosis various meningitis. Int J Biol Med Res 3(2):1644–1647

Jun HK, Kim TS, Sakai T (1991) Purification and characterization of extracellular adenosine deaminase from *Streptomyces* sp. J Ferment Bioeng 71:6–11

Lee G, Lee SS, Kay KY, Kim D, Choi S, Jun KH (2009) Isolation and characterization of a novel adenosine deaminase inhibitor, IADA-7, from *Bacillus* sp. J-89. J Enzyme Inhib Med Chem 24(1):59–64

Losey HC, Ruthenburg AJ, Verdine GL (2006) Crystal structure of *Staphylococcus aureus* tRNA adenosine deaminase TadA in complex with RNA. Nature Struct Mol Biol 13:153–159

Pospisilova H, Frebort I (2007) Aminohydrolases acting on adenine, adenosine and their derivatives. Biomed Pap Med Fac Univ Palacky Olomouc Czech Repub 151(1):3–10

Pospisilova H, Novak O, Frebortova J, Strnad M, Frebort I (2006) Oxidative and hydrolytic cleavage of cytokinin derivatives with biomedical and biotechnological potential. In: Abstracts of International Symposium of Fifth 21st Century COE "Towards Creating New Industries Based on Inter-Nanoscience". Awaji, Japan, pp 15–20

Saitou N, Nei M (1987) The neighbor-joining method: a new method for reconstructing phylogenetic trees. Mol Biol Evol 4:406–425

Sakai T, Jun H (1978) Purification and crystallization of adenosine deaminase in *Pseudomonas iodinum*. FEBS Lett 86:174–178

Sambrook J, Russel DW (2001) Rapid isolation of yeast DNA. Molecular cloning, a laboratory manual. Cold Spring Harbor Laboratory, New York, pp 631–632

Shanmugam S (2011) Optimization of synergistic parameters for atypical pterin deaminase activity from rattus norvegicus using response surface methodology Biochemical and Molecular Engineering XVII. Emerging Frontiers, Seattle

Srinivasa Rao K, Anand Kumar H, Rudresh BM, Srinivas T, Harish Bhat K (2010) A Comparative study and evaluation of serum adenosine deaminase activity in the diagnosis of pulmonary tuberculosis. Biomed Res 2:189–194

Strobel G, Daisy B, Castillo U, Harper J (2004) Natural products from endophytic microorganisms. J Nat Prod 67:257–268

Sugadev R, Kumaran D, Burley SK, Swaminathan S (2006) Crystal structure of an adenine deaminase. New York Structural Genomics Research Consortium (NYSGRC).

Tamura K, Nei M (1993) Estimation of the number of nucleotide substitutions in the control region of mitochondrial DNA in humans and chimpanzees. Mol Biol Evol 10(3):512–526

Tamura K, Dudley J, Nei M, Kumar S (2007) MEGA4: molecular evolutionary genetics analysis (MEGA) software version 4.0. Mol Biol Evol 24:1596–1599

Tamura K, Peterson D, Peterson N, Steker G, Nei M, Kumar S (2011) MEGA5: molecular evolutionary genetics analysis using maximum likelihood, evolutionary distance, and maximum parsimony methods. Mol Biol Evol 28(10):2731–2739

5

Enabling comparative modeling of closely related genomes: example genus *Brucella*

José P. Faria · Janaka N. Edirisinghe · James J. Davis · Terrence Disz ·
Anna Hausmann · Christopher S. Henry · Robert Olson · Ross A. Overbeek ·
Gordon D. Pusch · Maulik Shukla · Veronika Vonstein · Alice R. Wattam

Abstract For many scientific applications, it is highly desirable to be able to compare metabolic models of closely related genomes. In this short report, we attempt to raise awareness to the fact that taking annotated genomes from public repositories and using them for metabolic model reconstructions is far from being trivial due to annotation inconsistencies. We are proposing a protocol for comparative analysis of metabolic models on closely related genomes, using fifteen strains of genus *Brucella*, which contains pathogens of both humans and livestock. This study lead to the identification and subsequent correction of inconsistent annotations in the SEED database, as well as the identification of 31 biochemical reactions that are common to *Brucella*, which are not originally identified by automated metabolic reconstructions. We are currently implementing this protocol for improving automated annotations within the SEED database and these improvements have been propagated into PATRIC, Model-SEED, KBase and RAST. This method is an enabling step for the future creation of consistent annotation systems and high-quality model reconstructions that will support in predicting accurate phenotypes such as pathogenicity, media requirements or type of respiration.

Introduction

Since the first bacterial genome was sequenced in 1995 (Fleischmann et al. 1995), the number of genome sequences has grown exponentially (Lagesen et al. 2010). This increase in genomic data has demanded the improvements in high-throughput genome analysis tools that are widely being used today. It is now possible to automate the generation of annotations (Aziz et al. 2008) and initial draft metabolic models with minimal effort (Henry et al. 2010); however, the creation of accurate, high-quality models requires a substantial investment in mining phenotypic data (e.g., BioLog or RNAseq data) and an iterative reconciliation with the experimental data (Thiele and Palsson 2010).

The quality of the initial metabolic network reconstructions and their utility for formulating predictions depends on the quality and consistency of the annotations from which they were generated. If one attempts to compare the initial metabolic reconstructions for distinct organisms, a significant number of discrepancies in the resulting models are often found. However, isofunctional homologs must have the same annotations, so that they can be mapped to the same reactions in the models. Thus,

J. P. Faria · J. N. Edirisinghe · J. J. Davis · T. Disz ·
C. S. Henry · R. Olson · R. A. Overbeek · V. Vonstein
Mathematics and Computer Science Division, Argonne National
Laboratory, Argonne, IL, USA

A. Hausmann · R. A. Overbeek · G. D. Pusch · V. Vonstein
Fellowship for Interpretation of Genomes, Burr Ridge, IL, USA

M. Shukla · A. R. Wattam
Virginia Bioinformatics Institute, Virginia Tech, Blacksburg,
VA, USA

J. P. Faria
IBB-Institute for Biotechnology and Bioengineering, Centre of
Biological Engineering, University of Minho, Campus de
Gualtar, 4710-057 Braga, Portugal

J. N. Edirisinghe · J. J. Davis (✉) · C. S. Henry
Computation Institute, University of Chicago, Chicago, IL, USA
e-mail: jimdavis@uchicago.edu

improving annotation consistency and accuracy has become an issue of paramount importance.

In this report, we describe a broadly applicable protocol for improving the annotations and metabolic reconstructions for an entire genus. We demonstrate how this protocol has improved the annotations and metabolic reconstructions for genus *Brucella,* a group of intracellular facultative bacterial pathogens of humans and livestock. High-quality metabolic reconstructions and predictive metabolic models are available for several organisms, most notably model organisms such as *E. coli* (Orth et al. 2011) and *B. Subtilis* (Tanaka et al. 2013). A metabolic model for any *Brucella* strain has yet to be proposed. Since wet lab research with pathogenic organisms can be particularly challenging, this makes the development of predictive metabolic models for those organisms highly desirable. Maintaining annotation consistency among closely related genomes is the key step for enabling comparative modeling studies.

Results

Description of the protocol

Step 1. Genomes are chosen for analysis We have chosen fifteen genomes representing the major species, biovars and clades of the genus *Brucella* (Wattam et al. 2012) (Table 1).

Step 2. Potential mobile element proteins are identified and removed from consideration To find potential mobile element proteins we first identified repeat regions in each chromosome. BLASTN (Altschul et al. 1997) was used to compare each of the fifteen genomes against itself. Any DNA region (other than rRNA operons) occurring more than once in the genome with a nucleotide identity ≥90 % and a length ≥200 nucleotides was considered to be a repeat. Although there are many ways to identify mobile element proteins that could be substituted within this framework (e.g., Davis and Olsen 2011), for the purposes of this study, we define a potential mobile element protein as a one that overlaps a repeat region by at least 10 bp. All of the 15 *Brucella* genomes were then compared to the list of potential mobile element proteins using BLASTP, and matching proteins with ≥50 % identity over ≥80 % of the protein length were also considered to be potential mobile element proteins regardless of proximity to a repeat region. This resulted in the creation of 50 mobile element protein families, containing a total of 410 proteins. These proteins were excluded from subsequent steps due to their variability and because they are not currently used for metabolic model reconstructions.

Step 3. Families of core proteins are generated In order to find the core proteins, the remaining genes from each of the *Brucella* genomes were compared. Two proteins were placed in the same protein family if they were bi-directional best hits between a pair of genomes with >50 % identity and 80 % coverage, and the genes occurred within a conserved genomic context (Overbeek et al. 1999a, b). We considered the context of the matched pairs to be conserved if there were at least three pairs of bi-directional best hits co-occurring within a 10 Kb region. This resulted in 5,038 families (with two or more proteins) containing a total of 52,626 proteins. From these initial families we generated core protein families, which are defined as

Table 1 *Brucella* genomes used in this study with their SEED (Overbeek et al. 2005, 2013) and PATRIC (Gillespie et al. 2011; Wattam et al. 2013) identifiers, sizes, number of contigs and number of protein coding sequences (CDSs)

Genome name	PubSEED ID	PATRIC genome ID	Genome size (bp)	Number of contigs	Number of CDSs
Brucella abortus bv. 1 str. 9-941	262,698.4	15,061	3,286,445	2	3,413
Brucella canis ATCC 23365	483,179.4	25,663	3,312,769	2	3,394
Brucella ceti str. Cudo	595,497.3	28,239	3,389,269	7	3,578
Brucella ceti M13/05/1	520,460.3	83,544	3,337,230	22	3,367
Brucella melitensis bv. 1 str. 16 M	224,914.11	92,729	3,294,931	2	3,446
Brucella microti CCM 4915	568,815.3	92,249	3,294,931	2	3,374
Brucella neotomae 5K33	520,456.3	114,381	3,329,623	11	3,383
Brucella ovis ATCC 25840	444,178.3	136,990	3,275,590	2	3,499
Brucella pinnipedialis M292/94/1	520,462.3	74,143	3,373,519	15	3,356
Brucella sp. 83/13	520,449.3	75,385	3,153,851	20	3,152
Brucella inopinata BO1	470,735.4	109,945	3,366,774	55	3,361
Brucella inopinata-like BO2	693,750.4	146,994	3,305,941	174	3,276
Brucella sp. NVSL 07-0026	520,448.3	103,899	3,297,137	17	3,442
Brucella suis 1330	204,722.5	107,850	3,315,175	2	3,402
Brucella suis bv. 5 str. 513	520,489.3	73,489	3,323,676	19	3,316

families containing at most one protein from each genome, where 80 % of the genomes are represented in the family. Similar to Step 2, it would be possible to substitute other methods for finding orthologous genes at this step as well (e.g., Li et al. 2003).

Step 4. Annotation inconsistencies are removed The core protein families of the RAST-annotated *Brucella* genomes were compared and inconsistencies (defined as two or more family members having different annotations) were evaluated. We manually curated a total of 398 families containing 4,848 proteins. We defined two metrics to measure progress.

The first:

Given a protein family (i.e., from one of the 5,038 families we constructed), at what frequency has any given pair of proteins within the family been assigned precisely the same annotation by RAST (Overbeek et al. 2013)?

We report this property before and after manual cleanup, and compare our annotations to other public annotation resources (Table 2).

The second:

How many Brucella-universal-reactions have been assigned to each genome?

By universal reactions we mean the reactions that are present in all *Brucella* genomes used in this study. We chose this second metric to demonstrate that improvements in annotations lead to improvements in the metabolic reconstructions.

Step 5. Annotation and reaction database improvements are made based on metabolic network reconstructions Metabolic reconstructions were built for the fifteen *Brucella* genomes (Tables S1, S2), using the tools provided by DOE Systems Biology Knowledgebase (KBase) (http://kbase.us). Starting with the manually improved genomes, we focused on the reactions that were non-universal among the 15 *Brucella* strains. The annotations relating to these reactions were manually evaluated and corrected, if needed. This process was repeated.

The initial set of metabolic reconstructions from the original RAST annotations contained 1,011 *Brucella*-universal-reactions. The second set of reconstructions from the manually curated annotations (Step 4) contained 1,016, of which 20 were found to be new core reactions and 15 were removed from the set due to annotation errors. Finally, the third set, after using the metabolic reconstructions to guide the annotation cleanup, contained 1,047 *Brucella*-universal-reactions, of which 31 previously unrecognized core reactions were found.

Annotation improvements

To eliminate sequencing, annotation and modeling errors from true strain-specific differences, we manually examined the 86 non-universal reactions from the second set of metabolic reconstructions. This revealed problems with the automated assertion or omission of reactions in certain genomes (Table S3). We verified the absence of 39 reactions from the set of genomes and identified 31 cases of *Brucella*-universal-reactions that had not been identified in the first round of metabolic reconstruction. The leading cause for the omission of reactions was insufficient sequencing quality (e.g., frame shifts, incomplete ORFs at the end of contigs or stretches of low quality sequence) that resulted in gene-calling errors. We also found 16 annotation errors (outdated functional roles), errors in the reaction database (labeled as "functional role ambiguities" in Table S3) and one gene fusion.

More importantly, this process resulted in the identification of five unique non-universal reactions in the *Brucella inopinata* BO1 and *Brucella inopinata*-like BO2 strains. Those reactions are involved in rhamnose-containing glycan synthesis and confirm the findings for those strains reported in (Wattam et al. 2012). In addition, we proposed candidate proteins in all Brucella for the *N*-acetyl-L,L-diaminopimelate deacetylase, the missing step in the diaminopimelate pathway (DAP) of leucine biosynthesis. All Brucella non-universal reactions for each genome are provided in Tables S4 and S5.

Table 2 The consistency of annotations across different resources

Source	Number of pairs	Number of pairs inconsistently annotated	Percent of pairs inconsistently annotated
RefSeq	562,597,217	383,808,122	68.2
IMG	101,525,838	52,434,525	51.6
TrEMBL	112,735,194	46,284,849	41.1
SwissProt	803,819	42,429	5.3
SEED	271,622,566	9,056,551	3.3
Original RAST output	16,349,603	102,097	0.6
RAST after manual curation	16,349,603	47,504	0.3

For each protein in a *Brucella* protein family used in this study, all of the proteins with identical sequences were found in various databases and the percentage of pairs that were inconsistently annotated was computed. Annotations were collected from RefSeq (Pruitt et al. 2007), UniProt Knowledgebase (UniProtKB)(Apweiler et al. 2010), the Translated EMBL Nucleotide Sequence Data Library (TrEMBL) (Boeckmann et al. 2003), the Integrated Microbial Genomes (IMG) system (Markowitz et al. 2012) and the SEED (Overbeek et al. 2005, 2013)

Discussion

In this report, we have described a workflow for improving the annotations of an entire genus that utilizes metabolic reconstructions as a measure of annotation consistency. This has resulted in the production of an accurate and consistent collection of annotations and initial estimates of the metabolic network for the genus *Brucella*. By manual curation of 398 protein families (used in metabolic models) whose members had inconsistent annotations for isofunctional homologs, we have lowered the percentage of inconsistently annotated pairs of genes from 0.6 to 0.3 %. Those improvements have lead to changes in the metabolic reconstructions, generating a larger set of *Brucella*-universal-reactions and highlighting the real metabolic differences between organisms. We believe that knowledge of the real differences will be of importance when deciding on sets of "representative models" to portrait the entire genus. The "representative models" will aid in the research of less studied or newly sequenced strains.

With this work, we have demonstrated that the use of a controlled vocabulary for the annotation of genomes is a key for the construction of reaction networks and future predictive comparative models. The automated annotations provided by the RAST system and the SEED's controlled vocabulary (Overbeek et al. 2005, 2013) provide a good start, but annotation inconsistencies caused by sequencing and propagation errors have to be manually processed. This method was devised to reduce the workload of researchers who are trying to build models, but it also clearly exposes bottlenecks where future computational tools must be built that can meet and exceed the skill level of an expert human annotator.

This work has improved the annotations in the SEED and RAST (Overbeek et al. 2005, 2013) and the reaction databases in Model-SEED (Henry et al. 2010) and KBase by flagging ambiguities in current functional roles. It has also improved the *Brucella*-specific collections of protein families that are propagated to RAST and PATRIC, the PathoSystems Resource Integration Center (Gillespie et al. 2011; Wattam et al. 2013), which is dedicated to enabling bioinformatics research for bacterial pathogens and has particularly strong ties to the *Brucella* research community.

With this proof of concept, we plan to use this methodology to improve annotations of other conserved genera and extend it to less conserved phylogenetic groups and pave the way for comparative modeling.

Acknowledgments We thank Jean Jacques Letesson, Maite Iriarte, Stephan Köhler and David O'Callaghan for their input on improving specific annotations. This project has been funded by the United States National Institute of Allergy and Infectious Diseases, National Institutes of Health, Department of Health and Human Services, under Contract No. HHSN272200900040C, awarded to BW Sobral, and from the United States National Science Foundation under Grant MCB-1153357, awarded to CS Henry. J.P.F. acknowledges funding from [FRH/BD/70824/2010] of the FCT (Portuguese Foundation for Science and Technology) Ph.D. scholarship.

Conflict of interest The authors declare that they have no conflict of interest in the publication.

References

Altschul SF, Madden TL, Schaffer AA, Zhang J, Zhang Z, Miller W, Lipman DJ (1997) Gapped BLAST and PSI-BLAST: a new generation of protein database search programs. Nucleic Acids Res 25(17):3389–3402

Apweiler R, Martin MJ, O'Donovan C, Magrane M, Alam-Faruque Y, Antunes R, Barrell D, Bely B, Bingley M, Binns D et al (2010) The Universal Protein Resource (UniProt) in 2010. Nucleic Acids Res 38:D142–D148 (database issue)

Aziz RK, Bartels D, Best AA, DeJongh M, Disz T, Edwards RA, Formsma K, Gerdes S, Glass EM, Kubal M et al (2008) The RAST server: rapid annotations using subsystems technology. BMC Genomics 9:75.

Boeckmann B, Bairoch A, Apweiler R, Blatter MC, Estreicher A, Gasteiger E, Martin MJ, Michoud K, O'Donovan C, Phan I et al (2003) The SWISS-PROT protein knowledgebase and its supplement TrEMBL in 2003. Nucleic Acids Res 31(1):365–370

Davis JJ, Olsen GJ (2011) Characterizing the native codon usages of a genome: an axis projection approach. Mol Biol Evol 28(1):211–221

Fleischmann RD, Adams MD, White O, Clayton RA, Kirkness EF, Kerlavage AR, Bult CJ, Tomb JF, Dougherty BA, Merrick JM et al (1995) Whole-genome random sequencing and assembly of *Haemophilus influenzae* Rd. Science 269(5223):496–512

Gillespie JJ, Wattam AR, Cammer SA, Gabbard JL, Shukla MP, Dalay O, Driscoll T, Hix D, Mane SP, Mao C (2011) PATRIC: the comprehensive bacterial bioinformatics resource with a focus on human pathogenic species. Infect Immun 79(11):4286–4298

Henry CS, DeJongh M, Best AA, Frybarger PM, Linsay B, Stevens RL (2010) High-throughput generation, optimization and analysis of genome-scale metabolic models. Nat Biotechnol 28(9):977–982

Lagesen K, Ussery DW, Wassenaar TM (2010) Genome update: the 1000th genome—a cautionary tale. Microbiology 156(Pt 3):603–608.

Li L, Stoeckert CJ, Roos DS (2003) OrthoMCL: identification of ortholog groups for eukaryotic genomes. Genome Res 13(9):2178–2189

Markowitz VM, Chen IM, Palaniappan K, Chu K, Szeto E, Grechkin Y, Ratner A, Jacob B, Huang J, Williams P et al (2012) IMG: the integrated microbial genomes database and comparative analysis system. Nucleic Acids Res 40:D115–D122. (database issue)

Orth JD, Conrad TM, Na J, Lerman JA, Nam H, Feist AM, Palsson BO (2011) A comprehensive genome-scale reconstruction of *Escherichia coli* metabolism—2011. Mol Syst Biol 7:535.

Overbeek R, Fonstein M, D'Souza M, Pusch GD, Maltsev N (1999a) Use of contiguity on the chromosome to predict functional coupling. In Silico Biol 1(2):93–108

Overbeek R, Fonstein M, D'Souza M, Pusch GD, Maltsev N (1999b) The use of gene clusters to infer functional coupling. Proc Natl Acad Sci USA 96(6):2896–2901

Overbeek R, Begley T, Butler RM, Choudhuri JV, Chuang HY, Cohoon M, de Crecy-Lagard V, Diaz N, Disz T et al (2005) The subsystems approach to genome annotation and its use in the project to annotate 1000 genomes. Nucleic Acids Res 33(17):5691–5702.

Overbeek R, Olson R, Pusch GD, Olsen GJ, Davis JJ, Disz T, Edwards RA, Gerdes S, Parrello B, Shukla M, Vonstein V et al (2013) The SEED and the rapid annotation of microbial genomes using subsystems technology (RAST). Nucleic Acids Res. (database issue)

Pruitt KD, Tatusova T, Maglott DR (2007) NCBI reference sequences (RefSeq): a curated non-redundant sequence database of genomes, transcripts and proteins. Nucleic Acids Res 35:D61–D65 (database issue, pii:gkl842)

Tanaka K, Henry CS, Zinner JF, Jolivet E, Cohoon MP, Xia F, Bidnenko V, Ehrlich SD, Stevens RL, Noirot P (2013) Building the repertoire of dispensable chromosome regions in *Bacillus subtilis* entails major refinement of cognate large-scale metabolic model. Nucleic Acids Res 41(1):687–699.

Thiele I, Palsson B (2010) A protocol for generating a high-quality genome-scale metabolic reconstruction. Nat Protoc 5:93–121

Wattam AR, Inzana TJ, Williams KP, Mane SP, Shukla M, Almeida NF, Dickerman AW, Mason S, Moriyon I, O'Callaghan D et al (2012) Comparative genomics of early-diverging *Brucella* strains reveals a novel lipopolysaccharide biosynthesis pathway. mBio 3(5):e00246–12.

Wattam AR, Abraham D, Dalay O, Disz TL, Driscoll T, Gabbard JL, Gillespie JJ, Gough R, Hix D, Kenyon R (2013) PATRIC, the bacterial bioinformatics database and analysis resource. Nucleic Acids Res. (database issue)

6

Genotypic and phenotypic diversity of polyhydroxybutyrate (PHB) producing *Pseudomonas putida* isolates of Chhattisgarh region and assessment of its phosphate solubilizing ability

Toshy Agrawal · Anil S. Kotasthane ·
Renu Kushwah

Abstract A diverse and versatile spectrum of metabolic activities among isolates of fluorescent *Pseudomonas putida* indicates their adaptability to various niches. These polyhydroxybutyrate producing and phosphate solubilizing isolates showed a high level of functional and genetic versatility among themselves. One of the potential *P. putida* isolate P132 can contribute as a candidate agent for both biocontrol and PGPR applications. Identified as one of the most efficient PHB producer and phosphate solubilizer, in vitro detection of P132 showed the presence of genes for phenazine, pyrrolnitrin, pyoluteorin and 2,4 diacetylphloroglucinol along with polyhydroxyalkanoate.

Keywords Diversity · Phosphate solubilization · Polyhydroxybutyrate · Pseudomonas

Introduction

Pseudomonas putida are ubiquitous bacteria frequently present in water, in soils, and especially in the plant rhizosphere (Timmis 2002; Dos Santos et al. 2004). These aerobic, gram-negative Pseudomonads possess many traits

T. Agrawal and A.S. Kotasthane share equal contribution for the manuscript.

T. Agrawal (✉) · A. S. Kotasthane · R. Kushwah
Department of Plant Molecular Biology & Biotechnology, Indira Gandhi Krishi Vishwavidyalaya, Krishak Nagar, Raipur 492006, Chattisgarh, India
e-mail: toshy@rediffmail.com

A. S. Kotasthane
Department of Plant Pathology, Indira Gandhi Krishi Vishwavidyalaya, Krishak Nagar, Raipur 492006, Chattisgarh, India

that make them well suited as biocontrol and growth promoting agents (Weller 1988; Lemanceau 1992; Weller et al. 2002; Fravel 2005). A large number of secondary metabolites (Leisinger and Margraff 1979), growth hormones (Brown 1972), antibiotics (Fravel 1988, 2005; Weller et al. 2002) and chelating compounds such as siderophores (Leong 1986) are known to be released by these fluorescent Pseudomonads. Some of them may also be involved in the biodegradation of natural or man-made toxic chemical compounds (Holloway 1992; Ramos et al. 2009). *P. putida* show diverse spectrum of metabolic versatility and niche-specific adaptations (Rojo 2010; Wu et al. 2011).

There may be direct or indirect mechanisms of these rhizobacteria as plant growth promoters and biological control agents. Phosphate solubilization is one of the direct mechanisms (Rodríguez and Fraga 1999; Mayak et al. 2004; Shahzad et al. 2010) and production of antibiotics such as 2,4-diacetyl phloroglucinol (DAPG), phenazine, pyoluteorin and pyrrolnitrin against pathogenic fungi and bacteria is among indirect mechanisms of PGPR (Ramamoorthy et al. 2001; McSpadden Gardener B 2008). Apart from primary and secondary metabolite production, certain fluorescent Pseudomonads (especially *P. putida*) are suitable as whole-cell biocatalyzers for the production of several value-added industrial compounds such as biodegradable and biocompatible polyesters called polyhydroxyalkanoates (PHA) or polyhydroxybutyrates (PHB). It accumulates as discrete granules and is used as storage material for carbon and for reducing equivalents by *P. putida*. This property has been widely exploited for their targeted biosynthesis in this organism (Hoffmann and Rehm 2004). Different strains of *P. putida* such as *P. putida* KT2440, *P. putida* GPo1, *P. putida* S12, etc. have been investigated for its capacity to accumulate PHAs and

PHBs from different carbon sources (Durner et al. 2001; Hartmann et al. 2004; Kim et al. 2007; Meijnen et al. 2008). The pha gene cluster is responsible for the accumulation of PHAs and PHBs in *P. putida* (Chung et al. 2009; Vo et al. 2008; Wang and Nomura 2010).

PCR ribotyping has been used to characterize *Pseudomonas* spp. which concentrate on the analysis of the segments of the ribosomal genes and discriminate between isolates of same species by use of their chromosomal differences. The three main sets of repetitive elements used for typing purposes are the repetitive extragenic palindromic (REP) sequence, the enterobacterial repetitive intergenic consensus sequence (ERIC) and the BOX elements. Additionally, the availability of antibiotic gene sequences has enabled design of primers based on conserved regions for polymerase chain reaction (PCR) detection of antibiotic-producing bacteria (Raffel et al. 1996; Raaijmakers et al. 1997; Bangera and Thomashow 1999; McSpadden Gardener et al. 2001; de Souza and Raaijmakers 2003).

Multidisciplinary application of fluorescent Pseudomonads makes it significant and essential to study their phenotypic diversity along with genotypic variability. This will be helpful in designing strategies to use the indigenous isolates as bio-inoculants for biocontrol as well as plant growth promotion. This offers a viable substitute for the use of chemical inputs in agriculture. However, an effective biological control strain isolated from one region may not perform effectively in other soils or plants. Therefore, in an attempt to study the diversity of indigenous fluorescent Pseudomonads in Chhattisgarh, a large number of fluorescent Pseudomonads were isolated from the forest and agricultural soils, characterized and maintained in the Department of Plant Molecular Biology and Biotechnology, Indira Gandhi Krishi Vishwavidyalaya, Raipur. The purpose of present investigation was to assess the representative *P. putida* isolates for their polyhydroxybutyrate production and phosphate solubilizing ability using an array of in vitro assays; metabolite utilization tests and genotypic profiling were performed with species-specific ERIC primers and some antibiotic gene-specific primers. Species-specific primer was used for taxonomic affiliation of *P. putida*. The overall aim of the present investigation was thus to exploit and have a better understanding of beneficial activities of *P. putida* isolates.

Materials and methods

Bacterial isolates

The experimental material consisted of purified twenty-four isolates of *P. putida* isolated from soil samples of

Table 1 *Pseudomonas putida* isolates used in the present study

S. no.	Isolates	Origin/location
1	P2	Cowpea soil, IGKV Horticulture garden, Raipur
2	P3	Fallow land soil, IGKV Horticulture garden, Raipur
3	P7	Fenugreek soil, IGKV Horticulture, Raipur
4	P23	Termitorium soil, VIP Road, Raipur
5	P29	Termitorium soil, VIP Road, Raipur
6	P43	Abhanpur road
7	P45	Abhanpur road
8	P56	Bhatagaon
9	P59	Bhatagaon (degraded paddy straw)
10	P74	Chhati
11	P80	Darba
12	P123	Kanker forest-2
13	P130	Kanker forest-3
14	P132	Kanker forest-3
15	P144	Rice field, Kodebor
16	P150	Kurud
17	P163	Nursery, Raipur
18	P166	Purur
19	P174	Rice field, Rajiv Gandhi Marg, Raipur
20	P184	Arhar-rice field soil, Satpara
21	P187	Arhar-rice field soil, Satpara
22	P191	Forest soil, Raipur
23	P192	Forest soil, Raipur
24	P207	Bamboo soil, VIP road, Raipur

different geographical locations of Chhattisgarh as listed in Table 1. Both rhizospheric and non-rhizospheric soil samples were collected and used for isolation of fluorescent pseudomonads by adopting serial dilution method in King's B medium. After incubation at 28 °C for 2 days, fluorescent pseudomonad colonies from plates were identified under UV light (366 nm). Isolates were characterized on the basis of biochemical tests as per the procedures outlined in Bergey's Manual of Systematic Bacteriology (Sneath et al. 1986). Purified single colonies were further streaked onto KB agar plates to obtain pure cultures. The isolates were maintained in the culture collections of the Department of Plant Molecular Biology and Biotechnology, Indira Gandhi Krishi Vishwavidyalaya, Raipur, Chhattisgarh, India. Bacterial cultures were maintained at −80 °C on King's B broth (Himedia) containing 50 % (w/v) glycerol and revived on King's B slants as per requirement.

Phenotypic characterization of *P. putida* isolates

The phenotypic characterization of *P. putida* isolates were done on the basis of fluorescence on King's B (KB) medium, gelatin liquefaction, casein hydrolysis, lipolytic activity, nitrate reduction, growth at 4 and 42 °C, oxidase test, phenylalanine test and egg yolk medium test (Stanier et al. 1966; Holt et al. 1994). A rapid antibiotic sensitivity test was used to distinguish different species of fluorescent *Pseudomonas*. Antibiotic sensitivity studies were performed by the streak plate method of Bauer et al. (1966). Kanamycin and carbenicillin sensitivity was determined by incorporating 1 mg/ml of Kanamycin and 0.1 mg/ml of Carbenicillin, respectively, in King's B medium. *Pseudomonas* spp. showing positive growth on either of the antibiotic supplemented medium was resistant.

Hicarbohydrate™ kit was used to test carbon utilization profiles as described by the manufacturer (Himedia Laboratories, Mumbai, India). Cells were grown in King's medium B broth to reach density of 0.5 O.D. at 600 nm. An aliquot of 50 µl of this suspension was inoculated to each well of Hicarbohydrate™ kit, incubated at 30 °C for 24 h and the results were registered according to the instructions of the manufacturer. The experiment was done with three replicates.

Screening for polyhydroxybutyrate (PHB) production and its quantitative estimation

Pseudomonas putida isolates were screened for PHB accumulation qualitatively by following the viable colony method using Sudan Black B dye (Juan et al. 1998). Sterilized Nutrient agar (Himedia) supplemented with 1 % glucose was spot inoculated with the isolates and incubated at 30 °C for 24 h. Ethanolic solution (0.02 %) of Sudan Black B was spread over the colony and the plates were kept undisturbed for 30 min. Later, they were washed with ethanol (96 %) to remove the excess stain from the colony. The dark blue colored colony was taken as positive for PHB production.

The Sudan Black B positive isolates were subjected to quantification of PHB production as per the method of John and Ralph (1961). The bacterial cells containing the polymer were pelleted at 10,000 rpm for 10 min. and the pellet was washed with acetone and ethanol to remove the unwanted materials. The pellet was resuspended in equal volume of 4 % sodium hypochlorite and incubated at room temperature for 30 min. The whole mixture was again centrifuged and the supernatant discarded. The cell pellet containing PHB was again washed with acetone and ethanol. Finally, the polymer granules were dissolved in hot chloroform. The chloroform was filtered and to the filtrate, concentrated 10 ml hot H_2SO_4 was added. The addition of sulfuric acid converts the polymer into crotonic acid which is brown colored. The solution was cooled and the absorbance read at 235 nm against a sulfuric acid blank. By referring to the standard curve prepared using Poly[(*R*)-3-hydroxybutyric acid] (Sigma Aldrich, USA) by

following the method of Law and Slepecky (1969), the quantity of PHB produced by different bacterial isolates was determined.

Screening of phosphate solubilisation ability and its quantitative estimation

Qualitative screening of phosphate solubilising *P. putida* was performed on Pikovskaya agar medium (Himedia) containing tricalcium phosphate as a phosphate source and bromocresol purple (0.1 g/l) as a pH indicator for acidification (Vazquez et al. 2000). After incubation of fresh cultures of *P. putida* at 28 ± 2 °C for 48 h, phosphate solubilising isolates turned the media color from purple to yellow in the zones of acidification.

Quantitative estimation of phosphate solubilisation in Pikovskaya broth (Himedia) was performed according to the procedure of Murphy and Riley (1962). Fresh cultures of *P. putida* isolates were inoculated to 50 ml of Pikovskaya's broth and incubated at 28 ± 2 °C and 100 rpm. The amount of inorganic phosphate (Pi) released in the broth was estimated after 7 days of incubation in comparison with un-inoculated control. The broth culture was centrifuged at 10,000 rpm for 10 min to separate the supernatant from the bacterial growth and insoluble phosphate. To the 0.5 ml of the culture supernatant 5 ml of chloromolybdic acid was added and mixed thoroughly. Volume was made up to 10 ml with distilled water and 125 µl of chlorostannous acid was added to it. Immediately, the final volume was made up to 25 ml with distilled water and mixed thoroughly. After 15 min, the blue color developed was read in a spectrophotometer at 610 nm using a reagent blank. Corresponding amount of soluble phosphorous was calculated from standard curve of potassium dihydrogen phosphate (KH_2PO_4). Phosphate solubilizing activity was expressed in terms of tricalcium phosphate solubilization which in turn was measured by µg/ml of available orthophosphate as calibrated from the standard curve of KH_2PO_4.

16S rRNA gene amplification

Total genomic DNA from 24 *P. putida* isolates was extracted by the CTAB procedure (Ausubel et al. 1991) and used for amplification using various primers (Table 2). PCR primers designed from genes HI660468 (Pa49), HM067869 (Pa16S), HQ317190 (Pp16S), EF159157 (Pf16S) and AF869903 (Pf23S) were used to specifically distinguish species of *Pseudomonas* isolates by amplification of the nuclear rRNA gene cluster. These forward and reverse primers were designed using Batch primer3 software from following gene sequences: HI660468: Sequence 49 from Patent WO2010127969 of *Pseudomonas*

Table 2 Details of PCR primers used in the present study

S. no.	Primers	Gene/reference	Sequence(5'–3')	Length	Expected product size (bp)
1	Pa49-F	HI660468	TCTTCCGCCTGTTCAATTACCCGA	24	448
	Pa49-R		AATACCTTGGCCACCTTGTTCAGC	24	
2	Pa16S-F	HM067869	AGAGGGTGGTGGAATTTCCTGTGT	24	586
	Pa16s-R		TACCGACCATTGTAGCACGTGTGT	24	
3	Pp16S-F	HQ317190	ACCGACAGAATAAGCACCGGCTAA	24	364
	Pp16S-R		AAGAGTTCAAGACTCCCAACGGCT	24	
4	Pf16S-F	EF159157	TCCCTATCGATTGATCCGGCTTCT	24	250–260
	Pf16S-R		TTTAGATGGTGGAGCCAAGGAGGA	24	
5	Pf23S-F	AF869903	ACGCTTTCTTTAAAGGGTGGCTGC	24	400–420
	Pf23S-R		TCTATCCATGGGCAGGTTGAAGGT	24	
6	ERIC-F	de Bruijn (1992)	AAGTAAGTGACTGGGGTGAGCG	22	
	ERIC-R		TATAAGCTCCTGGGGATTCAC	21	
7	PhaJ1-F	Polyhydroxyalkanoate	AAGGCCGAGTACAAGAAGTCCGTT	24	240–250
	PhaJ1-R		TCACCGGTTTCTGGAAGCTCATCT	24	
8	PHZ1	Phenazine	GGCGACATGGTCAACGG	17	1,400
	PHZ2	Delaney et al. (2001)	CGGCTGGCGGCGTATAT	17	
9	PCA2a	Phenazine	TTGCCAAGCCTCGCTCCAAC	20	1,400
	PCA3B	Raaijmakers et al. (1997)	CCGCGTTGTTCCTCGTTCAT	20	
10	B2BF	2,4 Diacetyl phloroglucinol	ACCCACCGCAGCATCGTTTATGAGC	25	~470 or 629
	BPR4	McSpadden Gardener et al. (2001)	CCGCCGGTATGGAAGATGAAAAAGTC	26	
11	PrnAF	Pyrrolnitrin	GTGTTCTTCGACTTCCTCGG	20	1,050
	PrnAR	de Souza and Raaijmakers (2003)	TGCCGGTTCGCGAGCCAGA	19	
12	phlA-1f	2,4 Diacetyl phloroglucinol	TCAGATCGAAGCCCTGTACC	20	418
	phlA-1r	Rezzonico et al. (2003)	GATGCTGTTCTTGTCCGAGC	20	
13	plt1	Pyoluteorin	ACTAAACACCCAGTCGAAGG	20	~440
	plt2	Mavrodi et al. (2001)	AGGTAATCCATGCCCAGC	18	
14	PrnCf	Pyrrolnitrin	CCACAAGCCCGGCCAGGAGC	20	~720
	PrnCr	Mavrodi et al. (2001)	GAGAAGAGCGGGTCGATGAAGCC	23	
15	Phl2a	2,4 Diacetyl phloroglucinol	GAGGACGTCGAAGACCACCA	20	~745
	Phl2b	Raaijmakers et al. (1997)	ACCGCAGCATCGTGTATGAG	20	
16	PltBf	Pyoluteorin	CGGAGCATGGACCCCCAGC	19	~700 or 900
	PltBr	Mavrodi et al. (2001)	GTGCCCGATATTGGTCTTGACCGAG	25	

aeruginosa, HM067869 of *Pseudomonas aeruginosa* strain GIM 32 16S ribosomal RNA gene, partial sequence, HQ317190 of *P. putida* strain DYJL49 16S ribosomal RNA gene, partial sequence, EF159157 of *Pseudomonas fluorescens* strain TNAUA2 16S ribosomal RNA gene and 16S–23S ribosomal RNA intergenic spacer, partial sequence and AF369903 of *Pseudomonas fluorescens* 23S ribosomal RNA gene, partial sequence procured from NCBI database. PCR was carried in 20 µl reaction mixture containing 1× assay buffer (10 mM Tris–HCl at pH 9.0, 50 mM KCl, 2.5 mM MgCl$_2$), 0.1 mM each dNTP mix, 1 µM both forward and reverse primers, 60–90 ng of template DNA and 0.5 U Taq DNA polymerase (Axygen) in a programmable thermo cycler (M/s Biorad Laboratories India Pvt. Ltd) according to the following thermo-cycling

conditions: 95 °C for 5 min, 35 cycles of 1 min at 95 °C, 60 °C for 1 min, 72 °C for 1 min and final elongation step at 72 °C for 7 min.

ERIC-PCR-based genotypic analysis

ERIC primer sequences were used in PCR to detect differences in the number and distribution of these bacterial repetitive sequences in the bacterial genome. ERIC-PCR was carried out using the primer sequences ERIC-F (5'AAGTAAGTGACTGGGGTGAGCG3') and ERIC-R (5'TATAAGCTCCTGGGGATTCAC3') as described by de Bruijn (1992). ERIC-PCR was carried in 20 µl reaction mixture containing 1× assay buffer (10 mM Tris–HCl at pH 9.0, 50 mM KCl, 2.5 mM MgCl$_2$), 0.1 mM each dNTP mix, 1 µM both forward and reverse primers, 60–90 ng of template DNA and 1 U Taq DNA polymerase (Axygen) in a programmable thermo cycler (M/s Biorad Laboratories India Pvt. Ltd) according to the following thermo-cycling conditions: 94 °C for 3 min, 45 cycles of 45 s at 94 °C, 53 °C for 1 min, 72 °C for 1 min and final elongation step at 72 °C for 8 min.

In vitro detection of antibiotic-producing *P. putida* isolates using gene-specific primers

Primers (Imperial Life Sciences) for the different PCR-based screening of genes that encode for antibiotics are detailed in Table 2. Preparation of bacterial templates for detecting antibiotic-producing genes was carried out as described by Wang et al. (2001) and Rezzonico et al. (2003). PCR amplification of primers PhaJ1F-R, PHZ1-2, PCA2a-3B, B2BF-BPR4, PrnAF-R, phlA-1f-r, plt1-2, PrnCf-r, Phl2a-2b and PltBf-r was carried out in 20 µl reaction mixtures containing 3 µl of lysed bacterial suspension, 1X assay buffer (10 mM Tris–HCl at pH 9.0, 50 mM KCl, 2.5 mM MgCl$_2$), 0.4 mM dNTPs, 1 µM of each primer and 1 U of Taq DNA polymerase (Axygen). Amplification was performed in a programmable thermo cycler (M/s Biorad Laboratories India Pvt. Ltd). The cycling program for PCA2a-3B, PrnAF-R, plt1-2, PrnCf-r, PHZ1-2, Phl2a-2b, PhaJ1F-R, and PltBf-r included an initial denaturation at 95 °C for 3 min followed by 35 cycles of 95 °C for 1 min, 62 °C (for PCA2a-3B, PrnAF-R, plt1-2, PrnCf-r, PHZ1-2)/52 °C (for Plt1-2)/60 °C (for Phl2a-2b, PhaJ1F-R)/65 °C (for B2BF-BPR4) for 1 min, 72 °C for 1 min, and then a final extension at 72 °C for 5 min. However, the cycling program for phlA-1f-r included an initial denaturation at 94 °C for 5 min followed by 35 cycles of 94 °C for 30 s, 62 °C for 30 s, 72 °C for 45 s, and then a final extension at 72 °C for 5 min.

The amplification products were electrophoresed in a 1 % (w/v) agarose gel with 1× TBE buffer at 80 V at room temperature, stained with ethidium bromide and photographed under UV light by Biorad Gel-Documentation as well as on 5 % native polyacrylamide gel (visualized by silver staining).

Statistical analysis

All the experiments were conducted in three completely randomized replicates. On the basis of data derived from the carbon source utilization profiles, a matrix with binary code composing positive (1) and negative (0) values was made. SIMQUAL program was used to compute the symmetric matrix in the form of average taxonomic distances. Sequential, agglomerative, hierarchical and nested (SAHN) clustering was used for the cluster analyses.

Phenogram was constructed from the similarity matrix by the un-weighted pair group with mathematical averages (UPGMA) using NTSYS-pc2.02a (Exeter software, New York, USA) numerical taxonomy and multivariate analysis system. Similar method was followed to construct dendrogram using binary data of ERIC primer-based PCR amplification of *P. putida* isolates.

Replicated data of quantitative estimation of PHB production and P solubilization of all the 24 *P. putida* isolates were subjected to statistical analysis using WASP (Web Agri Stat Package) software (http://icargoa.res.in/wasp/index.php). Critical difference at 0.05 level of significance was calculated for the observed values along with average and standard deviation. Duncan's test controls the Type I comparison wise error rate and as per Duncan's grouping mean values with the same letter are not significantly different.

Results and discussion

Phenotypic characterization of *P. putida* isolates

Isolates were characterized on the basis of biochemical and antibiotic sensitivity tests. All the isolates of *P. putida* were positive for cytochrome oxidase. Isolates showed variability for traits such as gelatin liquefaction, casein hydrolysis, lipolytic activity, nitrate reduction and antibiotic sensitivity test. Of the 24 *P. putida*, 9 isolates (37.5 %) showed proteolytic activity (casein hydrolysis) by inducing clear zones around the cells on skim milk agar medium, 7 isolates (29.17 %) showed lipolytic activity, 14 isolates (58.33 %) were negative for nitrate test and 10 isolates (41.67) gave positive result for nitrate test (of which 6 isolates P23, P43, P59, P80, P144 and P174 were positive before addition of zinc and 4 isolates P187, P191, P192 and P207 showed positive response after addition of zinc). Only three isolates P56, P130 and P191 were positive for phenylalanine test as indicated by appearance of green color after addition of few drops of 10 % aq. ferric chloride to the cultures grown in phenyl alanine amended medium. Lecithinase production was observed as opaque precipitate around colonies of four *P. putida* isolates viz. P59, P80, P123 and P166 resulting in lecithin positive result in egg yolk medium. Blazevic et al. (1973) reported some diagnostic tests for differentiation of *P. fluorescens* and *P. putida* isolates. *P. aeruginosa* is the only fluorescent pseudomonads that can grow at 42 °C whereas *P. fluorescens* grow at 4 °C. In the present investigation none of the isolates showed growth at 4 °C and 42 °C. A shortened gelatin test can differentiate *P. fluorescens* (positive) from *P. putida* (negative). Present result correlated with this fact except the isolates P2, P3, P45, P144, P192 and P207

Table 3 Differential phenotypic characteristics revealed by twenty-four *P. putida* isolates

Tests	P2	P3	P7	P23	P29	P43	P45	P56	P59	P74	P80	P123
1	−	−	−	−	−	−	−	−	−	−	−	−
2	+	+	+	+	+	+	+	−	+	+	+	+
3	−	−	−	−	−	−	−	−	−	−	−	−
4	−	−	−	−	−	−	−	−	−	−	+	−
5	+	+	+	+	+	+	+	+	+	+	+	+
6	+	+	+	+	+	+	+	−	+	+	+	+
7	−	−	−	−	−	−	−	−	−	−	−	−
8	−	−	−	+	−	−	−	−	−	−	+	−
9	+	+	+	+	+	+	+	−	+	+	+	+
10	−	−	−	−	−	−	−	−	±	−	±	−
11	±	−	−	+	−	−	−	−	+	+	+	+
12	+	+	+	+	+	+	+	−	+	+	+	+
13	−	−	−	−	−	−	−	−	−	−	−	−
14	−	−	−	−	−	−	−	−	−	−	−	−
15	−	−	−	−	−	−	−	−	−	−	−	−
16	−	−	±	−	−	−	−	−	−	−	−	−
17	−	−	−	−	−	−	−	−	−	−	−	−
18	−	−	−	−	−	−	−	−	−	−	−	−
19	−	−	−	−	−	−	−	−	−	−	−	−
20	−	−	−	−	−	−	−	−	−	−	−	−
21	−	−	−	−	−	−	−	−	−	−	−	−
22	−	−	−	−	−	−	−	−	−	−	−	−
23	−	−	−	−	−	−	−	−	−	−	−	−
24	−	−	−	−	−	−	−	−	−	−	−	−
25	−	−	−	−	−	−	−	−	−	−	−	−
26	−	−	−	−	−	−	−	−	−	−	−	−
27	−	−	−	−	−	−	−	−	−	−	−	−
28	−	−	−	−	−	−	−	−	−	−	−	−
29	+	+	±	+	+	+	+	−	+	+	+	+
30	−	−	−	−	−	−	−	−	−	−	−	−
31	+	+		+	+	+	+	+	+	+	+	+
32	−	−	−	−	−	−	−	−	−	−	−	−
33	+	+	+	+	+	+	+	+	+	+	+	+
34	+	+	+	−	+	+	+	+	−	−	+	+
35	−	−	−	−	−	−	−	−	−	−	−	−
36	−	+	−	+	+	−	−	−	−	−	−	−
37	+	+	−	−	−	−	+	−	−	−	−	−
38	−	−	−	+	−	+	−	−	−	−	−	+
39	+	+	+	+	+	+	+	+	+	+	+	+
40	−	−	−	−	−	−	−	+	−	−	−	−
41	−	−	−	−	−	−	−	−	L+	−	L+	L+
42	−	−	−	−	−	−	−	−	−	−	−	−
43	−	−	−	+	−	+	−	−	+	−	+	−
44	−	−	−	NA	−	NA	−	−	−	−	NA	−
45	R	R	R	R	R	R	R	R	R	R	R	R
46	S	S	S	S	S	S	S	R	S	S	S	S

Table 3 continued

Tests	P130	P132	P144	P150	P163	P166	P174	P184	P187	P191	P192	P207
1	−	−	−	−	−	−	−	−	−	−	−	−
2	+	+	+	+	+	+	+	+	+	+	+	+
3	−	−	−	−	−	−	−	−	−	−	−	−
4	−	−	−	−	−	−	−	−	−	+	−	−
5	+	+	+	+	+	+	+	+	+	+	+	+
6	+	+	+	+	+	+	+	+	+	+	+	+
7	−	−	−	−	−	−	−	−	−	−	−	−
8	−	−	−	−	−	−	−	−	−	−	−	−
9	+	+	+	+	+	+	+	+	+	+	+	+
10	−	−	−	−	−	−	−	−	−	−	−	−
11	+	+	+	+	+	+	+	+	+	+	+	+
12	+	+	+	+	+	+	+	+	+	+	+	+
13	−	−	−	−	−	−	−	−	−	−	−	−
14	−	−	−	−	−	−	−	−	−	−	−	−
15	−	−	−	−	−	−	−	−	−	−	−	−
16	−	−	−	−	−	−	−	−	−	−	−	−
17	−	−	−	−	−	−	−	−	−	−	−	−
18	−	−	−	−	−	−	−	−	−	−	−	−
19	−	−	−	−	−	−	−	−	−	−	−	−
20	−	−	−	−	−	−	−	−	−	±	−	−
21	−	−	−	−	−	−	−	−	−	±	−	−
22	−	−	−	−	−	−	−	−	−	+	−	−
23	−	−	−	−	−	−	−	−	−	−	−	−
24	−	−	−	−	−	−	−	−	−	−	−	−
25	−	−	−	−	−	−	−	−	−	−	−	−
26	−	−	−	−	−	−	−	−	−	−	−	−
27	−	−	−	−	−	−	−	−	−	−	−	−
28	−	−	−	−	−	−	−	−	−	−	−	−
29	+	+	+	±	+	+	+	+	+	+	+	+
30	−	−	−	−	−	−	−	−	−	−	−	−
31	+	+	+	+	+	+	+	+	+	+	+	+
32	−	−	−	−	−	−	−	−	−	−	−	−
33	+	+	+	+	+	+	+	+	+	+	+	+
34	+	+	+	+	+	+	+	−	+	+	+	+
35	−	−	−	−	−	−	−	−	−	−	−	−
36	−	−	+	−	−	+	−	−	+	+	+	+
37	−	−	+	−	−	−	−	−	+	−	+	+
38	−	−	+	−	−	−	−	−	+	+	+	+
39	+	+	+	+	+	+	+	+	+	+	+	+
40	+	−	−	−	−	−	−	−	−	+	−	−
41	−	−	−	−	−	L+	−	−	−	−	−	−
42	−	−	−	−	−	−	−	−	−	−	−	−
43	−	−	+	−	−	−	+	−	NA	NA	NA	NA
44	−	−		−	−	−	NA	−	+	+	+	+
45	R	R	R	R	R	R	R	R	R	R	R	R
46	S	S	S	S	S	S	R	S	S	S	S	S

Different tests: 1, lactose; 2, xylose; 3, maltose; 4, fructose; 5, dextrose; 6, galactose; 7, raffinose; 8, trehalose; 9, melibiose; 10, sucrose; 11, L-arabinose; 12, mannose; 13, inulin; 14, sodium gluconate; 15, glycerol; 16, salicin; 17, dulcitol; 18, inositol; 19 sorbitol; 20, mannitol; 21, adonitol; 22, arabitol; 23, erythritol; 24, α-methyl-D-mannoside; 25, rhamnose; 26, cellobiose; 27, melezitose; 28, α-methyl-D-mannoside; 29, xylitol; 30, ONPG; 31, esculin hydrolysis; 32, D-arabinose; 33, citrate utilization; 34, malonate utilization; 35, sorbose; 36, casein hydrolysis; 37, gelatin hydrolysis; 38, lipase test; 39, oxidase test; 40, phenyl alanine test; 41, egg yolk reaction; 42, growth at 4 °C/42 °C; 43, nitrate test (before adding Zn); 44, nitrate test (after adding Zn); 45, carbenicillin sensitivity; 46, kanamycin sensitivity. +, Positive reaction; −, negative reaction; ±, partially positive; L+, lecithinase positive; R, resistant; S, susceptible; NA, not applicable

which liquefies gelatin. *P. fluorescens* and *P. putida* are very sensitive to low levels of kanamycin and resistant to carbenicillin, a pattern just the opposite of that obtained with *P. aeruginosa*. All the isolates were resistant to antibiotic carbenicillin and sensitive to kanamycin. However, isolates P56 and P174 were tolerant to both the antibiotics (Table 3). Several strains within the family *Pseudomonadaceae* such as *P. putida* S12 show significant intrinsic resistance to multiple antibiotics (Kieboom and de Bont 2001). Blazevic et al. (1973) further suggested that the shortened test for nitrate reduction, then, together with the marked sensitivity to kanamycin and resistance to carbenicillin would provide a rapid means of accurately identifying *P. fluorescens* and *P. putida* and separating them from *P. aeruginosa*. However, rare nitrate-negative *P. aeruginosa* or rare nitrate-positive *P. fluorescens* or *P. putida* should not be misidentified using both of these characteristics.

All the *P. putida* isolates utilized xylose, dextrose, galactose, melibiose, mannose, xylitol, esculin and citrate but exhibited varying degree of utilization towards

other carbon sources such as fructose, trehalose, L-arabinose, arabitol and malonate. These isolates did not utilize lactose, maltose, raffinose, sucrose, inulin, sodium gluconate, glycerol, salicin, dulcitol, inositol, sorbitol, mannitol, adonitol, arabitol, erythritol, α-methyl-D-gluconate, rhamnose, cellobiose, melezitose, α-methyl-D-mannoside, ONPG, D-arabinose and sorbose. Isolate P56 did not utilize xylose, galactose, melibiose, mannose and xylitol which were utilized by all other 23 *P. putida* isolates (Table 3). Numerical analysis of phenotypic characteristics revealed polymorphism among *P. putida* isolates. All 24 *P. putida* isolates were grouped into 3 major phenons at 0.76 similarity coefficient level (Fig. 1). The similarity coefficient range among 24 *P. putida* isolates was 0.39–1.00. Phenons 1, 2 and 3 consist a total of 17, 4 and 2 isolates, respectively. Isolate P56 did not fall into any of the phenogram and revealed all together distinct identity. Differential utilization of carbon sources by isolates of *P. putida* as identified by Hi-carbohydrate[TM] kit test may play an important role in adapting to a variety of

Fig. 1 Phenogram of 24 *P. putida* isolates based on their carbon source utilization profiles

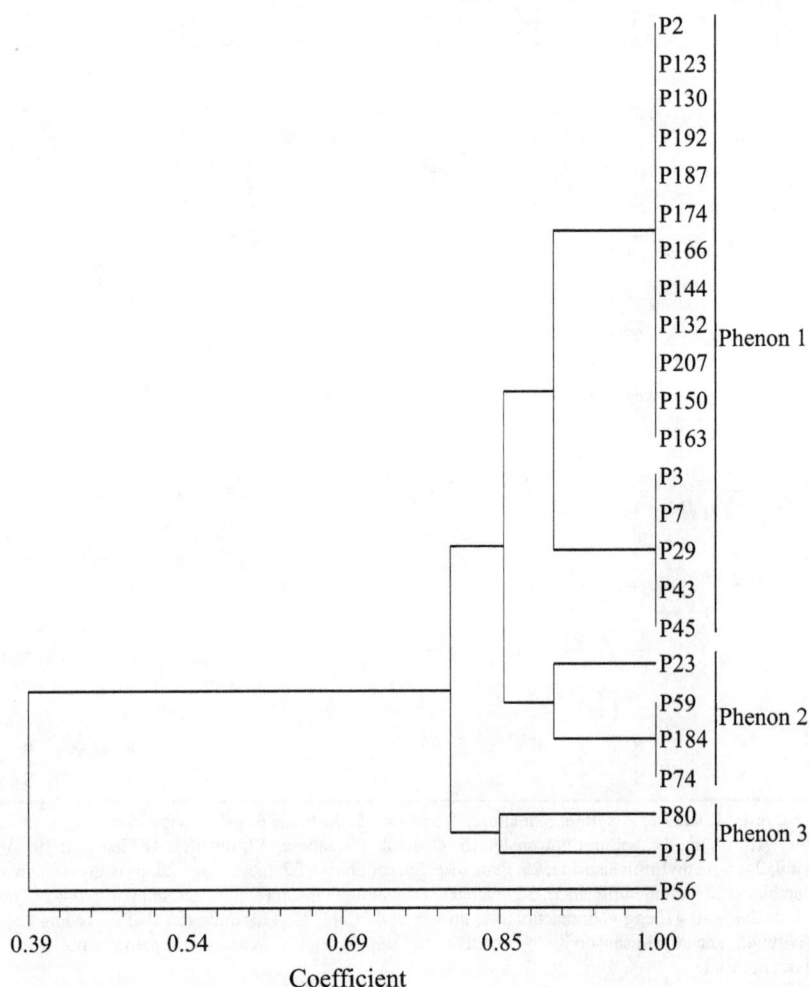

crop plants and soil types. Xylose being second to glucose in natural abundance is a promising candidate substrate for bacterial growth (Beall et al. 1991). Carbohydrates serve as primary substrate for the synthesis of many important metabolites and commercial products by microorganisms. Meur et al. (2012) reported that sequential feeding of relatively cheap carbohydrates such as xylose is a practical way to achieve more cost-effective medium-chain-length (mcl) PHA production. Reduction in cost can be achieved using two kinds of carbon sources, one for biomass production and the other for synthesis of PHA. Metabolite utilization diversity is also important because changes in their composition may affect the patterns and activities of rhizobacterial populations which are dependent upon rhizospheric nutrients for growth. Broad spectrum carbon source utilization among *P. putida* isolates in the present study may help in developing and designing stimulators for specific application.

Screening of polyhydroxybutyrate (PHB) producers and its quantification

Biodegradable and biocompatible polyesters such as polyhydroxyalkanoates (PHA) have potential pharmaceutical values (Takahashi et al. 1994). In an alkaline environment *P. putida* have been reported to produce medium-chain-length (R)-3-hydroxyalkanoates (Wang et al. 2007). In the present investigation all the 24 *P. putida* isolates gave positive result for PHB accumulation in Sudan Black B qualitatively screening test in glucose-supplemented nutrient agar medium. Madison and Huisman (1999) have also reported that these biopolymers are accumulated as inclusions (PHA granules) in the bacterial cytoplasm in response to inorganic nutrient limitations, generally, when the microbes are cultured in the presence of an excess carbon source. However, there was significant difference in quantitative analysis of all the isolates which ranged from 6.17 to 14.35 mg/ml. Variation was significant at both 0.01 and 0.05 levels. The lowest observed value was for isolate P130 and the highest was for isolate P2. Isolates P43, P56 and P187 also produced significantly higher amounts of PHB as compared to other isolates (Table 4). Expenditures for large-scale production of PHA are almost evenly divided among carbon source, fermentation process and separation process (Sun et al. 2007; Elbahloul and Steinbüchel 2009). Therefore screening for carbohydrate utilization of *P. putida* isolate may help in identifying candidate isolate which dwells upon cheaper carbon sources. Our work reports that all the 24 putida isolates utilized relatively cheaper carbohydrates such as xylose, dextrose, galactose, melibiose and mannose. PHB could be synthesized from a cheaper raw material xylose in

Table 4 PHB production and inorganic phosphate solubilization of *P. putida* isolates

S. no.	Isolates	PHB production (mg/ml)**	Phosphate solubilized (µg/ml)**
1	P2	$14.35^a \pm 0.45$	$57.38^j \pm 11.85$
2	P3	$10.09^{d,e,f,g} \pm 0.20$	$358.95^g \pm 25.38$
3	P7	$10.39^{c,d,e,f,g} \pm 1.19$	$651.95^{a,b} \pm 29.62$
4	P23	$9.89^{e,f,g} \pm 0.92$	$687.11^a \pm 19.95$
5	P29	$9.51^{g,h} \pm 0.67$	$566.12^{c,d} \pm 45.42$
6	P43	$13.48^b \pm 0.70$	$607.11^{b,c,d} \pm 5.81$
7	P45	$7.98^{h,i} \pm 0.82$	$260.56^h \pm 44.63$
8	P56	$12.27^b \pm 0.69$	$659.62^{a,b} \pm 54.61$
9	P59	$11.59^{b,c,d} \pm 0.73$	$478.11^f \pm 31.26$
10	P74	$11.42^{b,c,d,e,f} \pm 0.79$	$686.12^a \pm 21.38$
11	P80	$11.38^{b,c,d,e,f} \pm 1.02$	$607.11^{b,c,d} \pm 22.78$
12	P123	$6.96^{i,j} \pm 0.69$	$564.11^{c,d} \pm 49.65$
13	P130	$6.17^j \pm 1.15$	$626.45^{a,b} \pm 30.33$
14	P132	$11.22^{b,c,d,e,f} \pm 0.64$	$660.23^{a,b} \pm 46.99$
15	P144	$11.48^{b,c,d,e} \pm 0.72$	$594.67^{b,c,d} \pm 8.02$
16	P150	$10.20^{d,e,f,g} \pm 0.78$	$384.05^g \pm 14.21$
17	P163	$6.81^{i,j} \pm 0.76$	$582.68^{c,d} \pm 27.82$
18	P166	$8.78^{g,h} \pm 0.95$	$554.18^{d,e} \pm 35.60$
19	P174	$9.35^{g,h} \pm 0.88$	$494.89^{e,f} \pm 8.33$
20	P184	$9.73^{f,g} \pm 0.70$	$489.19^{e,f} \pm 14.40$
21	P187	$12.15^b \pm 1.09$	$653.98^{a,b} \pm 35.32$
22	P191	$10.02^{d,e,f,g} \pm 0.53$	$685.66^a \pm 23.55$
23	P192	$11.98^{b,c} \pm 1.14$	$548.23^{d,e} \pm 55.48$
24	P207	$8.94^{g,h} \pm 0.74$	$165.12^i \pm 46.84$
	Control		$30.50^j \pm 6.36$
	CV	8.039	6.417
	CD (0.05)	1.695	66.910

Values are average of 3 replications; values after $\pm$ represents standard deviation

CV coefficient of variance, *CD* critical difference

** Values are significant at 1 and 5 % levels; As per Duncan's grouping means with the same letter are not significantly different

Pseudomonas pseudoflava and *P. cepacia* up to 22 % (w/w) and 50 % (w/w), respectively (Bertrand et al. 1990; Young et al. 1994). Meur et al. (2012) also tested the growth of an engineered *P. putida* KT2440 strain in xylose to reduce the substrate cost for PHA production.

Since isolates of *P. putida* are versatile and robust in catabolizing a broad range of compounds and resist adverse environmental conditions their metabolic repertoires funneled resources can be channeled towards PHA and PHB synthesis (Meijnen et al. 2009; Ciesielski et al. 2010; Poblete-Castro et al. 2012). The twenty-four *P. putida* used in the present investigation vary to some extent in their phenotypic behavior creating a broad range of industrial application possibilities. Its fast growth, high biomass yield

Fig. 2 PCR amplification of Pseudomonads using designed 16sRNA-based primers. **a** *P. putida* isolates amplified with primers derived from gene HQ317190 (Pp16S) generating ~390 bp bands in all the 24 isolates. **b** Representative *P. aeruginosa* isolates amplified with primers derived from gene HQ317190 (Pp16S) generating ~600 bp bands in all the 24 isolates. **c** *P. putida* isolates amplified with primers derived from gene HM067869 (Pa16S) generating ~360 bp bands in all the 24 isolates. **d** Representative *P. aeruginosa* isolates amplified with primers derived from gene HM067869 (Pa16S) generating ~600 bp bands in all the 24 isolates. Representative *P. aeruginosa* isolates were used to show differential amplification of the two species *P. putida* and *P. aeruginosa*

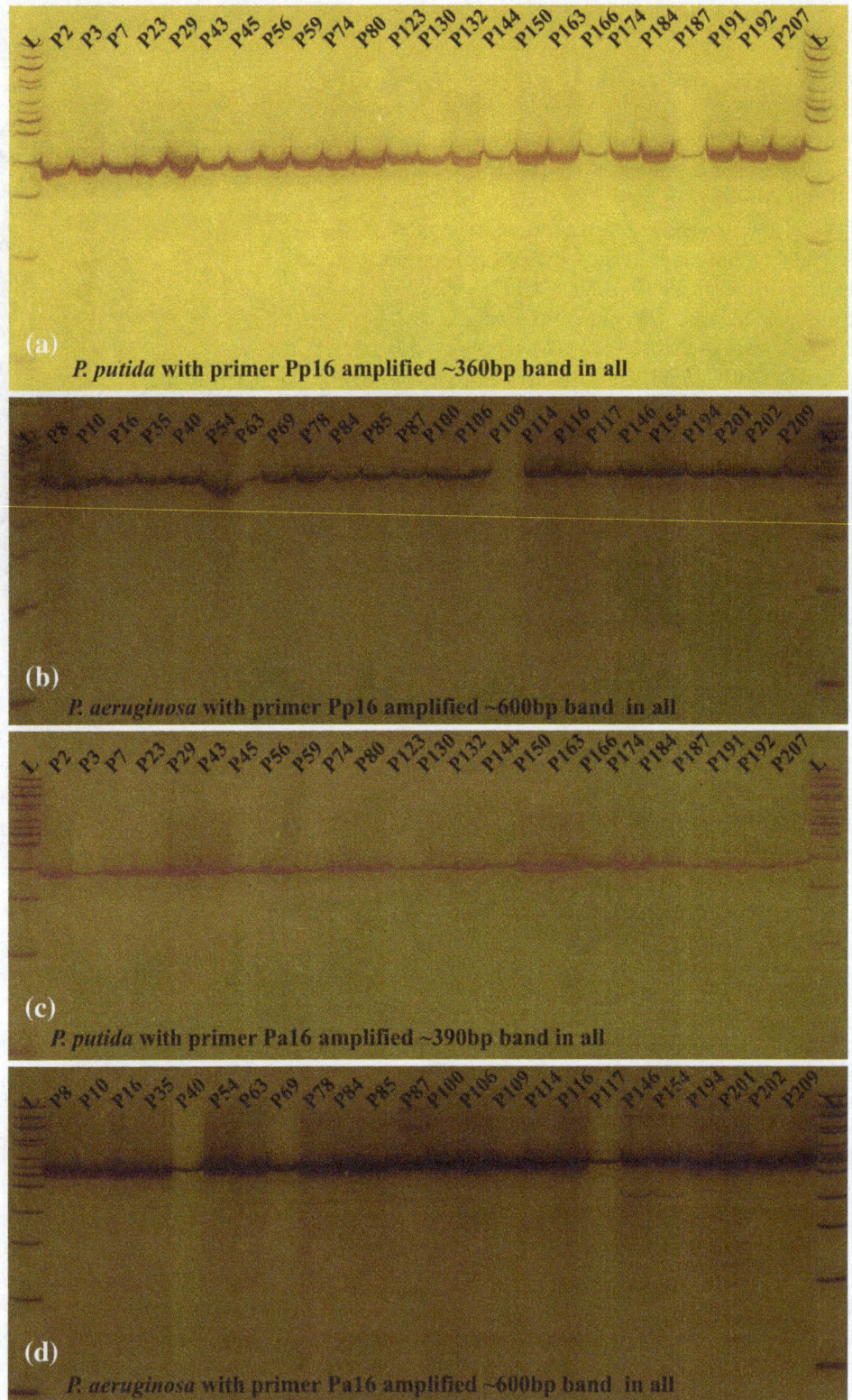

(a) *P. putida* with primer Pp16 amplified ~360bp band in all

(b) *P. aeruginosa* with primer Pp16 amplified ~600bp band in all

(c) *P. putida* with primer Pa16 amplified ~390bp band in all

(d) *P. aeruginosa* with primer Pa16 amplified ~600bp band in all

and low maintenance demands are among key features for successful industrial application. With the current lifestyle, need of an hour is the production of eco-friendly plastic materials such as polyhydroxyalkanoic acids (PHAs) by rational, efficient and sustainable use of natural resources. The outcomes of the present study can be exploited for selection of potential PHB producer *P. putida* isolate for commercial application.

Screening of phosphate solubilizers and its quantification

Phosphorus frequently is the least accessible macronutrient in many ecosystems and its low availability is often limiting to plant growth (Raghothama 1999). In vitro phosphate solubilization efficacy of *P. putida* isolates as performed on Pikovskaya agar by acidification showed positive results for all the 24 isolates tested. All the 24 isolates were capable of differentially utilizing tricalcium phosphate in both agar plate and broth assays. Quantitative estimation of soluble phosphate concentrations in Pikovskaya's broth was expressed as µg/ml and it varied significantly from 57.38 to 687.11 µg/ml. Variation was significant at both 0.01 and 0.05 levels. The lowest value was observed for isolate P2 and highest for isolate P23. Phosphate solubilization by isolates P23, P74 and P191 was significantly highest among all the other isolates. All the rhizospheric isolates of *P. putida* showed variable phosphate solubilizing potential with P7, P23, P56, P74, P132, P187 and P191 being the best P solubilizers among all other 24 isolates releasing more than 650 µg/ml inorganic phosphate (Table 4). These candidate isolates can be used as microbial inoculants to improve soil fertility by releasing bound phosphorus thereby increasing the crop yield potential. The production of organic acids and acid phosphatases plays a major role in the mineralization of organic phosphorous in soil. Stimulation of different crops by plant growth promoting *P. putida* isolate(s) with potential phosphate solubilization ability may help in exploiting large reserves of phosphorus present in most agricultural soils. Inoculation of plants by a target pseudomonas at a much higher concentration than that normally found in soil is necessary because the numbers of several phosphate solubilizing bacteria already present in soil are not high enough to compete with other bacteria commonly established in the rhizosphere. However, study of ecological roles of these characterized phosphate solubilizers in soil is necessary for sustainable agricultural practices and commercial applications. Several *Pseudomonas* species have been reported among the most efficient phosphate solubilizing bacteria and as important bio-inoculants due to their multiple biofertilizing activities of improving soil nutrient status, secretion of plant growth regulators and suppression of soil-borne pathogens (Rodríguez and Fraga 1999; Gulati et al. 2008; Vyas et al. 2009). Genes from potential phosphate solubilizer *P. putida* identified in the present investigation can be further exploited to study genetic regulation governing the mineral phosphate solubilization trait, which has an otherwise very less known information.

16S rRNA gene amplification and ERIC-PCR-based genotypic analysis

PCR amplification of 16S ribosomal RNA with primers designed from genes HI660468 (Pa49), HM067869 (Pa16S), HQ317190 (Pp16S), EF159157 (Pf16S) and AF869903 (Pf23S) resulted in specific distinguishing amplification products with only two primers from genes HM067869 (Pa16S) and HQ317190 (Pp16S). All 24 isolates resulted in positive reaction with primers designed from genes HM067869 (Pa16S) and HQ317190 (Pp16S). Primer from gene HQ317190 (Pp16S), amplified ∼360 bp band in all the *P. putida* isolates and ∼600 bp band in all the *P. aeruginosa* isolates. However primer from gene

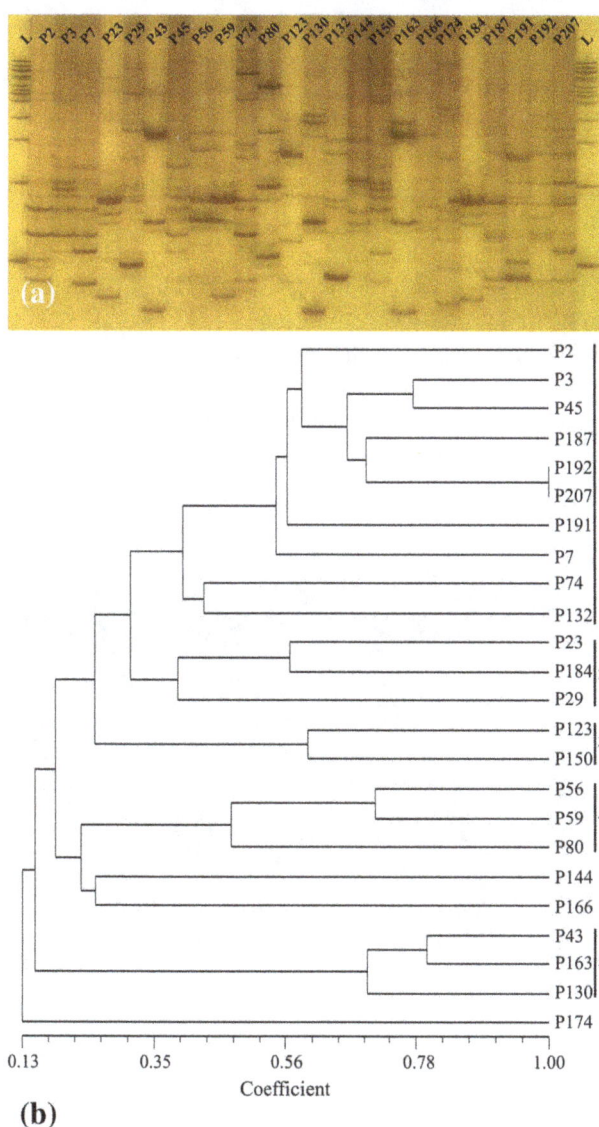

Fig. 3 ERIC-PCR-based genotypic analysis of 24 *P. putida* isolates. **a** PCR amplification of 24 *P. putida* isolates generated through ERIC primer. **b** Dendrogram of 24 *P. putida* isolates generated by binary matrix derived from ERIC amplicons

HM067869 (Pa16S) amplified ∼390 bp band in all the *P. putida* isolates and ∼600 bp band in all the *P. aeruginosa* isolates (Fig. 2a–d). Here *P. aeruginosa* isolates were used as a check to differentiate the two species using afore-mentioned primers. Absence of amplification with primers designed from genes EF159157 (Pf16S) and AF869903 (Pf23S) of *P. fluorescens* proves that none of the isolates used in the present investigation belonged to the species *P. fluorescens*.

However, a high level of polymorphism was seen in PCR of 24 *P. putida* isolates with ERIC primer. In the present study, the ERIC primer sequence was used in PCR to detect differences in the number and distribution of this bacterial repetitive sequence in the isolates of *P. putida* genomes. The number of bands after PCR amplification of isolates with ERIC primer varied from 5 to 16 with molecular weights between 50 and 1,500 bp (Fig. 3a). The most characteristic products of ERIC amplification for *P. putida* isolates were 310, 200, 180, 165 and 145 bp observed in 13, 16, 16, 13 and 13 out of 24 *P. putida* isolates, respectively. The genetic similarity among 24 *P. putida* isolates ranged from 0.13 to 1.00 (Fig. 3b). The similarity data obtained with ERIC primer identified two major clusters, one of which had only a single *P. putida* isolate P174. Overall the cluster analysis based on the pair-wise coefficient sim-ilarity with UPGMA of ERIC-PCR resulted into 5 dis-tinct genomic clusters at similarity coefficient 0.48, viz. groups 1, 2, 3, 4 and 5 consisting of ten (P2, P3, P45, P187, P192, P207, P191, P7, P74, P132), three (P23,

P184, P29), two (P123, P150), three (P56, P59, P80) and three (P43, P163, P130) isolates, respectively. Iso-lates P144 and P166 did not fall into any group. All the isolates exhibited their high degree of genetic variability and distributed into different clusters. This resulted in resolving microdiversity among *P. putida* isolates and significant levels of genomic heterogeneity between strains within and between sites, respectively. Grouping does not appear to be based on geographic origin. The ERIC-PCR fingerprints showed wide variations due to high degree of DNA heterogeneity over all the 24 iso-lates of *P. putida*. ERIC-PCR confirmed differences in repetitive elements dispersion in *Pseudomonas* genomes and a high degree of genetic variability among phos-phate solubilizing *P. putida* isolates. Bacterial isolates with similar biochemical property having approximately common genetic content exhibit molecular diversity. Similar results have been observed by other workers also (McSpadden Gardener 2008; Naik et al. 2008; Kaluzna et al. 2010; Charan et al. 2011).

In vitro detection of antibiotic-producing *P. putida* isolates using gene-specific primers

The results of the PCR analysis with primer PhaJ (Poly-hydroxyalkanoate gene) indicated that a DNA fragment approximately 250 bp in size was obtained in all *P. putida* isolates except P56 (Fig. 4). However, PCR analysis with primers PCA2a-3B (phenazine) and PltBf-r (Pyoluteorin) produced DNA fragments of size 1,400 bp and 800 bp,

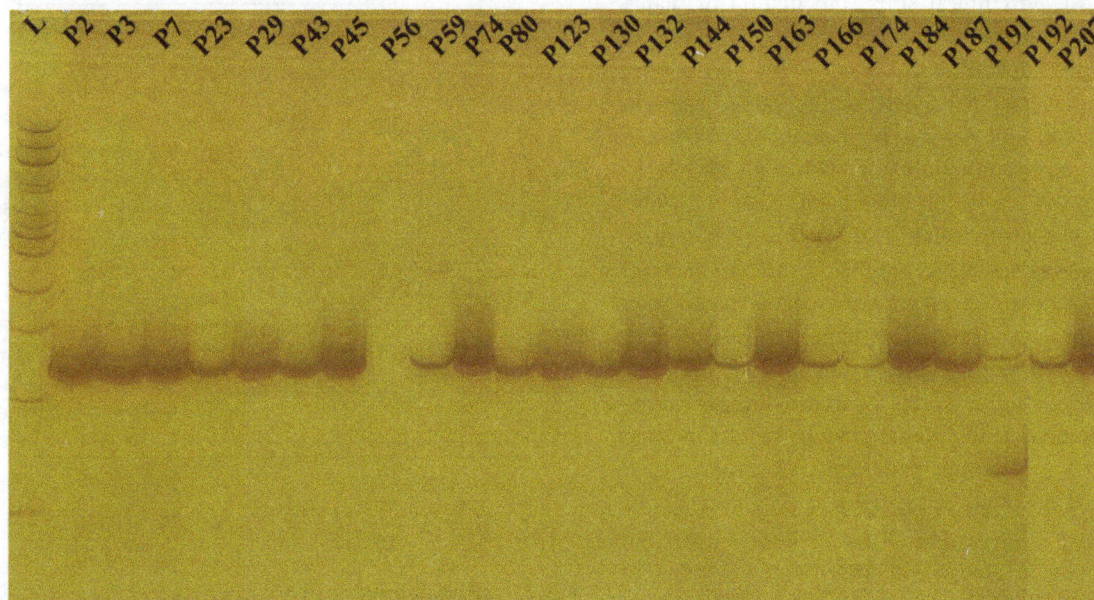

Fig. 4 PCR amplification of 24 *P. putida* isolates generated through PhaJ primer showing amplification of ∼250 bp in all the isolates except P56

respectively, in *P. putida* isolate P132 only; all other isolates showed negative results with these primers, i.e., these primers did not yield a PCR product. Pyoluteorin primer PltBf-r amplified another specific 700 bp band in *P. putida* isolates P80 and P132. It may be concluded that these genes were absent in other isolates. Primer plt1-2 (Pyrrolnitrin) amplified ~450 bp product in three isolates P56, P132 and P144 (Fig. 5a). Primer PrnAF-R (Pyrrolnitrin) amplified ~1,000 bp fragment in all the isolates except P7, P45, P130, P132, P144, and P150 (Fig. 5b). 2,4 Diacetylphloroglucinol gene-specific primer Phl2a-2b amplified an expected 750 bp fragment in isolates P56 and P132 only; faint bands were observed in P163 and P166 also (Fig. 5c). 2,4 Diacetylphloroglucinol primer B2BF-BPR4 amplified an expected DNA fragment of approximately 629 bp in three *P. putida* isolates P56, P132 and P174. However DNA fragment of 350 bp was observed in isolates P130, P144, P163, P174, P184 and P187 (Fig. 5d).

Results of PCR analysis with primers of different antibiotic genes showed that *P. putida* isolate P132 isolated from Kanker forest contained genes for phenazine, pyrrolnitrin, pyoluteorin and 2,4 diacetylphloroglucinol along with polyhydroxyalkanoate gene. Isolate P56 amplified 2,4 diacetylphloroglucinol gene with primers Phl2a-2b and B2BF-BPR4 and pyrrolnitrin gene with primers plt1-2 and PrnAF-R but amplification with primer specific for polyhydroxyalkanoate gene (PhaJ) was absent in it. The presence of antibiotic genes can be sued as a suitable marker for screening and selection of bacteria with potential biocontrol activity, in vitro and in situ conditions. However, it may not be necessary that biosynthetic genes for all the antibiotics may be present in all the *Pseudomonas* spp or isolates. Similar results have been reported by Zhang et al. (2006) who used 30 different PCR primers to identify antibiotic-related genes in previously isolated bacteria exhibiting good biocontrol activity. *Pseudomonas* spp. DF41 did not show amplification with primers specific for antibiotic biosynthetic genes encoding PCA, pyrrolnitrin, pyoluteorin and 2,4-DAPG, or for the zwittermicin A self-resistance gene.

It is often difficult and laborious to isolate and identify antibiotic-producing strains from natural environments. However, PCR detection can be a quick alternative to it which depends on the grouping of isolates based on different antibiotic-related genes present. In the present study an attempt was made to characterize *P. putida* isolates for the presence of biosynthetic genes involved in production of different types of antibiotics and related compounds by utilizing multiple primer sets. Significant phenotypic and genotypic differences have been reported during last decades in indigenous populations of fluorescent *Pseudomonas* spp.

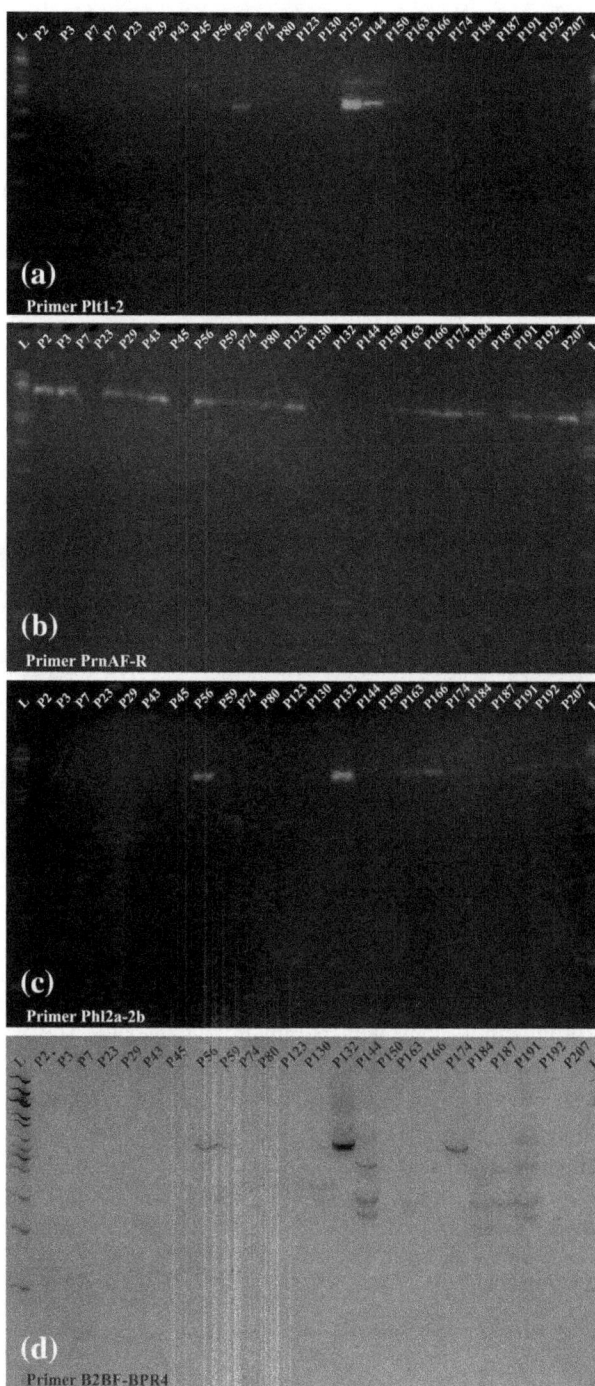

Fig. 5 PCR amplification of 24 *P. putida* isolates generated through primer. **a** plt1-2 (Pyrrolnitrin) amplified ~450 bp product in three isolates P56, P132 and P144. **b** PrnAF-R (Pyrrolnitrin) amplified ~1,000 bp fragment in all the isolates except P7, P45, P130, P132, P144, and P150. **c** Phl2a-2b amplified an expected 750 bp fragment in isolates P56 and P132 only; faint bands were observed in P163 and P166 also. **d** B2BF-BPR4 showing amplification of ~629 bp in isolates P56, P132 and P174 and ~350 bp in isolates P130, P144, P163, P174, P184 and P187

isolated from the rhizoplane, rhizosphere soil, or nonrhizo-sphere soil (Lemanceau et al. 1995; Raaijmakers et al. 1997). The genotypic diversity that exists within antibiotic producer *P. putida* isolates can be exploited to improve the rhizosphere competence and biocontrol activity of intro-duced rhizobacteria. However to manipulate the behavior and activity of biocontrol agents it is necessary to under-stand the influence of several abiotic and biotic factors on expression of antibiotic biosynthesis.

Conclusion

The present work provides an insight into the nutritional, biochemical and genetic versatility of 24 *P. putida* iso-lates. There is variability among isolates for PHB pro-duction and release of inorganic phosphate which can be further exploited to select potential isolate for industrial, biocontrol and plant growth promoting applications. All the isolates (except P56) were amplified with polyhy-droxyalkanoate gene-specific primer PhaJ but the vari-ability in PHB production may be because of the fact that synthesis of PHB is pathway-dependent and environment-dependent. Also it is a multi-gene-dependent process which necessitates requirement for more number of primers for screening. One of the potential *P. putida* isolate P132 can contribute as a candidate agent for both biocontrol and PGPR applications. Identified as one of the most efficient PHB producer and phosphate solubilizer, in vitro detection of P132 showed the presence of genes for phenazine, pyrrolnitrin, pyoluteorin and 2,4 diac-etylphloroglucinol along with polyhydroxyalkanoate. The diversity and adaptation of different isolates in terms of their respective environments provide further insights into their potential applications for the bioremediation of contaminated environments.

Acknowledgements Authors acknowledge the financial support provided by the Indira Gandhi Krishi Vishwavidyalaya, Raipur for carrying out this work.

Conflict of interest The authors declare that they have no conflict of interest.

References

Ausubel FM, Brent R, Kingston RE, Moore DD, Seidman JG, Smith JA, Struhl K (eds) (1991) Current protocols in molecular biology. Greene Publishing Associates, New York

Bangera MG, Thomashow LS (1999) Identification and characteriza-tion of a gene cluster for synthesis of the polyketide antibiotic 2,4-diacetylphloroglucinol from *Pseudomonas fluorescens* Q2-87. J Bacteriol 181:3155–3163

Bauer AW, Kirby WMM, Sherris JC, Turck M (1966) Antibiotic susceptibility testing by a standardized single disk method. Am J Clin Pathol 45:93–496

Beall DS, Ohta K, Ingram LO (1991) Parametric studies of ethanol-production from xylose and other sugars by recombinant *Escherichia coli*. Biotechnol Bioeng 38(3):296–303

Bertrand JL, Ramsay BA, Ramsay JA, Chavarie C (1990) Biosyn-thesis of poly-betahydroxyalkanoates from pentoses by *Pseudo-monas pseudoflava*. Appl Environ Microbiol 56(10):3133–3138

Blazevic DJ, Koepcke MH, Matsen JM (1973) Incidence and Identification of *Pseudomonas fluorescens* and *Pseudomonas putida* in the Clinical Laboratory. Appl Microbiol 25(1):107–110

Brown ME (1972) Seed and root bacterization. Ann Rev Phytopathol 12:181–197

Charan RA, Reddy VP, Reddy PN, Reddy SS (2011) Assessment of genetic diversity in *Pseudomonas fluorescens* using PCR based methods. Bioremediat Biodivers Bioavailab 5(1):10–16

Chung A, Liu Q, Ouyang SP, Wu Q, Chen GQ (2009) Microbial production of 3-hydroxydodecanoic acid by pha operon and fadBA knockout mutant of *Pseudomonas putida* KT2442 harboring tsB gene. Appl Microbiol Biotechnol 83(3):513–519

Ciesielski S, Pokoj T, Klimiuk E (2010) Cultivation-dependent and -independent characterization of microbial community produc-ing polyhydroxyalkanoates from raw glycerol. J Microbiol Biotechnol 20(5):853–861

Delaney SM, Mavrodi D, Bonsall VRF, Thomashow LS (2001) *phzO*, a gene for biosynthesis of 2-hydroxylated phenazine compounds in *Pseudomonas aureofaciens* 30-84. J Bacteriol 183:318–327

De Bruijn FJ (1992) Use of repetitive (repetitive extragenic, palindromic and enterobacterial repetitive intergenic concensus) sequences and the polymerase chain reaction to fingerprint the genomes of *Rhizobium meliloti* isolates and other soil bacteria. Appl Environ Microbiol 58:2180–2187

de Souza JT, Raaijmakers JM (2003) Polymorphisms within the *prnD* and *pltC* genes from pyrrolnitrin and pyoluteorin producing *Pseudomonas* and *Burkholderia* spp. FEMS Microbiol Ecol 43:21–34

Dos Santos VA, Heim S, Moore ER, Stratz M, Timmis KN (2004) Insights into the genomic basis of niche specificity of *Pseudo-monas putida* KT2440. Environ Microbiol 6:1264–1286

Durner R, Zinn M, Witholt B, Egli T (2001) Accumulation of poly[(R)]-3-hydroxyalkanoates] in *Pseudomonas oleovorans* during growth in batch and chemostat culture with different carbon sources. Biotechnol Bioeng 72(3):278–288

Elbahloul Y, Steinbüchel A (2009) Large–scale production of poly(3-hydroxyoctanoic acid) by *Pseudomonas putida* GPo1 and a simplified downstream process. Appl Environ Microbiol 75:643–651

Fravel DR (1988) Role of antibiosis in the biocontrol of plant diseases. Annu Rev Phytopathol 26:75–91

Fravel DR (2005) Commercialization and implementation of biocon-trol. Annu Rev Phytopathol 43:309–335

Gulati A, Rahi P, Vyas P (2008) Characterization of phosphate–solubilizing fluorescent pseudomonads from the rhizosphere of seabuckthorn growing in the cold deserts of Himalayas. Curr Microbiol 56:73–79

Hartmann R, Hany R, Geiger T, Egli T, Witholt B, Zinn M (2004) Tailored biosynthesis of olefinic medium-chain-length poly[(R)-3-hydroxyalkanoates] in *Pseudomonas putida* GPo1 with improved thermal properties. Macromol 37(18):6780–6785

Hoffmann N, Rehm BHA (2004) Regulation of polyhydroxyalkanoate biosynthesis in *Pseudomonas putida* and *Pseudomonas aeruginosa*. FEMS Microbiol Lett 237(1):1–7

Holloway B (1992) *Pseudomonas* in the late twentieth century. In: Galli E, Silver S, Witholt B (eds) *Pseudomonas* molecular biology and biotechnology. American Society for Microbiology, Washington, DC, pp 1–8

Holt JG, Krieg NR, Sneath PHA, Staley JT, Williams ST (1994) Bergey's manual of determinative bacteriology. Williams & Wilkins, Baltimore

John HL, Ralph AS (1961) Assay of poly-β-hydroxybutyric acid. J Bacteriol 82:33–36

Juan ML, Gonazalez LM, Walker GC (1998) A novel screening method for isolating exopolysaccharide deficient mutants. Appl Environ Microbiol 64:4600–4602

Kaluzna M, Ferrante P, Sobiczewski P, Scortichini M (2010) Characterization and genetic diversity of *Pseudomonas syringae* from stone fruits and hazelnut using repetitive-PCR and MLST. J Plant Pathol 92(3):781–787

Kieboom J, de Bont JAM (2001) Identification and molecular characterization of an efflux system involved in *Pseudomonas putida* S12 multidrug resistance. Microbiol 147:43–51

Kim DY, Hyung WK, Moon GC, Young HR (2007) Biosynthesis, modification, and biodegradation of bacterial medium-chain-length polyhydroxyalkanoates. J Microbiol 45(2):87–97

Law J, Slepecky RA (1969) Assay of poly-β-hydroxybutyric acid. J Bacteriol 82:52–55

Leisinger T, Margraff R (1979) Secondary metabolites of the fluorescent Pseudomonads. Microbiol Rev 43(3):422–442

Lemanceau P (1992) Beneficial effects of rhizobacteria on plants: example of fluorescent *Pseudomonas* spp. Agronomie 12:413–437

Lemanceau P, Corberand T, Gardan L, Latour X, Laguerre G, Boeufgras JM, Alabouvette C (1995) Effect of two plant species, flax (*Linum usitatissimum* L.) and tomato (*Lycopersicon esculentum* Mill.), on the diversity of soil borne populations of fluorescent pseudomonads. Appl Environ Microbiol 61:1004–1012

Leong J (1986) Siderophores : their biochemistry and possible role in the biocontrol of plant pathogens. Annu Rev Phytopathol 24:187–209

Madison LL, Huisman GW (1999) Metabolic engineering of poly (3-hydroxyalkanoates): from DNA to plastic. Microbiol Mol Biol Rev 63:21–53

Mavrodi OV, McSpadden Gardener BB, Mavrodi DV, Bonsall RF, Weller DM, Thomashow LS (2001) Genetic diversity of *phlD* from 2,4-diacetylphloroglucinol-producing fluorescent *Pseudomonas* spp. Phytopathol 91:35–43

Mayak S, Tirosh T, Glick BR (2004) Plant growth-promoting bacteria that confer resistance to water stress in tomato and pepper. Plant Sci 166:525–530

McSpadden Gardener BB (2008) Diversity and ecology of biocontrol *Pseudomonas* spp. in agricultural systems. Phytopathol 97(2):221–226

McSpadden Gardener BB, Mavrodi DV, Thomashow LS, Weller DM (2001) A rapid polymerase chain reaction-based assay characterizing rhizosphere populations of 2,4-diacetylphloroglucinol-producing bacteria. Phytopathol 91:44–54

Meijnen JP, De Winde JH, Ruijssenaars HJ (2008) Engineering *Pseudomonas putida* S12 for efficient utilization of D-xylose and L-arabinose. Appl Environ Microbiol 74(16):5031–5037

Meijnen JP, De Winde JH, Ruijssenaars HJ (2009) Establishment of oxidative D-xylose metabolism in *Pseudomonas putida* S12. Appl Environ Microbiol 75(9):2784–2791

Meur SL, Zinn M, Egli T, Thöny-Meyer L, Ren Q (2012) Production of medium-chain-length polyhydroxyalkanoates by sequential feeding of xylose and octanoic acid in engineered *Pseudomonas putida* KT2440. BMC Biotechnol 12:53

Murphy J, Riley JP (1962) A modified single solution method for determination of phosphate in natural waters. Anal Chem Acta 27:31–36

Naik PR, Raman G, Narayanan KB, Sakthivel N (2008) Assessment of genetic and functional diversity of phosphate solubilizing fluorescent pseudomonads isolated from rhizospheric soil. BMC Microbiol 8:230–235

Poblete-Castro I, Becker J, Dohnt K, dos Santos V, Wittmann C (2012) Industrial biotechnology of *Pseudomonas putida* and related species. Appl Microbiol Biotechnol 93(6):2279–2290

Raaijmakers JM, Weller DM, Thomashow LS (1997) Frequency of antibiotic-producing *Pseudomonas* spp. in natural environments. Appl Environ Microbiol 63:881–887

Raffel SJ, Stabb EV, Milner JL, Handelsman J (1996) Genotypic and phenotypic analysis of zwittermicin A-producing strains of *Bacillus cereus*. Microbiol 142:3425–3436

Raghothama KG (1999) Phosphate acquisition. Annu Rev Plant Physiol Mol Biol 50:665–693

Ramamoorthy V, Viswanathan R, Raguchander T, Prakasam V, Samiyappan R (2001) Induction of systemic resistance by plant growth promoting rhizobacteria in crop plants against pests and diseases. Crop Prot 20:1–11

Ramos JL, Krell T, Danield C, Segura A, Duque E (2009) Responses of *Pseudomonas* to small toxic molecules by a mosaic of domains. Curr Opin Microbiol 12:215–220

Rezzonico F, Moënne-Loccoz Y, Défago G (2003) Effect of stress on the ability of a *phlA*-based quantitative competitive PCR assay to monitor biocontrol strain *Pseudomonas fluorescens* CHA0. Appl Environ Microbiol 69:686–690

Rodríguez H, Fraga R (1999) Phosphate solubilizing bacteria and their role in plant growth promotion. Biotechnol Adv 17:319–339

Rojo F (2010) Carbon catabolite repression in *Pseudomonas*: optimizing metabolic versatility and interactions with the environment. FEMS Microbiol Rev 34:658–684

Shahzad SM, Khalid A, Arshad M, Kalil-ur-Rehman (2010) Screening rhizobacteria containing ACC-deaminase for growth promotion of chickpea seedlings under axenic conditions. Soil Environ 29(1):38–46

Sneath PHA, Mair NS, Elisabeth Sharpe M, Holt JG (1986) Bergey's manual of systematic bacteriology. Williams and Wilkins, Baltimore

Stanier RY, Palleroni NJ, Doudoroff M (1966) The aerobic pseudomonads: a taxonomic study. J Genet Microbiol 43:159–271

Sun Z, Ramsay JA, Guay M, Ramsay BA (2007) Fermentation process development for the production of medium-chain-length poly-3-hyroxyalkanoates. Appl Microbiol Biotechnol 75:475–485

Takahashi K, Murakami T, Kamata A, Yumoto R, Higashi Y, Yata N (1994) Pharmacokinetic analysis of the absorption enhancing action of decanoic acid and its derivatives in rats. Pharm Res 11:388–392

Timmis KN (2002) *Pseudomonas putida*: a cosmopolitan opportunist par excellence. Environ Microbiol 4:779–781

Vazquez PG, Holguin ME, Puente A, Cortes L, Bashan Y (2000) Phosphate-solubilising microorganisms associated with the rhizosphere of mangroves in a semiarid coastal lagoon. Bio Fert Soils 30:460–468

Vo TM, Kwang-Woo L, Young-Mi J, Yong-Hyun L (2008) Comparative effect of overexpressed *phaJ* and *fabG* genes supplementing (R)-3-hydroxyalkanoate monomer units on biosynthesis of mcl-polyhydroxyalkanoate in *Pseudomonas putida* KCTC1639. J Biosci Bioeng 106(1):95–98

Vyas P, Rahi P, Gulati A (2009) Stress tolerance and genetic variability of phosphate-solubilizing fluorescent *Pseudomonas* from the cold deserts of the trans-Himalayas. Microb Ecol 58:425–434

Wang Q, Nomura CT (2010) Monitoring differences in gene expression levels and polyhydroxyalkanoate (PHA) production in *Pseudomonas putida* KT2440 grown on different carbon sources. J Biosci Bioeng 110(6):653–659

Wang C, Ramette A, Punjasamarnwong P, Zala M, Natsch A, Moënne-Loccoz Y, D'efago G (2001) Cosmopolitan distribution of phlD-containing dicotyledonous crop associated biocontrol pseudomonads of worldwide origin. FEMS Microbiol Ecol 37:105–116

Wang L, Armbruster W, Jendrossek D (2007) Production of medium chain-length hydroxyalkanoic acids from *Pseudomonas putida* in pH stat. Appl Microbiol Biotechnol 75:1047–1053

Weller DM (1988) Biological control of soilborne plant pathogens in the rhizosphere with Bacteria. Annu Rev Phytopathol 26:379–407

Weller DM, Raaijmakers J, McSpadden Gardener BB, Thomashow LS (2002) Microbial populations responsible for specific soil suppressiveness to plant pathogens. Annu Rev Phytopathol 40:309–348

Wu X, Monchy S, Taghavi S, Zhu W, Ramos JL, vander Lelie D (2011) Comparative genomics and functional analysis of niche-specific adaptation in *Pseudomonas putida*. FEMS Microbiol Rev 35:299–323

Young FK, Kastner JR, May SW (1994) Microbial production of poly-betahydroxybutyric acid from D-xylose and lactose by *Pseudomonas cepacia*. Appl Environ Microbiol 60(11):4195–4198

Zhang Y, Fernando WGD, de Kievit TR, Berry C, Daayf F, Paulitz TC (2006) Detection of antibiotic-related genes from bacterial biocontrol agents with polymerase chain reaction. Can J Microbiol 52:476–481

7

Specific oligonucleotide primers for detection of endoglucanase positive *Bacillus subtilis* by PCR

S. Ashe · U. J. Maji · R. Sen ·
S. Mohanty · N. K. Maiti

Abstract A polymerase chain reaction (PCR) assay was developed for discrimination of *Bacillus subtilis* from other members of *B. subtilis* group as well as rapid identification from environmental samples. Primers ENIF and EN1R from endoglucanase gene were used to amplify a1311 bp DNA fragment. The specificity of the primers was tested with seven reference strains and 28 locally isolated strains of endoglucanase positive *Bacillus* species. The PCR product was only produced from *B. subtilis*. The results demonstrated high specificity of two oligonucleotides for *B. subtilis*. This species-specific PCR method provides a quick, simple, powerful and reliable alternative to conventional methods in the detection and identification of *B. subtilis*. To our knowledge this is the first report of a *B. subtilis* specific primer set.

Keywords Endoglucanase gene · PCR · Specific primers · *B. subtilis*

Introduction

Genus *Bacillus* is a Gram-positive, spore-forming, fermentative, aerobic and rodshaped bacteria. Several species of this group are non-pathogenic, simple to cultivate and secrete enzymes such as proteases, amylases and cellulases that are useful for a number of industrial applications (Arbige et al. 1993). The *Bacillus subtilis* group contains the closely related taxa *Bacillus subtilis* subsp. *subtilis* (Smith et al.

1964; Nakamura et al. 1999), *Bacillus licheniformis* (Skerman et al. 1980), *Bacillus amyloliquefaciens* (Priest et al. 1987), *Bacillus atrophaeus (*Nakamura 1989), *Bacillus mojavensis* (Roberts et al. 1994), *Bacillus vallismortis* (Roberts et al. 1996*), Bacillus subtilis* subsp. *spizizenii* (Nakamura et al. 1999). Classical identification methods based on biochemical tests or fatty acid methyl ester profiling were laborious and hence not applicable for the purpose of a rapid screening. These taxa can be differentiated from one another by fatty acid composition analysis, restriction digest analysis and DNA–DNA hybridization analysis, but are quite difficult to differentiate by phenotypic characteristics (Roberts et al. 1994; Nakamura et al. 1999). The use of the 16S rRNA sequence as a target for genetic detection was therefore considered. Numerous *Bacillus* species described so far have been found to display rather conserved 16S rRNA sequences compared to other genera. Thus the use of this taxonomic marker is sometimes inadequate for species definition according to generally accepted criteria (Stackebrandt and Goebel 1994). Such unusual similarities exist for members of the '*Bacillus* 16S rRNA groupI', including *B. subtilis*, which displays 99.3 % similarity at the 16S rRNA level to *B. atrophaeus* and 98.3 % to *B. licheniformis* and *B. amyloliquefaciens* (Ash et al. 1991). In the present study, it has been shown that specific primers for detection of endoglucanase gene could be used for identification of *B. subtilis*.

Materials and methods

Bacterial strains and culture conditions

A total 35 *Bacillus* strains were used in this study (Table 1). Of 12 *B. subtilis*, ten strains were isolates from pond sediments. For *Bacillus cereus,* one ATCC strain and nine pond

S. Ashe · U. J. Maji · R. Sen · S. Mohanty · N. K. Maiti (✉)
Division of Fish Health Management, Central Institute of
Freshwater Aquaculture, Kaushalyaganga,
Bhubaneswar 751002, Orissa, India
e-mail: maitink@yahoo.co.in

Table 1 Bacterial strains used

Species	Total no. of strains	Source	Accession number of sediment isolates
Bacillus subtilis	14	ATCC 11,774, ATCC 6,051 and 12 pond sediment	GQ214130, GQ21413 HQ388810–HQ388813 JX438679–JX438684
Bacillus cereus	10	ATCC 13,061 and 9 pond sediment	GQ214131 HQ388814–HQ388817 JX438685–JX438788
Bacillus pumilus	6	ATCC 14,884 and 5 pond sediment	HQ388808 JX438694–JX438697
Bacillus megaterium	1	ATCC 9,885	
Bacillus thuringiensis	1	ATCC 10,792	
Bacillus licheniformis	1	ATCC 13,061	
Bacillus amyloliquefaciens	2	Pond sediment	JX438692–JX438693

Pond sediment isolates were confirmed by 16s rDNA sequencing

sediment isolates were tested. Five pond isolates that had been classified as *Bacillus pumilus* and two as *B. amyloliquefaciens* were also included in the test. For *Bacillus megaterium*, *B. licheniformis* and *Bacillus thuringiensis*, ATCC strains were analysed. All the pond sediment isolates were identified by 16S rDNA sequencing and available in our laboratory.

Soil samples

24 Soil samples collected from agriculture field and fish culture ponds were also included.

Primers

The endoglucanase gene sequences (EC, 3.2.1.4) of *B. subtilis* were retrieved from GenBank nucleotide database and were aligned using Clustal W (1.82) Multiple Alignment Program. Two sets of primers EN1F (103–124 bp) 5′-CCAGTAGCCAAGAATGGCCAGC-3′, EN1R (1,413–1,393 bp) 5′-GGAATAATCGCCGCTTTG TGC-3′) were designed by analyzing the conserved regions of the aligned sequences.

DNA isolation

The total genomic DNA was extracted from bacterial suspension (after 12 h incubation in LB) using DNA extraction kit (Merck Bioscience, India) following the manufacture's instruction. Soil genomic DNA was extracted by using ultra clean soil DNA isolation kits (MoBio, USA).

Polymerase chain reaction

The PCR reaction mixtures (50 μl) contained, dNTPs each 100 μmol; 1X PCR buffer (10 mM Tris Cl, 50 mM KCl,

1.5 mM $MgCl_2$ and 0.01 % gelatin); each primer 10 pmol; *Taq* DNA polymerase (NEB) 0.75U and bacterial DNA 100 ng. The touch down PCR in a volume of 50 μl was carried out with initial denaturation of 94 °C for 5 min followed by ten cycles of touch down program (94 °C for 30 s, 70 °C for 20 s and 74 °C for 45 s, followed by a 1 °C decrease of the annealing temperature every cycle). After completion of the touch down program, 25 cycles were subsequently performed (94 °C for 30 s, 60 °C for 20 s and 74 °C for 45 s) and ending with a 10 min extension at 74 °C. PCR reactions were run on a 1.5 % agarose gel in 1X TBE.

Cloning and sequencing

Band was excised from the gel and PCR product was purified by using the QIAquick gel purification kit according to the manufacturer's instructions (QIAGEN, Germany). The purified PCR product was cloned in pGEM®-T Easy vector following manufacturer's protocol (promega) and transformed into DH5α cells. Sequencing of the positive clones were done by Sanger method using 96 capillary high through put sequencer; ABI 3,730 XL (Xcelris, India) with T7 and SP6 universal primer.

Results and discussion

BlastN seach of endoglucanase gene of *B. subtilis* (accession numbers HM470252.1, AF355629.1 and CP002906.1) revealed 93 % similarity with *B. amyloliquefaciens*, *B. megaterium*, *B. pumilus* and *B. licheniformis* 90 % with *B. subtilis* subsp. *spizizenii* and 98–99 % with *Geobacillus stearothermophilus* and *Paenibacillus campinasensis*. Based on multiple alignments of endo-β-1,4-glucanase genes, a specific consensus motif was identified in the endoglucanase gene of *B. subtilis, G. stearothermophilus* and *P. campinasensis*. Two PCR primers, EN1F and EN1R, were chosen that were predicted to specifically amplify a 1,311 bpDNA fragment of the *B. Subtilis, G. stearothermophilus* and *P. campinasensis*. The Genbank database (NCBI) search for complimentary sequences revealed 100 % homology between the primers and the gene encodes endo-β-1,4-glucanase of *B. subtilis* as well as *G. stearothermophilus* and *P. campinasensis*. No homologous sequences were found for other members of genus *Bacillus* indicating an excellent specificity of the primers for *B. subtilis*.

As expected, the test turned out to be positive only for *B. subtilis* among the different species of Genus *Bacillus* PCR amplification with genomic DNA isolated from in vitro cultured *B. subtilis* resulted in a reproducible amplification of 1,311 bp product with primer combinations EN1F/EN1R. To determine the sensitivity of PCR, endpoint titration with serial dilutions of genomic DNA isolated from the standard strain of *B. subtilis* was carried out and positive results obtained as little as 500 ficogram of DNA (Fig. 1).

To assess the range of specificity of the PCR test, a number of endoglucanase positive *Bacillus* species were assayed. Given the considerable number of species established to date under *B. subtilis* group, our choice to assess the range of specificity was restricted to *B. subtilis* subsp. *subtilis* that was representative of the *B. subtilis* group.

To test the specificity of the amplified products, control experiments were performed under the same conditions with DNA from different members of *B. subtilis* group as well as *B. cereus* group. The test found to be positive only for *B. Subtilis* (Fig. 2). It is noteworthy that the species detected as positive with this test are very close from a

taxonomic point of view when phylogenetic tree was constructed on the basis of endoglucanase gene sequences of different species of Genus *Bacillus* (Fig. 3). Attempt to detect *B. subtilis* directly from soil samples collected from agriculture field and fish culture ponds were successful, out of 24 soil samples collected from fish pond and agricultural field ten samples were positive for amplification. Cloning and sequencing confirmed the amplicon to be endoglucanase gene of *B. subtilis*. However, after enrichment of negative soil samples on TSB 20 % increased positivity rate was obtained by PCR, demonstrating that the initial

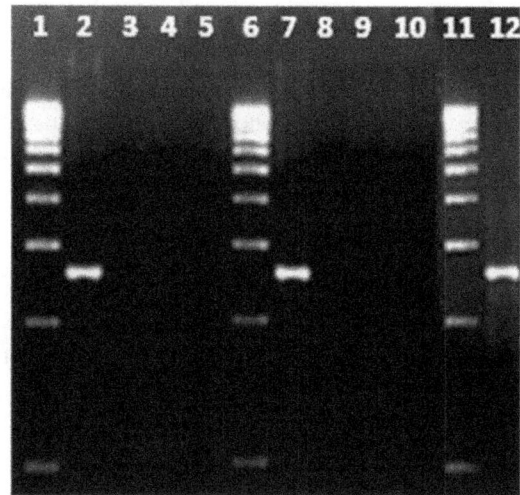

Fig. 2 PCR amplification for endoglucanase gene in different *Bacillus* spp. *Lane 1* size marker (500 bp ladder); *lanes 2–5 B. subtilis* ATCC-6,051, *B. cereus*-ATCC 13,061, *B. pumilus* ATCC-14,884, *B. megaterium* ATCC-9,885; *lane 6* size marker (500 bp ladder); *lanes 7–10 B. subtilis* ATCC-11,774, *B. thuringiensis* ATCC-10,792, *B. licheniformis* ATCC-13,061, *B. amyloliquefaciens* CF8; *lane 11* size marker (500 bp ladder); *lane 12 B. subtilis* C11B1

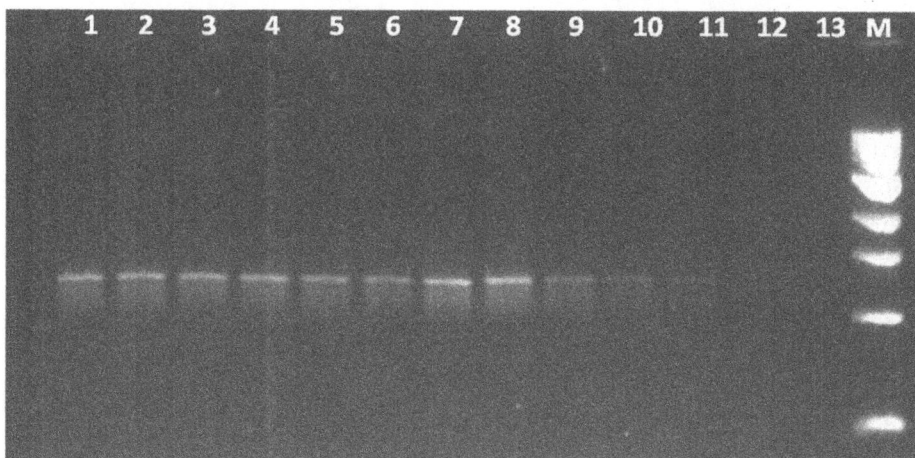

Fig. 1 Limit of detection of endoglucanase gene in different concentration of DNA. *Lane 1* 100 ng, *lane 2* 50 ng, *lane 3* 10 ng, *lane 4* 5 ng, *lane 5* 1 ng, *lane 6* 100 pg, *lane 7* 50 pg, *lane 8* 10 pg, *lane 9* 5 pg, *lane 10* 1 pg, *lane 11* 500 fg, *lane 12* 100 fg, *lane 13* negative control, *lane M* size marker (1 kb ladder, NEB)

Fig. 3 Phylogenetic tree of
Bacillus spp. based on
endoglucanase gene sequence
data

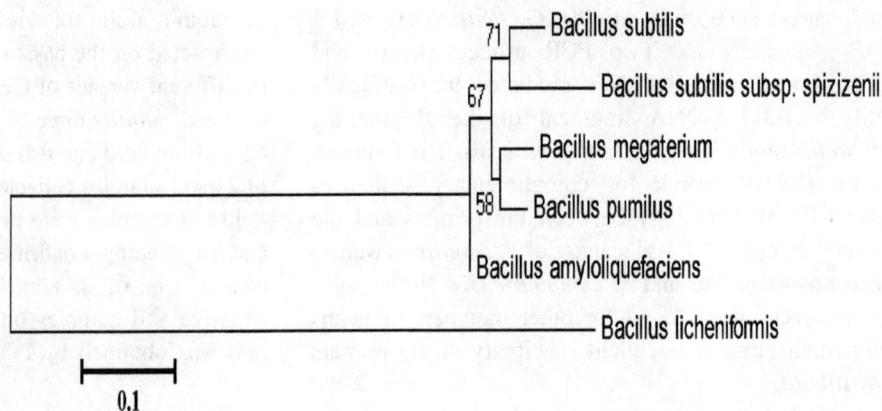

concentration of *B. subtilis* was at a proportion below the detection limit. The primers not only differentiated *B. subtilis* from other species but also differentiated at subspecies level as expected product size could not be predicted from *B. subtilis* subsp. *spizizenii* by primer blast (NCBI).

In this report, a PCR method has been established for identification of *B. subtilis*. Endo-β-1,4-glucanase gene has been chosen to design primers that could be useful for identification and direct detection of *B. subtilis* from environmental samples. Detection of *B. subtilis* has been shown to be specific, although the primers showed specificity for *G. stearothermophilus* and *P. campinasensis* in primer blast, and predicted amplicon to be 1,311 bp. However, *B. subtilis* could be differentiated from *G. stearothermophilus* and *P. campinasensis* by sequencing of the pcr product. 16S rRNA gene sequence analysis is the most commonly used method for identifying bacteria or for constructing bacterial phylogenetic relationships (Woese 1987; Vandamme et al. 1996; Joung and Cote 2002); however, its usefulness is limited because of the high percentage of sequence similarity between closely related species (Ash et al. 1991; Marti'nez-Murcia et al. 1992; Christensen et al. 1998). The use of protein-encoding genes as phylogenetic markers is now a common approach (Yamamoto and Harayama 1998; Ko et al. 2004; Chelo et al. 2007). Detailed investigations have demonstrated that sequences from protein-encoding genes can accurately predict genome relatedness and may replace DNA–DNA hybridization for species identification and delineation in the future (Stackebrandt et al. 2002; Zeigler 2003). Wang et al. (2007) clearly showed that in the *B. subtilis* group, within which species differentiation is very difficult, core genes such as gyrB allow differentiation on genetic basis. Compared to other genera, *Bacillus* species are having conserved 16s rRNA sequences and are difficult to identify at species level using this marker (Stackebrandt and Goebel 1994).

Conclusion

In the present study, the demonstrated specificity of the oligonucleotides used as PCR primers and results of experiments with soil samples provide the basis to develop a diagnostic assay for identification and detection *B. subtilis* form environment samples. As the primers used in this study have been found to be specific to endoglucanase gene of *G. stearothermophilus* and *P. campinasensis*, we suggest to use these primers as supplementary PCR assay to 16s rRNA sequencing for identification of *B. subtilis*.

Acknowledgments The financial help received from AMAAS nodal centre ICAR, New Delhi, India for carrying out the work is dully acknowledged.

Conflict of interest None.

References

Arbige MV, Bulthuis BA, Schultz J, Crabb D (1993) Fermentation of *Bacillus*. In: Sonenshein AL, Hoch JA, Losick R (eds) *Bacillus subtilis* and other gram-positive bacteria: biochemistry, physiology, and molecular genetics. American Society for Microbiology, Washington, DC, pp 871–895

Ash C, Farrow JA, Dorsch M, Stackebrandt E, Collins MD (1991) Comparative analysis of *Bacillus anthracis*, *Bacillus cereus*, and related species on the basis of reverse transcriptase sequencing of 16S rRNA. Int J Syst Bacteriol 41:343–346

Chelo IM, Ze'-Ze' L, Tenreiro R (2007) Congruence of evolutionary relationships inside the Leuconostoc–Oenococcus–Weissella clade assessed by phylogenetic analysis of the 16S rRNA gene, dnaA, gyrB, rpoC and dnaK. Int J Syst Evol Microbiol 57:276–286

Christensen H, Nordentoft S, Olsen JE (1998) Phylogenetic relationships of *Salmonella* based on rRNA sequences. Int J Syst Bacteriol 48:605–610

Joung KB, Cote JC (2002) Evaluation of ribosomal RNA gene restriction patterns for the classification of *Bacillus* species and related genera. J Appl Microbiol 92:97–108

Ko KS, Kim JW, Kim JM, Kim W, Chung SI, Kim IJ, Kook YH (2004) Population structure of the *Bacillus cereus* group as determined by sequence analysis of six housekeeping genes and the plcR gene. Infect Immun 72:5253–5261

Li-T Wang, Lee F-L, Tai C-J, Kasai H (2007) Comparison of gyrB gene sequences, 16S rRNA gene sequences and DNA–DNA hybridization in the *Bacillus subtilis* group. Int J Syst Evol Microbiol 57:1846–1850

Martı́nez-Murcia AJ, Benlloch S, Collins MD (1992) Phylogenetic interrelationships of members of the genera *Aeromonas* and *Plesiomonas* as determined by 16S ribosomal DNA sequencing: lack of congruence with results of DNA–DNA hybridization. Int J Syst Bacteriol 42:412–421

Nakamura LK (1989) Taxonomic relationship of black-pigmented *Bacillus subtilis* strains and a proposal for *Bacillus atrophaeus* sp. *nov*. Int J Syst Bacteriol 39:295–300

Nakamura LK, Roberts MS, Cohan FM (1999) Relationship of *Bacillus subtilis* clades associated with strains 168 and W23: aproposal for *Bacillus subtilis* subsp. *subtilis* subsp. nov. and *Bacillus subtilis* subsp. *spizizenii* subsp. nov. Int J Syst Bacteriol 49:1211–1215

Priest FG, Goodfellow M, Shute LA, Berkeley RCW (1987) *Bacillus amyloliquefaciens* sp. nov., nom. rev. Int J Syst Bacteriol 37:69–71

Roberts MS, Nakamura LK, Cohan FM (1994) *Bacillus mojavensis* sp. *nov.*, distinguishable from *Bacillus subtilis* by sexual isolation, divergence in DNA sequence, and differences in fatty acid composition. Int J Syst Bacteriol 44:256–264

Roberts MS, Nakamura LK, Cohan FM (1996) *Bacillus vallismortis* sp. *nov.*, a close relative of *Bacillus subtilis*, isolated from soil in Death Valley, California. Int J Syst Bacteriol 46:470–475

Skerman VBD, McGowan V, Sneath PHA (1980) Approved lists of bacterial names. Int J Syst Bacteriol 30:225–420

Smith NR, Gibson T, Gordon RE, Sneath PH (1964) Type cultures and proposed neotype cultures of some species in the genus Bacillus. J Gen Microbiol 34:269–272

Stackebrandt E, Goebel BM (1994) Taxonomic note: a place for DNA–DNA reassociation and 16S rRNA sequence analysis in the present species definition in bacteriology. Int J Syst Bacteriol 44:846–849

Stackebrandt E, Frederiksen W, Garrity GM, Grimont PAD, Kampfer P, Maiden MCJ, Nesme X, Rossello-Mora R, Swings J et al (2002) Report of the ad hoc committee for the reevaluation of the species definition in bacteriology. Int J Syst Evol Microbiol 52:1043–1047

Vandamme P, Pot B, Gillis M, De Vos P, Kersters K, Swings J (1996) Polyphasic taxonomy, a consensus approach to bacterial systematics. Microbiol Rev 60:407–438

Woese CR (1987) Bacterial evolution. Microbiol Rev 51:221–271

Yamamoto S, Harayama S (1998) Phylogenetic relationships of *Pseudomonas putida* strains deduced from the nucleotide sequences of gyrB, rpoD and 16S rRNA genes. Int J Syst Bacteriol 48:813–819

Zeigler DR (2003) Gene sequences useful for predicting relatedness of whole genomes in bacteria. Int J Syst Evol Microbiol 53:1893–1900

8

Genetic diversity and gene differentiation among ten species of Zingiberaceae from Eastern India

Sujata Mohanty · Manoj Kumar Panda · Laxmikanta Acharya · Sanghamitra Nayak

Abstract In the present study, genetic fingerprints of ten species of Zingiberaceae from eastern India were developed using PCR-based markers. 19 RAPD (Rapid Amplified polymorphic DNA), 8 ISSR (Inter Simple Sequence Repeats) and 8 SSR (Simple Sequence Repeats) primers were used to elucidate genetic diversity important for utilization, management and conservation. These primers produced 789 loci, out of which 773 loci were polymorphic (including 220 unique loci) and 16 monomorphic loci. Highest number of bands amplified (263) in *Curcuma caesia* whereas lowest (209) in *Zingiber cassumunar*. Though all the markers discriminated the species effectively, analysis of combined data of all markers resulted in better distinction of individual species. Highest number of loci was amplified with SSR primers with resolving power in a range of 17.4–39. Dendrogram based on three molecular data using unweighted pair group method with arithmetic mean classified all the species into two clusters. Mantle matrix correspondence test revealed high matrix correlation in all the cases. Correlation values for RAPD, ISSR and SSR were 0.797, 0.84 and 0.8, respectively, with combined data. In both the genera wild and cultivated species were completely separated from each other at genomic level. It also revealed distinct genetic identity between species of *Curcuma* and *Zingiber*. High genetic diversity documented in the present study provides a baseline data for optimization of conservation and breeding programme of the studied zingiberacious species.

S. Mohanty (✉) · M. K. Panda · L. Acharya · S. Nayak
Centre of Biotechnology, Siksha 'O' Anusandhan University, Bhubaneswar 751003, Odisha, India
e-mail: sujatamohantyils@gmail.com

Keywords Zingiberaceae · Genetic diversity · RAPD · ISSR · SSR

Introduction

Zingiber and *curcuma* are two interesting genus of Zingiberaceae mostly contain spice plants with very high medicinal value. Among them few species are being cultivated but majority of them are wild in nature. Thus, these species are being depleted from nature due to extensive collection and habitat destruction. Strategic management and planning for their conservation is still far away from reality. Some species are still unknown due to their unavailability and extensive use by the tribal people for medicine as well as spice. The knowledge of genetic variability is a pre requisite to study the evolutionary history of a species, as well as for other studies like intraspecific variation, genetic resource conservation etc. (Islam et al. 2007). Hence, genetic diversity and gene differentiation through molecular marker analysis are essential for their taxonomic relationship evaluation, conservation and sustainable utilization.

For proper conservation programme it is essential to characterize the plants genetically. Number of molecular markers is being regularly used for studying genetic relations, population genetics, genetic characterizations in different plant groups and cultivars. The molecular markers are not influenced by the external environmental factor unlike that of morphological markers and hence accurately testify the genetic relationship between and among plant groups. Molecular markers like RAPD, ISSR and SSR are being used regularly for genetic diversity assessment as a thorough knowledge of the level and distribution of genetic variation is essential for conservation (Dreisigacker et al.

2005; Sharma et al. 2008; Naik et al. 2010; Das et al. 2011). PCR-based DNA fingerprinting techniques like RAPD, ISSR and SSR are proven to be very informative and cost-effective techniques in many plant species as these primers do not require prior knowledge of a species genetics (Williams et al. 1990; Zeitkiewicz et al. 1994; Lee et al. 2007).

Many workers have been reported the genetic diversity among zingiber and curcuma species (Das et al. 2011; Jatoi et al. 2006; Syamkumar and Sasikumar 2007) but the studied species are area specific based on their availability in that region. Also, less is known about the genetic relationship among cultivated and wild species which is the main reason for extinction of wild but important species having future drug yielding potential. Studies have been attempted based on morphological, biochemical and anatomical characterization in *Zingiber* and *Curcuma* species (Jiang et al. 2006; Zhou et al. 2007; Paramasivam et al. 2009), but relying on morphological and biochemical characters it has its own limitations as they are always not completely represent the genetic structure (Noli et al. 1997). Molecular profiling of zingiberaceous species is still in an emerging stage. Reports were restricted to specific species and species within a single genus. Thus the present work is an attempt to study the genetic relationship existing within and among two different genera i.e., *Zingiber* and *Curcuma*.

The objective of the present study is to examine the level of genetic diversity among and within two populations of Zingiberaceae using DNA-based molecular markers like RAPD, ISSR and SSR to give baseline knowledge for optimization of combining strategies for efficient management of both the population.

Materials and methods

Plant material and DNA extraction

The present investigation deals with ten species of Zingiberaceae from eastern India and was collected from both Odisha and West Bengal. Among them *Z. officile* and *C. longa* are cultivated varieties and rest species were collected from wild habitat. They were maintained in the greenhouse of Centre of Biotechnology of Siksha 'O' Anusandhan University. Genomic DNA was isolated from frozen leaf samples by grinding with a mortar, pestle in extraction buffer (0.1 M Tris–HCl pH 7.5, 0.25 M NaCl, 25 mM EDTA pH 8.0, 10 % CTAB), and incubated at 65 °C for 10 min. It was thrice treated with phenol: chloroform: isoamyl alcohol (25:24:1) for removal of non-nucleic acid compounds. DNA was precipitated using isopropanol and resuspended in 100 μl of 10 mM Tris, pH

8.0 with 10 μg RNaseA. The quality and quantity of the DNA was determined with a Thermo Scientific UV–Vis spectrophotometer. The sample DNA was diluted as 25 ng μl^{-1} for RAPD and ISSR analysis.

Molecular marker analysis

Nineteen RAPD primers (Operon Technologies, Alemada, USA), eight ISSR and eight SSR primers (Bangalore Genei, Bangalore, India) were used for PCR amplification. Based on previous results, good resolution and reproducibility ability, all RAPD, ISSR and ISSR primers were selected out of several primers utilized during screening. In RAPD, PCR was performed at an initial temperature of 94 °C for 5 min for complete denaturation. The second step consisted of 42 cycles having three ranges of temperature, i.e., at 92 °C for 1 min for denaturation of template DNA, at 37 °C for 1 min for primer annealing, and at 72 °C for 2 min for primer extension, followed by running the samples at 72 °C for 7 min for complete polymerization. For ISSR, the same temperature profile was followed, but the primer annealing temperature was set at 5 °C lower than the melting temperature. The PCR products obtained from RAPD were analyzed in 1.5 % agarose gel whereas the ISSR products were analyzed in 2 % agarose gel stained with ethidium 15-μl of PCR products from RAPD and ISSR markers were combined with 2-μl of a loading buffer (0.4 % Bromo-phenol Blue, 0.4 % xylene cyanole and 5 ml of glycerol) and were analyzed directly on 1.5 % agarose gels in 0.5X TBE buffer. Electrophoresis was done for about 3 h at 60 volts. 100 bp ladder (MBI Fermentas) was used to compare the molecular weights of amplified products. Visualization of the amplified bands was done by UV transilluminator (Bgenei, Bangalore, India).

The detection of microsatellite polymorphism was performed using seven SSR markers from the eight SSR markers characterized by Lee et al. (2007). The SSR amplification condition was as follows: an initial hot start and denaturing step at 93 °C for 3 min followed by 35 cycles of a 1 min denaturation at 93 °C, a 1 min annealing at 55 °C, and a 1 min primer elongation at 72 °C. A final extension step at 72 °C for 5 min was performed. The PCR amplified products were resolved in a 12 % non-denaturing polyacrylamide gel. Electrophoresis was done for about two and half hours at 120 volts. 50 bp ladder (MBI Fermentas) was used to compare the molecular weights of amplified products.

Data analysis

Bands generated by all markers were compiled into a binary data matrix based on the presence (1) or absence

(0) of the selected band. The Primer Index was calculated from the polymorphic index. Polymorphic index of polymorphic information content (PIC) was calculated as per the formula of Roldan-Ruiz et al. (2000). Jaccard's coefficient of genetic similarity was calculated using the binary data matrix between all accessions. Similarity coefficient values estimated were used to construct a dendrogram (cluster diagram) using the method of unweighted pair group with arithmetic averages (UP-GMA) and principal co-ordinate analysis (PCA) was also carried out using NTSYSpc ver.2.2 software (Rohlf 1997).

Results

RAPD polymorphism

Twenty-five random decamer oligonucleotide primers were used for the present work and out of them 19 primers were selected and proved to be informative. A total of 317 bands amplified, were all polymorphic in nature. Among all the primers highest number of bands (29) were amplified in primer OPA4 which were in the range of 320–2,100 bp. Lowest number of bands (3) was amplified in primer OPD12 in the range of 850–1,031 bp. Average number of bands per primer was found to be 16.7. The largest amplicons (>3,000 bp) were amplified with OPA7, OPC2, OPD18 and OPAF5 primer and the smallest (200 bp) with OPA18, OPN16, and OPAF5 primer (Table 1) (Fig. 1). Among all the polymorphic bands amplified 111 numbers of bands were found to be unique. Highest number of unique bands (12) was amplified in primer OPA7 in and lowest (2) in primer OPD7, OPD12, respectively. The resolving power of the primers was varied from 11.27 to 0.72 where the primer OPD20 had maximum resolution power (11.27) and the primer OPD12 had minimum resolution power (0.727). However, the RAPD primer index was maximum (8.264) in case of primer OPA4 and minimum (0.628) in primer OPD12. The RPI value revealed that the primer OPA4 was the best for species segregation at nuclear DNA level. Jaccard's coefficient showed that the species *C. amada* and *C. aromatica* were most closely related with a similarity value of 0.29 and *Z. cassumunar* and *C. amada* were most remotely placed with a similarity coefficient of 0.132. Between any two species the average similarity coefficient was calculated to be 0.121.

ISSR polymorphism

Out of 10 ISSR primers, 8 primers resulted in the amplification of 147 fragments of which 3 were monomorphic bands and 144 were polymorphic bands which again includes 36 unique bands. The bands were amplified in the range of 200–3,000 bp (Fig. 1). The primer $(GACA)_4$ produced maximum number of bands (22), while the primer $(GTG)_5$ resulted in amplification of 15 loci only (Table 1). Maximum number of polymorphic bands (22) was found in primer $(GACA)_4$ and minimum (15) bands amplified in primer $(GTG)_5$. Monomorphic bands were found in primer $(GTG)_5$, $(GAC)_5$ and $(GGA)_4$ whereas unique bands were found in all primers. Among these ISSR primers, maximum resolving power was obtained for $(GGA)_4$ (18.364) and the minimum was for $T(GA)_9$ (10.545) with an average of 13.42. Maximum primer index was calculated for $(GACA)_4$ and minimum for $(GTG)_4$ with an average of 6.47. Jaccard's coefficient showed that the species *C. caesia* and *C. aromatica* were most closely related with a similarity value 0.67. Between any two species the average similarity coefficient was calculated at 0.311.

SSR polymorphism

Eight SSR primers were used for the present work. The primer combinations had amplified 325 loci among which 13 were found to be monomorphic in nature and the rest were polymorphic (Table 2) (Fig. 1). From the 312 polymorphic bands amplified, 73 were found to be unique. Maximum number of 52 bands was resolved for the primer CBT04 and the minimum 21 for CBT09. The average number of bands amplified for primer was 40.62. Bands resolved between 3,000 and 20 bp were taken into consideration for the present investigation. Maximum number of polymorphism was found in primer CBT04 (98 %) and minimum in primer CBT07 (91 %). The resolving power was maximum for the primer CBT04 (39.455) and minimum for primer CBT09 (17.455) with an average of 27.97. However, the primer index was found to be highest in the primer CBT04 (20.231) and lowest in primer CBT09 (7.074) with an average of 13.61. Jaccard's coefficient showed that the species *Z. rubens* and *Z. zerumbet* were most closely related with a similarity value 0.476 and *C. caesia* and *Z. cassumunar* were most remotely placed with a similarity coefficient of 0.17. Average similarity coefficient was calculated at 0.296.

Combined marker analysis

Total number of bands amplified with all markers in ten species is 2317 (Fig. 2) with an average of 231.7 per species. Highest number of bands amplified (263) in *Curcuma caesia* whereas lowest (209) in *Zingiber cassumunar*. Average number of bands amplified in all

Table 1 Details of RAPD and ISSR analysis in 10 species of Zingiberaceae

Markers	Primer	Primer sequence	Approx fragment Size (bp)	Total bands	Unique bands	Resolving power	Primer index
RAPD	OPA4	5′AATCGGGCTG3′	320 to 2,100	29	10	10.72	8.26
	OPA7	5′GAAACGGGTG3′	300 to >3,000	21	12	7.09	5.45
	OPA9	5′GGGTAACGCC3′	600 to 2,100	12	6	4.00	3.07
	OPA18	5′AGGTGACCGT3′	200 to 2,250	25	8	9.81	7.33
	OPC2	5′GTGAGGCGTC3′	500 to >3,000	22	8	10.36	7.07
	OPC5	5′GATGACCGCC3′	900 to 2,500	14	4	5.81	4.36
	OPC11	5′AAAGCTGCGG3′	450 to 2,100	17	3	9.27	6.11
	OPD3	5′GTCGCCGTCA3′	400 to 1,950	13	6	4.90	3.66
	OPD7	5′TTGGCACGGG3′	400 to 2,200	17	2	9.09	6.34
	OPD8	5′GTGTGCCCCA3′	500 to 2,800	18	10	6.90	4.69
	OPD12	5′CACCGTATCC3′	850 to 1,031	3	2	0.72	0.62
	OPD18	5′GAGAGCCAAC3′	450 to >3,000	16	4	6.90	4.86
	OPD20	5′ACCCGGTCAC3′	400 to 3,000	24	6	11.27	7.90
	OPN4	5′GACCGACCCA3′	500 to 2,200	16	3	8.54	5.58
	OPN16	5′AAGCGACCTG3′	200 to 2,100	18	11	6.54	4.72
	OPN18	5′GGTGAGGTCA3′	400 to 1,800	14	3	5.81	4.26
	OPAF5	5′CCCGATCAGA3′	200 > 3,000	15	5	6.90	4.36
	OPAF14	5′GGTGCGCACT3′	400 to 3,000	13	4	4.90	3.80
	OPAF15	5′CACGAACCTC3′	550 to 2,000	10	4	3.45	2.71
	Total		–	317*	111	–	–
	Mean		–	16.7	5.8	6.99	5
	Range		200 to >3,000	3–29	2–12	0.72–11.27	0.62–8.26
ISSR	SPS 1	(GAC)5	200 to 1,200	16**	2	14.72	4.92
	SPS 2	(GTGC)4	250 to 2,000	17	7	10.90	5.52
	SPS 3	(GACA)4	300 to 3,000	22	4	16.54	8.16
	SPS 4	(AGG)6	325 to 1,700	21	6	13.27	7.20
	SPS 5	(GA)9T	350 to 2,000	19	2	12.00	6.64
	SPS 6	T(GA)9	325 to 1,400	17	6	10.54	6.67
	SPS 7	(GTG)5	250 to 1,200	15**	6	11.09	4.79
	SPS 8	(GGA)4	300 to 1,800	20**	3	18.36	7.93
	Total		–	147	36	–	–
	Mean		–	18.37	4.5	13.42	6.47
	Range		200 to >3,000	15–22	2–7	10.54–18.36	4.79–8.16

* All are polymorphic bands, ** contain one monomorphic band, rest are all polymorphic

species is 16, 18 and 30 with RAPD, ISSR and SSR, respectively. Jaccard similarity showed that the species *C. aromatica* and *C. amada* were closely related with a similarity value of 0.4. Most distantly placed species were *C. aromatica* and *Z. cassumunar* with a similarity value of 0.132. The average similarity was found to be 0.19 (Table 3).

The dendrogram constructed through SHAN clustering and UPGMA analysis using Jaccard's similarity coefficient by RAPD, ISSR and SSR divided it into two clusters consisting *Zingiber* and *Curcuma* species. The dendrogram obtained presents two main clusters with ten species. The first cluster was again divided into two sub-clusters, the

first having species *Z. chrysanthum*, *Z. clarkei* and *Z. rubens* whereas the second subcluster was consisting of *Z. officinale* cv. Suprava, *Z. zerumbet* and *Z. cassumunar*, respectively. The second main cluster consists of *C. aromatica*, *C. longa*, *C. amada* and *C. caesia*. The two clusters had bootstrap value of 100 which proved its stability. The two subclusters were also stable having bootstrap value 72 and 88, respectively. Mantle matrix correspondence test revealed high matrix correlation in all the cases. Correlation values for RAPD, ISSR and SSR were 0.797, 0.84 and 0.8, respectively, with combined data. ISSR and RAPD exhibited very good correlation value ($r = 0.7$) with each other.

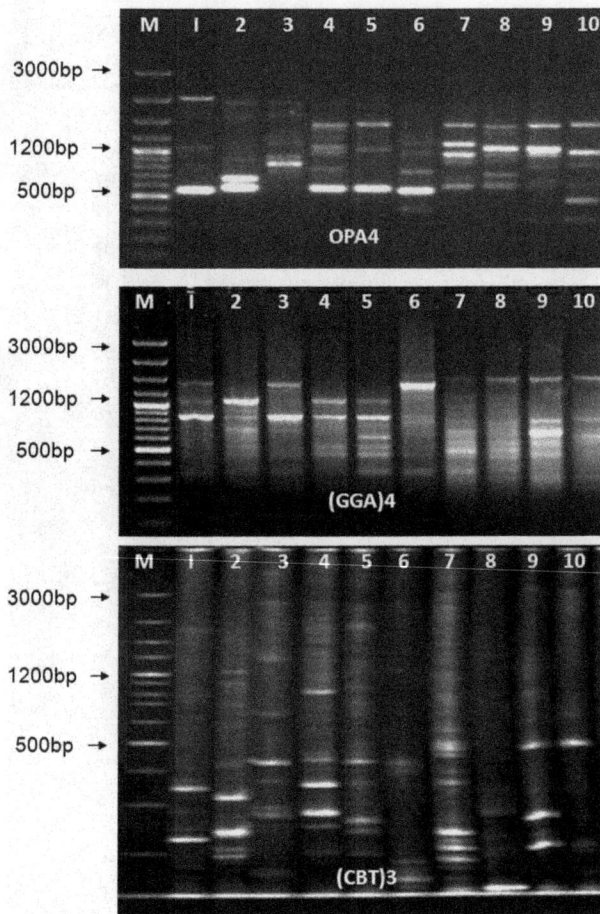

Fig. 1 RAPD, ISSR and SSR banding pattern in 10 species of Zingiberaceae with primer OPA4,(GGA)4 and (CBT)3 (M = gene ruler 100 bp ladder. Lane 1–10 represents, *Z. officinale*, *Z. rubens*, *Z.zerumbet*, *Z. chrysanthum*, *Z. clarkii*, *Z. cassumunar*, *C. longa*, *C. amada*, *C. aromatica*, *C. caesia*

Discussion

Accurate identification and characterization of different germplasm resources is important for species identification, cultivar development, certification and breeder's right's protection (Hale et al. 2006). With the advent of molecular biology techniques, DNA-based markers very efficiently augment morphological, cytological, and biochemical characters in germplasm characterization, varietal identification, clonal fidelity testing, assessment of genetic diversity, validation of genetic relationship, phylogenetic and evolutionary studies, marker-assisted selection, and gene tagging. Owing to plasticity, ubiquity and stability, DNA markers are easier, efficient and less time consuming, especially in perennials where morphological markers are few. The relatively easy to use, low-cost, and highly accurate nature of the polymerase chain reaction (PCR)—

based technologies such as RAPD, ISSR, AFLP, and microsatellites are widely appreciated. Although work on morphological characterization of *Zingiberaceous species* has been attempted, its molecular characterization is still in a nascent stage except for some genetic fidelity studies of micropropagated plants and isozyme-based characterization (Sasikumar 2005; Nayak et al. 2005; Mandal et al. 2007).

All the species taken in the present investigation were segregated into two different groups. Interestingly, all species under *Curcuma* and *Zingiber* genus were exclusively grouped under the respective genera. Within the *Zingiber* genus *Z. cassumunar* was found to be isolated from rest of the *Zingiber* species which may be due to its very rare distribution in wild. *Z. officinale* was separated to a distinct clade as it was the only cultivated species. Similarly, *C. longa* was separated from others as it was also a cultivated species.

Chen et al. (1999) used RAPD polymorphism to differentiate within and among *Curcuma wenyujin*, *C. sichuanensis*, and *C. aromatica*. Kress et al. (2002) studied the phylogeny of the gingers (*Zingiberaceae*) based on DNA sequences of the nuclear internal transcribed spacer (ITS) and plastid *matK* regions. Our result was in agreement with their report that the Zingiber and Curcuma genus were grouped separately. ISSR and SSR markers showed similar grouping with average similarity of 0.311 and 0.296, respectively, whereas RAPD showed less average similarity among species. This could be due to higher polymorphism observed in the above marker. The combined data showed less similarity among species in comparison to SSR and ISSR which might be due to more number of loci observed with RAPD marker. Similar clustering pattern was reflected when correlation was calculated with respect to pooled data versus RAPD, ISSR and SSR markers (0.797, 0.84 and 0.8, respectively).This showed that all the markers were equally effective in characterizing the studied species of Zingiberaceae. A phylogenetic analysis of the tribe Zingibereae (Zingiberaceae) was performed by Ngamriabsakul et al. (2003) using nuclear ribosomal DNA (ITS1, 5.8S, and ITS2) and chloroplast DNA (trnL [UAA] 5′ exon to trnF [GAA]). The study indicated that tribe Zingibereae is monophyletic with two major clades, the *Curcuma* clade, and the *Hedychium* clade. The genera *Boesenbergia* and *Curcuma* are apparently not monophyletic.

Application of PCR-based molecular markers like RAPD, ISSR, SSR and AFLP had been proved by many workers in molecular characterization of different species (Mukerjee et al. 2003; Dikshit et al. 2007; Fu et al. 2008; Lu et al. 2009).

Table 2 Details of SSR analysis in 10 species of Zingiberaceae

Primer	Primer sequence	Approx fragment Size(bp)	Total bands	Polymorphic Bands	Pol (%)	Monomorphic bands	Unique bands	Resolving power	SSR primer index
CBT-02	F:TCCTCCCTCCCTTCGCCCACTG R:CGATGTTCGCCATGGCTGCTCC	180 to 2,500	47	44	93.6	3	7	38.18	16.82
CBT-03	F:ATCAGCAGCCATGGCAGCGAC R:AGGGGATCATGTGCCGAAGGC	100 to 3,000	47	46	97.8	1	4	35.63	18.44
CBT-04	F:ACCCTCTCCGCCTCGCCTCCTC R:CTCCTCCTCCTGCGACCGCTCC	<100 to 2,900	52	51	98.1	1	5	39.45	20.23
CBT-05	F:CTCTGTCTCCTCCCCCGCGTCG R:TCAGCTTCTGGCCGGCCTCCTC	<100 to 3,000	49	48	97.9	1	13	36.72	16.13
CBT-06	F:GCCTCGAGCATCATCATCAG R:ATCAACCTGCACTTGCCTGG	<100 to 3,000	42	40	95.2	2	20	18.36	10.51
CBT-07	F:CGATCCATTCCTGCTGCTCGCG R:CGCCCCCATGCATGAGAAGAG	<100 to >3,000	34	31	91.2	3	15	18.36	8.33
CBT-08	F:CAGCAGATTTTTGCTCCG R:GTCGCGTTCGTGGAAAT	<100 to >3,000	33	32	96.9	1	6	19.63	11.37
CBT-09	F:AGGGGGCAGTGGAGAG R:ACGTTCCTGCACTTGACG	<100 to 1,600	21	20	95.2	1	3	17.45	7.07
Total		–	325	312	–	13	73	–	–
Mean		–	40.62	39	95.7	1.6	9.1	27.97	13.6
Range		<100 to >3,000	21–52	20–51	91–98	1–3	3–20	17.4–39.4	7.07–20.23

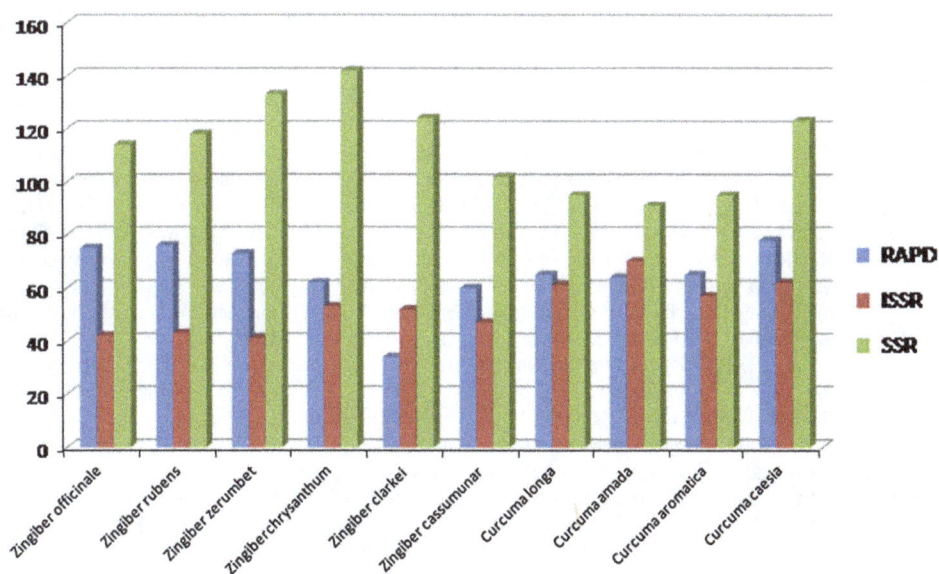

Fig. 2 Total number of bands amplified in ten species of zingiberaceae with RAPD, ISSR and SSR primers

In this report, three PCR-based markers have been used to characterize 10 important species of Zingiberaceae. The uniqueness of the specific bands amplified in different species of Zingiberaceae would help in certification of species as well as in protection of breeder's right. A correct phylogeny can be established among all constituent taxa of Zingiberaceae through molecular characterization. The potential of the technique will be further realized to fullest extent for identification and tagging of important and novel gene in different taxa, unexplored yet, thus facilitating the conservation and improvement of desired taxa of Zingiberaceae.

Table 3 Jaccard's similarity coefficient among 10 species of Zingiberaceae for RAPD, ISSR and SSR

	Z. officinale	Z. rubens	Z. zerumbet	Z. chrysanthum	Z. clarkei	Z. cassumunar	C. longa	C. amada	C. aromatica	C. caesia
Z. officinale	1.0000									
Z. rubens	0.1754	1.0000								
Z. zerumbet	0.2183	0.1863	1.0000							
Z. chrysanthum	0.1787	0.2487	0.2081	1.0000						
Z. clarkei	0.2012	0.2197	0.2012	0.3019	1.0000					
Z. cassumunar	0.1949	0.1463	0.2139	0.2094	0.2025	1.0000				
C. longa	0.1539	0.1577	0.1914	0.1991	0.1848	0.1691	1.0000			
C. amada	0.1429	0.2036	0.1781	0.1644	0.2270	0.1461	0.2757	1.0000		
C. aromatica	0.1652	0.1741	0.1333	0.1614	0.2174	0.1324	0.2796	0.4091	1.0000	
C. caesia	0.1458	0.1542	0.1447	0.1717	0.1553	0.1342	0.2793	0.3139	0.3810	1.0000

Acknowledgments　We are thankful to Prof. Manoj Ranjan Nayak, President, Siksha O Anusandhan University for his encouragement and support. The research is funded by Department of Biotechnology, Govt. of India.

Conflict of interest　The authors declare that they have no conflict of interest.

References

Chen Y, Bai S, Cheng K, Zhang S, Nian LZ et al (1999) Random amplified polymorphic DNA analysis on *Curcuma wenuujin* and *C. sichuanensis*. Zhongguo ZhongYao Za Zhi 24:131–133

Das A, Kesari V, Satyanarayana VM, Parida A, Rangan L et al (2011) Genetic relationship of Curcuma species from Northeast India using PCR-based markers. Mol Biotechnol 49:65–70

Dikshit HK, Jhang T, Singh NK, Koundal KR, Bansal KC, Chandra N, Tickoo JL, Sharma TR et al (2007) Genetic differentiation of Vigna species by RAPD URP and SSR markers. Biol Plant 519(3):451–457

Dreisigacker S, Zhang P, Warburton ML, Skovmand B, Hoisington D, Melchinger AE (2005) Genetic diversity among and within CIMMYT wheat landrace accessions investigated with SSRs and implications for plant genetic resources management. Theor Appl Genet 101(1–2):653

Fu X, Ning G, Gao L, Bao M et al (2008) Genetic diversity of *Dianthus* accessions as assessed using two molecular marker systems (SRAPs and ISSRs) and morphological traits. Sci Hort 117:263–270

Hale AL, Farnham MW, Menz MA (2006) Effectiveness of PCR-based markers for differentiating elite broccoli inbreds. J Am Soc Hortic Sci 131:418–423

Islam MA, Meister A, Schubert V, Kloppstech K, Esch E et al (2007) Genetic diversity and cytogenetic analysis in *Curcuma zedoaria* (Christm.) Roscoe. from Bangladesh. Genet Resour Crop Evol 54:149–156

Jatoi SA, Kikuchi A, Yi SS, Naing KW, Yamanaka S, Junko A et al (2006) Use of SSR markers as RAPD markers for genetic diversity analysis in Zingiberaceae. Breed Sci 56:107–111

Jiang H, Xie Z, Koo HJ, Mclaughhn SP, Timmermann BN, Gang DR et al (2006) Metabolic profiling and phylogenetic analysis of medicinal *Zingiber* species tools for authentication of gingers (*Zingiber officinale*). Phytochemistry 67:1673–1685

Kress WJ, Prince LM, Williams KJ (2002) The phylogeny and a new classification of the gingers (*Zingiberaceae*): evidence from molecular data. Am J Bot 89:1682–1696

Lee SY, Fai WK, Zakaria M, Ibrahim H, Othman RY, Gwag JG, RAO VR, Park YJ et al (2007) Characterization of polymorphic microsatellite markers, isolated from ginger (*Zingiber officinale* Rosc.). Mol Ecol Notes 7:1009–1011

Lu X, Liu L, Gong Y, Song X, Zhu X et al (2009) Cultivar identification and genetic diversity analysis of broccoli and its related species with RAPD and ISSR markers. Sci Hort 122(4):645–648

Mandal AB, Thomas VA, Elanchezhian R (2007) RAPD pattern of *Costus spaciosus* Koen ex. Retz., an important medicinal plant from the Andaman and Nicober Islands. Curr Sci 93(3):369–373

Mukerjee AK, Acharya LK, Mattagajasingh I, Panda PC, Mohapatra T, Das P et al (2003) Molecular characterization of three *Heritiera* species using AFLP markers. Biol Plant 47:445–448

Naik PK, Alam MA, Singh H, Goyal V, Parida S, Kalia S, Mohapatra T et al (2010) Assesment of genetic diversity through RAPD, ISSR and AFLP markers in Podophyllum hexandrum: a medicinal herb from northeastern Himalayan region. Physiol Mol Biol Plant 16(2):145–148

Nayak S, Naik PK, Acharya LK, Mukherjee AK, Panda PC, Das P (2005) Assessment of genetic diversity among 16 promising cultivars of ginger using cytological and molecular markers. Z Naturforsch 60c:485–492

Ngamriabsakul C, Newman MF, Cronk QCB (2003) The phylogeny of tribe *Zingibereae* (*Zingiberaceae*) based on its (nrDNA) and *trn*l–f (cpDNA) sequences. Edinb J Bot 60:483–507

Noli E, Cont S, Maccaferi M, Sanguineti MC et al (1997) Molecular characterization of Tomato cultivars. Seed Sci Technol 27:1–10

Paramasivam M, Poi R, Banerjee H, Bandopadhyay A et al (2009) High performance thin laye chromatographic method for quantitative determination of curcuminoids in Curcuma longa germplasm. Food Chem 113:640–644

Rohlf FJ (1997) NTSYS-pc Numerical taxonomy and multivariate analysis system. Version 2.02e. Exeter Software, Setauket, New York

Roldan-Ruiz I, Dendauw J, Van Bockstaele E, Depicker A, De Loose M (2000) AFLP markers reveal high polymorphic rates in ryegrasses (*Lolium* spp.). Mol Breed 6:125–134

Sasikumar B (2005) Genetic resources of Curcuma: diversity characterization and utilization. Plant Genet Resour Conser Util 3:230–251

Sharma A, Namedo AG, Mahadik KR (2008) Molecular markers: net prospects in plant genome analysis. Pharmacogn Rev 2(3):23–34

Syamkumar S, Sasikumar B (2007) Molecular marker based genetic diversity analysis of curcuma species from India. Sci Hortic 112:235–241

Williams JKF, Kubelik AR, Livak KG, Rafalki JA, Tingey SV et al (1990) DNA polymorphisms amplified by arbitrary primers are useful as genetic markers. Nucl Acids Res 18:6531–6535

Zeitkiewicz E, Rafalski A, Labuda D et al (1994) Genome finger printing by simple sequence repeat (SSR)-anchored PCR amplification. Genomics 20:176–183

Zhou X, Zhangwan L, Liang G, Zhub J, Wang D, Cai Z et al (2007) Analysis of volatile components of *C. sichanesis* X. X. Chen by gas chromatography-mass spectrometry. J Pharm Biomed Anal 43:440–444

9

Molecular characterization of an endophytic *Phomopsis liquidambaris* CBR-15 from *Cryptolepis buchanani* Roem. and impact of culture media on biosynthesis of antimicrobial metabolites

H. C. Yashavantha Rao · Parthasarathy Santosh · Devaraju Rakshith · Sreedharamurthy Satish

Abstract An endophytic fungus *Phomopsis liquidambaris* CBR-15, was isolated from *Cryptolepis buchanani* Roem. (Asclepiadaceae) and identified by its characteristic culture morphology and molecular analysis of the ITS region of rDNA and intervening 5.8S rRNA gene. The impact of different culture media on biosynthesis of antimicrobial metabolites was tested by disc diffusion assay. Polyketide synthase gene (PKS) of the endophytic fungus was investigated using three pairs of degenerate primers LC1–LC2c, LC3–LC5c and KS3–KS4c by PCR. TLC-bioautography method was employed to detect the antimicrobial metabolites. Antimicrobial metabolites fractionated with ethyl acetate extract showed significant antimicrobial activity against the test bacteria and fungi. Biosynthesis of antimicrobial metabolites was optimum as depicted by zone of inhibition from ethyl acetate extract cultured in potato dextrose broth. Strain CBR-15 was identified as *Phomopsis liquidambaris* and PKS genes of the fungus were amplified with LC3–LC5c and KS3–KS4c sets of degenerate primers. These findings suggest that endophytic *P. liquidambaris* CBR-15 harbor iterative type I fungal PKS gene domain which indicates the biosynthetic potential of endophytic fungi as producers of natural antimicrobial metabolites. The study also demonstrates the utilization and optimization of different culture media which best supports for the biosynthesis of the antimicrobial metabolites from *P. liquidambaris*.

Keywords Endophytic fungus · *Phomopsis liquidambaris* · *Cryptolepis buchanani* · Polyketide synthase gene · Antimicrobial metabolites

The sequence data of this fungus is deposited in GenBank under the accession no. KF032029.

H. C. Yashavantha Rao · D. Rakshith · S. Satish (✉)
Department of Studies in Microbiology, University of Mysore, Manasagangotri, Mysore 570 006, Karnataka, India
e-mail: satish.micro@gmail.com

P. Santosh
Plant Biotechnology Division, Unit of Central Coffee Research Institute, Coffee Board, Manasagangotri, Mysore 570 006, Karnataka, India

Introduction

Natural products have been the major potential sources of chemical diversity while driving pharmaceutical discovery over the past century (Mishra and Tiwari 2011). Despite the present focus on synthetic products, natural products serve as continuing source of novel bioactive metabolites, retaining an immense impact on modern medicine (Wang et al. 2012). Microbial endophytes are viewed as an outstanding and unexplored source of novel bioactive natural products because many of them occupy literally millions of unique biological niches growing in a variety of unusual environments (Verma et al. 2007). Endophytic fungi are known to be as potential resources for producing bioactive compounds (Aly et al. 2010). They have proven to be promising sources of new and biologically active metabolites which are of interest for specific medicinal or agrochemical applications (Strobel and Daisy 2003). Infrequently, endophytic fungi capable of producing their host plant compounds have been discovered (Eyberger et al. 2006; Kusari et al. 2008, 2009a, b; Kusari and Spiteller 2011; Shweta et al. 2010). Based on knowledge of the chemistry and biology of endophytic fungi, the isolation of natural products can give us a platform to replace the existing synthetic drugs that provide resistance to pathogens and contaminate safe environment (Gond et al. 2012).

Polyketides have a great commercial interest for drug discovery and account for medicinal sales exceeding $20 billion per year (Cheng et al. 2009). They are large family of structurally diverse natural products found in plants, fungi and bacteria. The study of PKS gene in natural environments may provide important ecological insights, in addition to opportunities for antimicrobial drug development (Zhao et al. 2008). The production of antibiotics by filamentous fungi can be enhanced by genetic modification, mixed culture fermentation, immobilization of the cells, optimization of fermentation conditions or enzymes induction (Oyama and Kubota 1993; Ho et al. 2003). Even minor variations in the environment or nutrition have the potential to impact the quantity and diversity of fermentation products. As an initial step in media optimization, nutritional array could be applied to cognize the conditions in which they would be more apt to produce antibiotics or secondary metabolites, resulting in enriched biological activity (Bills et al. 2008).

Cryptolepis buchanani Roem. and Schult. belongs to the family Asclepiadaceae, a climbing tree which is widely used in folk medicine in Southeast Asia (Laupattarakasem et al. 2003). It also plays a great medicinal value in Ayurveda as anti-diarrheal, anti-inflammatory and blood purifier (Kaul et al. 2003). In view of this, *C. buchanani* was selected for the isolation of fungal endophytes.

Here, we report for the first time on incidence of endophytic fungus from *C. buchanani* Roem. which comprises KS domain of fungal PKS gene as indicators of bioactivity. The impact of different culture media on biosynthesis of antimicrobial metabolites from *P. liquidambaris* CBR-15 was evaluated. The endophytic fungus has been identified by molecular analysis of the ITS region of rDNA containing ITS1, ITS2 and the intervening 5.8S rRNA gene.

Materials and methods

Collection site and source of endophytic fungus

Mysore (12.3°N 76.6°E, elevation 754 m) is located in the Southern part of India which has an annual mean temperature of 30 °C with about 786 mm precipitation per annum. *C. buchanani* Roem. was selected for the present study from this region. The plant is located in the campus of Mysore University. Healthy asymptomatic leaf, bark and root samples were collected and brought to the laboratory in an icebox which was used to isolate endophytic fungus within 24 h of collection.

Isolation of endophytic fungus

Collected samples were washed thoroughly in running tap water followed by distilled water before processing. To eliminate the epiphytic microorganisms, all the samples were initially rinsed with 70 % ethanol for 2 min and surface sterilized by sodium hypochlorite (4 %) for 5 min and again rinsed with 70 % ethanol for 30 s. The samples were rinsed two times in sterile double distilled water and allowed for surface drying in sterile conditions. The plant materials were cut into small segments (5 mm size) and placed on water agar plates (distilled water, 1.5 % agar) amended with chloramphenicol (250 ppm) and incubated at 30 °C for 3–4 days to few weeks till the growth initiated. The hyphal tips, that emerged from the plant tissues were picked and maintained on PDA plates for further studies (Wang et al. 2012). The endophytic fungus used in the present study is maintained in the Department of Studies in Microbiology, University of Mysore, India (Voucher number: MGMB/DCC/02/2012).

Culture media

Four different culture media were used: Potato dextrose broth (PDB), Malt extract broth (MEB), Yeast extract sucrose broth (YSB) and Mycological broth (MCB).

Fermentation and extraction

The endophytic fungus was cultured in 1-l Erlenmeyer flasks containing 500 ml of each different culture media for 3 weeks at 25 °C under static conditions. The culture broth was then filtered to separate the culture filtrate and mycelium. Culture filtrate was blended thoroughly and centrifuged at 4,000 rpm for 5 min. Liquid supernatant was extracted with an equal volume of ethyl acetate thrice separately and this was evaporated to dryness under reduced pressure at 45 °C using rotary flash evaporator (Buatong et al. 2011). All the experiments were conducted in triplicates.

Test microorganisms

Gram-positive bacteria *Staphylococcus aureus* (MTCC 7443), *Bacillus subtilis* (MTCC 121), *Listeria monocytogenes* (MTCC 839). Gram-negative bacteria *Escherichia coli* (MTCC 7410), *Salmonella typhi* (MTCC 733), *Pseudomonas aeruginosa* (MTCC 7903), *Candida albicans* (MTCC 183) and *Fusarium verticillioides*.

Antimicrobial susceptibility testing

The determination of antimicrobial susceptibility testing was carried out by disc diffusion assay. The sterile discs

(5 mm) were impregnated with 20 µl (100 µg/disc) of ethyl acetate extracts obtained from different culture media. The discs impregnated with ethyl acetate extract were dried in laminar hood and placed on the surface of the media already seeded with test microorganisms in Petri plates. One control disc impregnated with only 20 µl of ethyl acetate was also placed for each test organism with a positive control. The plates were incubated at 37 ± 2 °C and room temperature (for test bacteria and fungi, respectively) and the diameter of the zone of inhibition was measured (Sadrati et al. 2013).

Statistical analyses

Statistical analysis of results was performed using IBM SPSS version 20 (2011). One-way ANOVA (analysis of variance) at value $p < 0.001$ followed by Tukey's Post Hoc test with $p < 0.05$ was used to determine the significant differences between the results obtained in each experiment.

Molecular characterization of the strain CBR-15

Isolation of genomic DNA

The endophytic fungus was cultured in potato dextrose broth for 7 days at 30 °C under shaking conditions and the resultant mycelium was harvested by vacuum filtration and stored at −70 °C. The chilled mycelia were ground with mortar and pestle under liquid nitrogen then transferred into an Eppendorf microcentrifuge tube with 1 ml of pre-warmed (65 °C) $2 \times$ CTAB extraction buffer (2 % w/v CTAB, 100 mM Tris–HCl, 1.4 M NaCl, 20 mM EDTA, 1 % β-mercaptoethanol, pH 8.0), and then incubated in a 65 °C water bath for 60 min with occasional gentle swirling. After centrifugation, the aqueous phase of the mixture containing the total DNA was reextracted with an equal volume phenol:chloroform:isoamyl alcohol (25:24:1). The residual phenol was removed with chloroform:isoamyl alcohol (24:1) twice. DNA in the aqueous phase was precipitated by adding 2 volume ethanol and 0.1 volume 3 M NaAc (pH 5.2) and then incubated at −20 °C overnight. The DNA pellet was washed with 70 % ethanol twice, and suspended in 50 µl of TE buffer (10 mM Tris–HCl, 1 mM EDTA, pH 8.0) (Kim et al. 2010).

PCR amplification of ITS region of rDNA

The ITS regions of the fungus were amplified by ITS primers, ITS1 (5′ TCCGTAGGTGAACCTGCGG 3′) and ITS4 (5′ TCCTCCGCTTATTGATATGC 3′) (White et al. 1990). The PCR amplification was carried out in 0.2 ml PCR tubes, using Master cycler personal (Eppendorf). The PCR reaction mixture (50 µl) contained 5 µl 10 × PCR buffer containing 15 mM $MgCl_2$, 5 µl 2 mM deoxynucleoside triphosphates mix (dNTPs mix), 2 µl of each primer (5 pmol/µl), 4 µl template DNA, 2 µl (1 U/ml) *Taq* polymerase and deionised water (30 µl). Thermal cycling conditions were as follows: initial denaturation (4 min at 95 °C), followed by 30 cycles of denaturation (94 °C for 50 s), annealing (51 °C for 1 min), and primer extension (72 °C for 1 min), followed by final extension for 10 min at 72 °C. Amplification products were electrophoretically resolved on 1.4 % (w/v) agarose gel containing ethidium bromide (0.5 µg/ml), using 1 × TAE buffer at 70 V (Bhagat et al. 2012).

Amplification of ketosynthase domain of fungal PKS gene of strain *Phomopsis iquidambaris* CBR-15

Three pairs of degenerate primers, LC1 and LC2c, LC3 and LC5c (Bingle et al. 1999), KS3 and KS4c (Nicholson et al. 2001), which are ketosynthase (KS) domain specific primers were used to amplify the KS domain sequence of the PKS genes of *P. liquidambaris* by PCR (Nicholson et al. 2001). PCR reactions (50 µl) contained approximately 4 µl genomic DNA template, 5 µl 10 × PCR buffer, 4 µl 2.5 mM of each dNTP, 3 µl of each primer, 1 µl of 2 U/µl *Taq* DNA polymerase and 30 µl deionised water. The thermal cycling program was as follows: 5 min at 94 °C; 34 cycles of 1 min at 94 °C, 1.5 min at 55 °C, 3 min at 72 °C and 10 min at 72 °C.

TLC-bioautography assay

The antimicrobial activity of ethyl acetate extract was investigated by thin layer chromatography (TLC) using the bioautographic agar overlay method (Valgas et al. 2007). 10 µl of ethyl acetate extract cultured in each different media was spotted on precoated TLC silica gel plates (TLC, ALUGRAM® SIL G/UV$_{254}$, Machereye-Nagel, Germany) in an optimized solvent system of chloroform and methanol (9:1). The developed TLC plates were observed under visible light and UV light at 254 nm and 365 nm, respectively. The developed TLC plates were air dried and UV sterilized for 30 min. The TLC plates were then encased in sterile Petri plates and overlaid with Brain heart infusion medium (for *S. aureus*), Mueller–Hinton medium (for *E. coli*) and Sabouraud dextrose medium (for *C. albicans*) containing 0.65 % agar incorporated with 1 mg ml^{-1} 2,3,5-triphenyl tetrazolium chloride (Sigma-Aldrich) inoculated with 1 % standardized microbial inocula. After 8 h of diffusion at 8 °C, the plates were incubated for 24 h at 37 °C for bacteria and for 48–72 h at 25 °C for fungi, then for fungi the upper agar was sprayed with [3-(4,5 dimethylthiazol-2-yl)-2,5 diphenyltetrazolium bromide] (MTT) (Sigma-Aldrich) 5 mg mL–1 which was

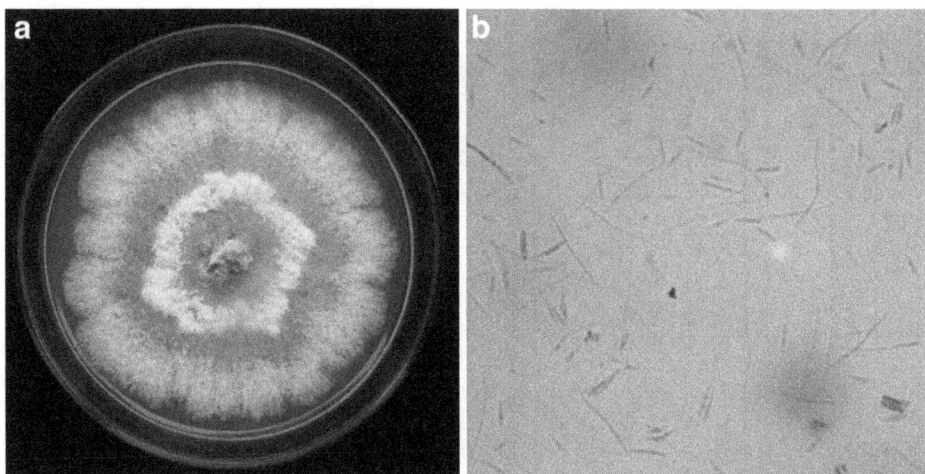

Fig. 1 a Colony morphology of *P. liquidambaris* CBR-15 on PDA and **b** microscopic features at ×40 magnification showing alpha and beta conidia

converted to a formazan dye by the test fungi. Inhibition zones were observed as clear spots against a red and purple background for bacteria and fungi, respectively. The areas of inhibition on the active spot were compared with the R_f value of the related spots on the reference TLC plate.

Results and discussion

Isolation of endophytic fungi and colony morphology endophytic fungal strain CBR-15

The endophytic fungus *P. liquidambaris* CBR-15 which is used in the present study was isolated from the leaf tissue of *C. buchanani*. Colony morphology on PDA, after 5 days at 25 °C, initially cottony, white to olive gray later on turns to gray to light brown and a radiating growth pattern, margin regular. Conidiomata eustromatic, black, spherical to irregular on the upper regions; Conidiophores: short to elongated, aseptate to septate and branched, Conidia are of two types: (a) alpha conidia hyaline, straight, aseptate, forming white to yellowish cirrhi; (b) beta conidia hyaline, filiform, straight or curved, aseptae morphology characters observed, were closely related with the genus *Phomopsis*, belongs to class coelomycetes (Fig. 1).

Impact of culture media on antimicrobial activity

Antimicrobial activity was determined by disc diffusion assay to assess the relative concentration of the active metabolites in ethyl acetate extract cultured in different media. The strain CBR-15 cultured on different media composition exhibited significantly distinct impacts on their antimicrobial activity. Different media compositions induced comparable antimicrobial responses in an individual culture media; however, minor variations were also observed. Interestingly, the optimal response in respect to

antimicrobial activity was observed in the ethyl acetate extract cultured in PDB when compared to rest of the media (Table 1). This may be due to the need for certain nutritional supplements, which may serve as precursors, for the biosynthesis of bioactive secondary metabolites in endophytic fungi (Tong et al. 2011). Gram-positive bacteria were more susceptible than Gram-negative bacteria. Ethyl acetate extract of *P. liquidambaris* cultured in PDB exhibited zone of inhibition 22.33 ± 0.33 and 21.00 ± 0.00 mm for *S. aureus* and *E. coli*, respectively. Extract from YSB medium exhibited moderate antimicrobial response followed by PDB (Table 1).

During disc diffusion assay, antimicrobial activity of ethyl acetate extracts cultured in different media were analyzed. The utilization of different mycological media as nutritional supplements can impact on the production of bioactive secondary metabolites. Application of multiple fermentation conditions is the desirable method that could enhance the probability of successful discovery of bioactive metabolites from a given strain (Bills et al. 2008). One such way to trigger the production of secondary metabolites is to vary the medium composition. The principle behind this method, named as one strain—many compounds (OSMAC) approach, is to expose the microorganism to other cultivating conditions than the standards used in laboratories (Fuchser and Zeeck 1997; Schiewe and Zeeck 1999; Hofs et al. 2000; Bills et al. 2008). Media composition, temperature, pH, culture vessel, aeration, cultivation time, light intensity can increase or reduce the production of the bioactive compounds by the strain (Bode et al. 2002; Siqueria et al. 2011). Yenn et al. (2012) reported anti-candidal activity of *Phomopsis* sp. ED2 cultured in yeast extract sucrose (YES) broth with aqueous extract of host plant. However, understanding of the exact mechanisms for the change in metabolic profile due to change in culture or fermentation conditions is usually not completely understood and

Table 1 Antimicrobial activity of ethyl acetate extract of endophytic *P. liquidambaris* CBR-15 fermented in different culture media against test microorganisms by disc diffusion assay (100 µg/disc)

Culture media[A]	Test microorganisms							
	Bacteria						Fungi	
	Gram-positive			Gram-negative			Yeast	Mold
	S. aureus	*B. subtilis*	*Listeria monocytogenes*	*S. typhi*	*E. coli*	*P. aeruginosa*	*C. albicans*	*F. verticillioides*
PDB	22.33 ± 0.33[b]	24.66 ± 0.33[b]	22.00 ± 0.57	20.00 ± 0.00[b]	21.00 ± 0.00[a]	14.66 ± 0.33[b]	17.66 ± 0.66[b]	20.33 ± 0.33[a]
MPY	16.33 ± 0.33[e]	18.33 ± 0.33[d]	14.33 ± 0.33[e]	14.66 ± 0.33[d]	18.33 ± 0.33[d]	12.00 ± 0.57[c]	12.66 ± 0.33[d]	17.00 ± 0.57[b]
YSB	20.33 ± 0.33[c]	20.33 ± 0.33[c]	18.66 ± 0.33[c]	17.33 ± 0.33[c]	20.00 ± 0.57b	12.33 ± 0.33[c]	14.66 ± 0.33[c]	17.33 ± 0.66[b]
MCB	18.33 ± 0.33[d]	18.66 ± 0.33[d]	16.33 ± 0.33	15.00 ± 0.00d	16.66 ± 0.33[d]	13.66 ± 0.33[bc]	13.33 ± 0.33[cd]	16.33 ± 0.33[b]
Gentamicin (C)	28.00 ± 0.00[a]	33.33 ± 0.33[a]	24.33 ± 0.33[a]	30.33 ± 0.33[a]	29.66 ± 0.33[a]	20.33 ± 0.33[a]	ND	ND
Nystatin (C)	ND	ND	ND	ND	ND	ND	21.00 ± 0.00[a]	22.00 ± 0.00[a]

Value represents diameter of zone of inhibition in mm. Data are means from three replicates ± SE and those representing similar superscripts in the appropriate columns are not significantly different (ANOVA, Tukey's HSD at $p \leq 0.05$). C—positive control; Gentamicin—10 µg/disc, Nystatin—100 µg/disc

ND not determined

[A] See "Materials and methods" for abbreviation

therefore difficult to predict (Bode et al. 2002). This work demonstrates that PDB serves as optimum culture media for the biosynthesis of antimicrobial metabolites which facilitates isolation and characterization of antimicrobial metabolites from *P. liquidambaris* (Fig. 2).

Molecular identification and amplification of ketosynthase domain of *Phomopsis* sp., PKS gene

The amplified ITS region of rDNA was sequenced and aligned with the ITS sequences of the different organisms retrieved from NCBI databases, using CLUSTAL W (Thompson et al. 1994). Dendrogram was generated using neighbor joining (NJ) plot and the boot strapping was carried out using 1,000 replications. The acquisition of ITS1-5.8S-ITS2 sequence data and NJ plot showed that the isolate belongs to *P. liquidambaris* CJBB25-20, KC895530 (Fig. 3) which is also an endophytic fungus isolated from *Saraca asoca* (http://www.ncbi.nlm.gov/nuccore/KC895530). The partial ITS sequence data of this fungus was deposited in GenBank, under accession no. KF032029.

Traditional bioprospecting of microbial endophytes recently initiated a genetic-based screening program of culturable endophytes to identify strains capable of producing bioactive secondary metabolites. Endophytic fungi have been genetically screened for the presence of PKS genes as indicators of bioactivity. In the present investigation, the genomic DNA of *P. liquidambaris* CBR-15 was amplified by LC3–LC5c and KS3–KS4c sets of degenerate primers (Fig. 4). Fungal PKSs are defined as iterative type I synthases and are classified into three groups based on the extent of reduction of the polyketide ring produced, namely non-reduced, partially reduced and highly reduced PKSs (Nicholson et al. 2001). To identify the presence of iterative type I PKS in *P. liquidambaris* CBR-15, LC1–LC2c, LC3–LC5c and KS3–KS4c sets of degenerate primers were used which are specific for the particular types of the non-reduced, partially reduced and highly reduced KS domains of endophytic fungal PKS gene, respectively (Lin et al. 2010). From this study, *P. liquidambaris* CBR-15 might be capable of producing bioactive polyketide metabolites which are partially or highly reduced in nature due the amplification by LC3-LC5c and KS3-KS4c set of degenerate primers. The DNA of endophytic fungal strains isolated from *Annona squamosa* was investigated for PKS gene by KS domain specific primers. All three KS domains were present in the strains belonging to the Diaporthales (Lin et al. 2010). Investigation on PKS diversities in natural environment appears as an addition to opportunities for the development of microbial drugs which may provide important ecological insights (Zhao et al. 2008).

Fig. 2 Antimicrobial activity of ethyl acetate extract of *P. liquidambaris* CBR-15 cultured in different media by disc diffusion assay against *E. coli* (*2a* and *2b*) and *B. subtilis* (*2c* and *2d*) where +*VE* positive control, −*VE* negative control and PDB,MPY, YSB, MCB are the different culture media (see "Materials and methods" for abbreviation) extract of *Phomopsis liquidambari* CBR-15

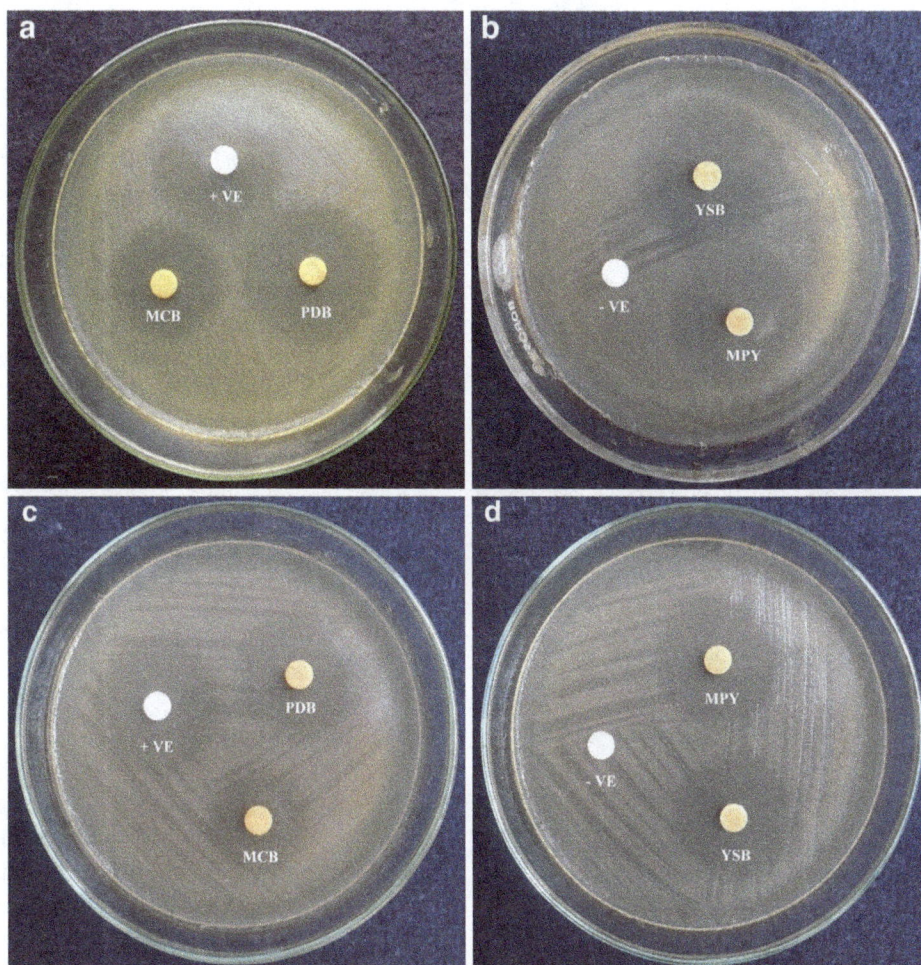

Fig. 3 ITS sequence-based Neighbor Joining tree of *Phomopsis* sp. isolates. A consensus NJ dendrogram with bootstrap values (1,000 replications) based on multiple sequence alignment. *Scale bar* indicated nucleotide substitutions per nucleotide position. * denotes the isolate obtained in the present study (accession no. KF032029)

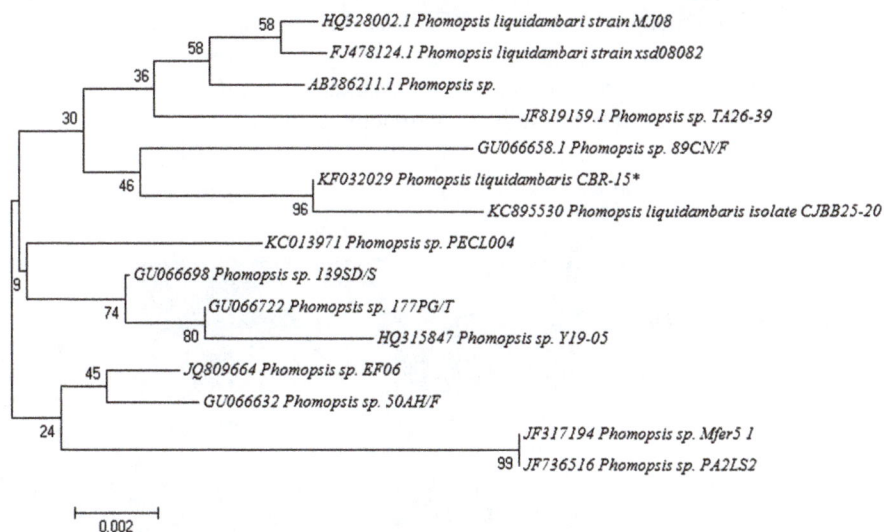

TLC-bioautography

Metabolite profiling of ethyl acetate extract inclined from PDB medium by TLC showed better resolution of metabolites compared to rest of the media. Two major spots were observed under UV light of 254 and 365 nm, respectively. On TLC, no comparable good resolution of metabolites was observed other than cultured in PDB. This implies that the availability of nutrient supplements in PDB medium enhanced the production of bioactive metabolites of

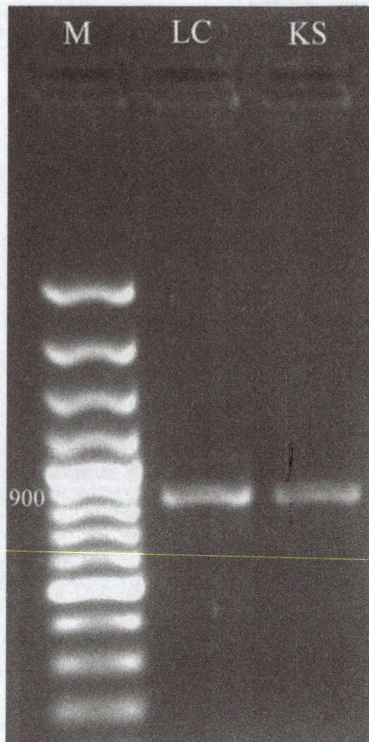

Fig. 4 PCR amplification of polyketide synthase gene (amplicon size about 900 bp) from *P. liquidambaris* CBR-15 by LC3–LC5c and KS3–KS4c pairs of degenerate primers. Lane: *M* 100 bp DNA ladder; *LC* LC3–LC5c degenerate primers; *KS* KS3–KS4c degenerate primers

P. liquidambaris CBR-15. Certain nutrients act as environmental factors, quantitatively and qualitatively affecting the production of antimicrobial metabolites (Tabbene et al. 2009).

In the bioautography assay, the ethyl acetate extract inclined from PDB culture filtrate exhibited antimicrobial activity by producing zone of inhibition at R_f value 0.56 against *S. aureus*, *E. coli* and *C. albicans* (Fig. 5). A mild

activity was also observed from YSB culture at similar R_f value against *S. aureus* and *C. albicans*, but it did not show any activity against *E. coli*. This indicates that the compound against *E. coli* can only be produced when the *Phomopsis liquidambaris* strain CBR-15 is cultured in PDB. A spot with a similar R_f was not observed in the ethyl acetate extract from MPY and MCB media. Yenn et al. (2012) reported anti-candidal activity of *Phomopsis* sp. ED2 cultured in yeast extract sucrose broth with addition of host extract, but in our study we reported the antimicrobial activity of *Phomopsis* sp. CBR-15 without host extract. The present study for the detection of antimicrobial compound by TLC is one of the simplest, economical and reproducible methods for drug discovery from natural products (Ahmed 2008; Hota 2010; Marston 2011; Patra et al. 2012). Further investigation is needed to characterize the antimicrobial metabolite.

Conclusions

Our finding implies that *P. liquidambaris* CBR-15, an endophytic fungus of *C. buchanani*, has antimicrobial properties and PDB is the best supporting media for the biosynthesis of antimicrobial metabolites. Bioactive natural compounds produced by endophytic fungi may provide new alternatives to address the problem of drug resistance development by human pathogens and multi-drug resistance microorganisms. The genome mining strategy employed here might assist strain prioritization for the isolation and characterization of antimicrobial metabolites with polyketide biosynthetic origin. This work is the first report on incidence of endophytic fungus inhabiting *C. buchanani* Roem. which comprises KS domain of fungal PKS gene as indicators of bioactivity.

Fig. 5 TLC-bioautography agar over lay assay of ethyl acetate extract cultured in PDB against **a** *S. aureus* and **b** *C. albicans*

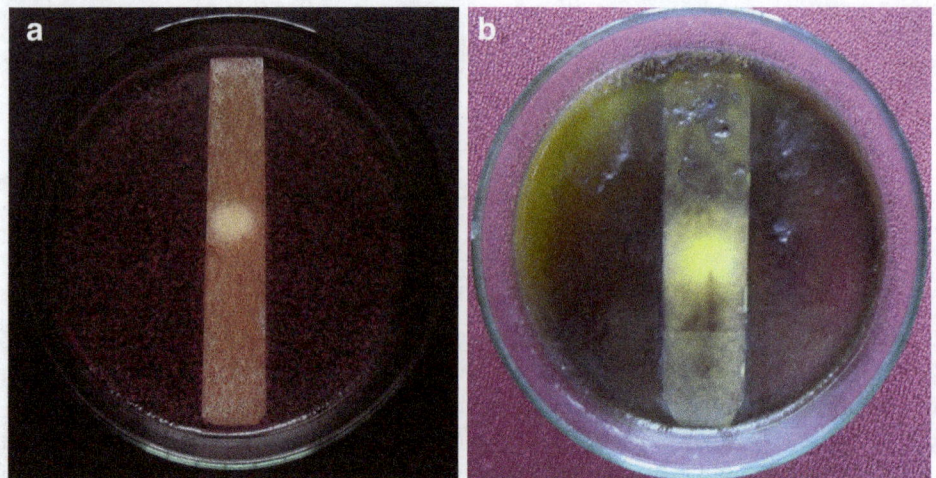

Acknowledgments This work was supported by University Grant Commission (UGC-Major Research Project) and Institution of Excellence, MHRD, New Delhi, India. Authors thank Department of Studies in Microbiology, University of Mysore for laboratory and instrumentation facilities.

Conflict of interest The authors declare that they have no conflict of interest.

References

Ahmed MKK (2008) Introduction to isolation, identification and estimation of lead compounds from natural products. In: Hiremath SR (ed) Textbook of industrial pharmacy. Orient Longman Private Ltd, Chennai, pp 345–379

Aly AH, Debbab A, Kjer J, Proksch P (2010) Fungal endophytes from higher plants: a prolific source of phytochemicals and other bioactive natural products. Fungal Divers 41(1):1–16

Bhagat J, Kaur A, Sharma M, Saxena AK, Chadha BS (2012) Molecular and functional characterization of endophytic fungi from traditional medicinal plants. World J Microbiol Biotechnol 28:963–971

Bills GF, Platas G, Fillola A, Jimenez MR, Collado J, Vicente F et al (2008) Enhancement of antibiotic and secondary metabolite detection from filamentous fungi by growth on nutritional arrays. J Appl Microbiol 104:1644–1658

Bingle LE, Simpson TJ, Lazarus CM (1999) Ketosynthase domain probes identify two subclasses of fungal polyketide synthase genes. Fungal Genet Biol 26:209–223

Bode HB, Bethe B, Hofs R, Zeeck A (2002) Big effects from small changes: possible ways to explore nature's chemical diversity. Chembiochem 3:619–627

Buatong J, Phongpaichit S, Rukachaisirikul V, Sakayaroj J (2011) Antimicrobial activity of crude extracts from mangrove fungal endophytes. World J Microbiol Biotechnol 27:3005–3008

Cheng YQ, Coughlin LM, Lim SK, Shen B (2009) Type I polyketide synthases that require discrete acyltransferases. Methods Enzymol 459:165–186

Eyberger AL, Dondapati R, Porter JR (2006) Endophyte fungal isolates from *Podophyllum peltatum* produce podophyllotoxin. J Nat Prod 69:1121–1124

Fuchser J, Zeeck A (1997) Aspinolides and aspinonene/aspyrone co-metabolites, new pentaketides produced by *Aspergillus ochraceus*. Liebigs Ann Recl 1997(1):87–95

Gond SK, Mishra A, Sharma VK, Verma SK, Kumar J, Kharwar RN et al (2012) Diversity and antimicrobial activity of endophytic fungi isolated from *Nyctanthes arbortristis*, a well-known medicinal plant of India. Mycoscience 53:113–121

Ho WH, To PC, Hyde KD (2003) Induction of antibiotic production of freshwater fungi using mix-culture fermentation. Fungal Divers 12:45–51

Hofs R, Walker M, Zeeck A (2000) Hexacyclinic acid, a polyketide from *Streptomyces* with a novel carbon skeleton. Angew Chem Int Ed 39:3259–3261

Hota D (2010) Evalution of plant extracts. In: Bioactive medicinal plants. Gene-Tech Books, New Delhi, pp 86–87

Kaul A, Bani S, Zutshi U, Suri KA, Satti NK, Suri OP (2003) Immunopotentiating properties of *Cryptolepis buchanani* root extract. Phytother Res 17:1140–1144

Kim JS, Seo SG, Jun KB, Kim JW, Kim SH (2010) Simple and reliable DNA extraction method for the dark pigmented fungus, Cercospora sojina. Plant Pathol J 26(3):289–292

Kusari S, Spiteller M (2011) Are we ready for industrial production of bioactive plant secondary metabolites utilizing endophytes? Nat Prod Rep 28:1203–1207

Kusari S, Lamshöft M, Zühlke S, Spiteller M (2008) An endophytic fungus from *Hypericum perforatum* that produces hypericin. J Nat Prod 71:159–162

Kusari S, Lamshöft M, Spiteller M (2009a) *Aspergillus fumigatus* Fresenius, an endophytic fungus from *Juniperus communis* L. Horstmann as a novel source of the anticancer pro-drug deoxypodophyllotoxin. J Appl Microbiol 107:1019–1030

Kusari S, Zühlke S, Spiteller M (2009b) An endophytic fungus from Camptotheca acuminate that produces camptothecin and analogues. J Nat Prod 72:2–7

Laupattarakasem P, Houghton PJ, Hoult JR, Itharat A (2003) An evaluation of the activity related to inflammation of four plants used in Thailand to treat arthritis. J Ethnopharmacol 85:207–215

Lin X, Jian Y, Zhong J, Zheng H, Su WJ, Qian XM et al (2010) Endophytes from the pharmaceutical plant, *Annona squamosa*: isolation, bioactivity, identification and diversity of its polyketide synthase gene. Fungal Divers 41:41–51

Marston A (2011) Thin-layer chromatography with biological detection in phytochemistry. Planar Chromatogr 13:2676–2683

Mishra BB, Tiwari VK (2011) Natural products: an evolving role in future drug discovery. Eur J Med Chem 46:4769–4807

Nicholson TP, Rudd BA, Dawson M, Lazarus CM, Simpson TJ, Cox RJ (2001) Design and utility of oligo nucleotide gene probes for fungal polyketide synthases. Chem Biol 8:157–178

Oyama M, Kubota K (1993) Induction of antibiotic production by protease in *Bacillus brevis* (ATCC8185). J Biochem 113:637–641

Patra JK, Gouda S, Sahoo SK, Thatoi HN (2012) Chromatography separation, ^{1}H NMR analysis and bioautography screening of methanol extract of *Excoecaria agallocha* L. from Bhitarkanika, Orissa, India. Asian Pac. J. Trop. Biomed 2(1):S50–S56

Sadrati N, Daoud H, Zerroug A, Dahamna S, Bouharati S (2013) Screening of antimicrobial and antioxidant secondary metabolites from endophytic fungi isolated from wheat (*Triticum durum*). J Plant Prot Res 53(2):128–136

Schiewe HJ, Zeeck A (1999) Cineromycins, γ-Butyrolactones and ansamycins by analysis of the secondary metabolite pattern created by a single strain *Streptomyces*. J Antibiot 52:635–642

Shweta S, Zühlke S, Ramesha BT, Priti V, Kumar PM, Ravikanth G, Spiteller M, Vasudeva R, Shaanker RU (2010) Endophytic fungal strains of Fusarium solani, from *Apodytes dimidiata* E. Mey. ex Arn (Icacinaceae) produce camptothecin, 10-hydroxycamptothecin and 9-methoxycamptothecin. Phytochemistry 71:117–122

Siqueria VMD, Conti R, Araujo JMD, Motta CMS (2011) Endophytic fungi from the medicinal plant *Lippia sidoides* Cham. and their antimicrobial activity. Symbiosis 53(2):89–95

Strobel G, Daisy B (2003) Bioprospecting for microbial endophytes and their natural products. Microbiol Mol Biol R 67(4):491–502

Tabbene O, Slimene IB, Djebali K, Mangoni ML, Urdaci MC, Limam F (2009) Optimization of medium composition for the production of antimicrobial activity by *Bacillus subtilis* B38. Biotechnol Prog 25(5):1267–1274

Thompson JD, Higgins DG, Gibson TJ (1994) CLUSTAL W: improving the sensitivity of progressive multiple sequence alignment through sequence weighting, position-specific gap penalties and weight matrix choice. Nucleic Acids Res 11(22):4673–4680

Tong WY, Darah I, Latiffah Z (2011) Antimicrobial activities of endophytic fungal isolates from medicinal herb *Orthosiphon stamineus* Benth. J Med Plants Res 5:831–836

Valgas C, Souza SM, Smania EFA, Smania AJ (2007) Screening methods to determine antibacterial activity of natural products. Braz J Microbiol 38:369–380

Verma VC, Gond SK, Kumar A, Kharwar RN, Strobel GA (2007) Endophytic mycoflora of bark, leaf, and stem tissues of *Azadirachta indica* A. Juss. (neem) from Varanasi (India). Microb Ecol 54:119–125

Wang LW, Xu G, Wang JY, Su ZZ, Lin FC, Zhang CL et al (2012) Bioactive metabolites from *Phoma* species, an endophytic

fungus from the Chinese medicinal plant *Arisaema erubescens*. Appl Microbiol Biotechnol 93:1231–1239

White TJ, Bruns T, Lee S, Taylor JW (1990) PCR Protocols: A Guide to Methods and Applications Amplification and direct sequencing of fungal genes for phylogenetics. In: Innis M, Gelfand DH, Sninsky JJ, White TJ (eds) Academic Press, San Diego, pp 315–322

Yenn TW, Lee CC, Ibrahimand D, Zakaria L (2012) Enhancement of anti-candidal activity of endophytic fungus *Phomopsis* sp. ED2, isolated from *Orthosiphon stamineus* Benth, by incorporation of host plant extract in culture medium. J Microbiol 50:581–585

Zhao J, Yang N, Zeng R (2008) Phylogenetic analysis of type I polyketide synthase and non-ribosomal peptide synthetase genes in Antarctic sediment. Extremophiles 12:97–105

10

Bamboo: an overview on its genetic diversity and characterization

Lucina Yeasmin · Md. Nasim Ali · Saikat Gantait ·
Somsubhra Chakraborty

Abstract Genetic diversity represents the heritable variation both within and among populations of organisms, and in the context of this paper, among bamboo species. Bamboo is an economically important member of the grass family Poaceae, under the subfamily Bambusoideae. India has the second largest bamboo reserve in Asia after China. It is commonly known as "poor man's timber", keeping in mind the variety of its end use from cradle to coffin. There is a wide genetic diversity of bamboo around the globe and this pool of genetic variation serves as the base for selection as well as for plant improvement. Thus, the identification, characterization and documentation of genetic diversity of bamboo are essential for this purpose. During recent years, multiple endeavors have been undertaken for characterization of bamboo species with the aid of molecular markers for sustainable utilization of genetic diversity, its conservation and future studies. Genetic diversity assessments among the identified bamboo species, carried out based on the DNA fingerprinting profiles, either independently or in combination with morphological traits by several researchers, are documented in the present review. This review will pave the way to prepare the database of prevalent bamboo species based on their molecular characterization.

Keywords *Bambusa* · Biodiversity · *Dendrocalamus* · Molecular marker

Abbreviation

AFLP	Amplified fragment length polymorphism
ISSR	Inter simple sequence repeat
RAPD	Randomly amplified polymorphic DNA
RFLP	Restriction fragment length polymorphism
SNP	Single nucleotide polymorphism
SSR	Simple sequence repeat

L. Yeasmin · Md. N. Ali (✉) · S. Chakraborty
Department of Agricultural Biotechnology, Faculty Centre for Integrated Rural Development and Management, School of Agriculture and Rural Development, Ramakrishna Mission Vivekananda University, Ramakrishna Mission Ashrama, Narendrapur, Kolkata 700103, India
e-mail: nasimali2007@gmail.com

S. Gantait
Department of Crop Science, Faculty of Agriculture, Universiti Putra Malaysia, 43400 Serdang, Selangor, Malaysia

S. Gantait
Department of Biotechnology, Instrumentation and Environmental Science, Bidhan Chandra Krishi Viswavidyalaya, Mohanpur, WB 741252, India

Introduction

Bamboo: Taxonomy

Bamboo, the fastest growing perennial, evergreen, arborescent plant is a member of the grass family (i.e., Poaceae) and constitutes a single subfamily Bambusoideae (Kigomo 1988). The Bambusoideae subfamily includes both herbaceous bamboo or Olyreae tribe and woody bamboos or Bambuseae tribe (Ramanayake et al. 2007). The Bambuseae tribe differs from the Olyreae on the basis of the presence of abaxial ligule (Zhang and Clark 2000; Grass Phylogeny Working Group 2001). The most recent classification systems have placed 67 genera of woody bamboos under nine subtribes, mainly depending on various floral characters (Dransfield and Widjaja 1995; Li 1997).

Area and distribution

The distribution of bamboos on planet earth extends from 51°N latitude in Japan (Island of Sakhalin) to 47°S latitude in South Argentina. A total number of 1,400 bamboo species are distributed worldwide. The bamboo can grow in an altitudinal range which extends from just above the sea level up to 4,000 m (Behari 2006). About 14 million hectares of the earth surface is covered by bamboos with 80 percent in Asia (Tewari 1992). The major species richness is found in Asia-pacific followed by South America, whereas the least number of species is found in Africa (Bystriakova et al. 2003). It has been reported that Europe has no native bamboo species (Liese and Hamburg 1987). According to FAO, total area under bamboo cultivation is 11,361 ha as in 2005, of which 1,754 ha is under private ownership. Herbaceous bamboo constitutes about 110 species which are mainly concentrated in the Neotropics of Brazil, Paraguay, Mexico and West Indies. The natural bamboo forest covers approximately 600,000 ha area across Brazil, Peru and Bolivia, which is known as "Tabocais" in Brazil and "Pacales" in Peru (Filgueiras and Goncalves 2004 cited in Das et al. 2008). The Bambuseae tribe includes about 1,290 species worldwide and constitutes three major groups (Das et al. 2008). The Paleotropical

woody bamboo is distributed in the tropical and subtropical regions of Africa, Madagascar, Sri Lanka, India, Southern Japan, Southern China and Oceania. The Neotropical woody bamboos are distributed in Southern Mexico, Argentina, Chile and West Indies. The north temperate woody bamboos are found in the North Temperate Zone and a small amount at a higher elevation of Madagascar, Africa, India and Sri Lanka (http://www.eeob.iastate.edu/bamboo/maps.html) (Fig. 1). Bamboo can thrive in hot, humid rainforests to cold resilient forests. It can tolerate as well as can grow in extreme temperature of about −20 °C. It also can tolerate excessive precipitation ranging from 32 to 50 inch. annual rainfall (Goyal et al. 2012).

Bamboo in India

India is the second richest country in bamboo genetic resources following China, ranking first in this aspect (Bystriakova et al. 2003). Several reports have been found regarding the species richness of bamboo in India. Bahadur and Jain (1983) reported about 113 bamboo species, whereas reports on the number of species varies from 102 (Ohrnberger 2002) to 136 (Sharma 1980). In India, 9.57 million ha which is about 12.8 % of the total forest area of the country is covered by bamboo plantation (Sharma

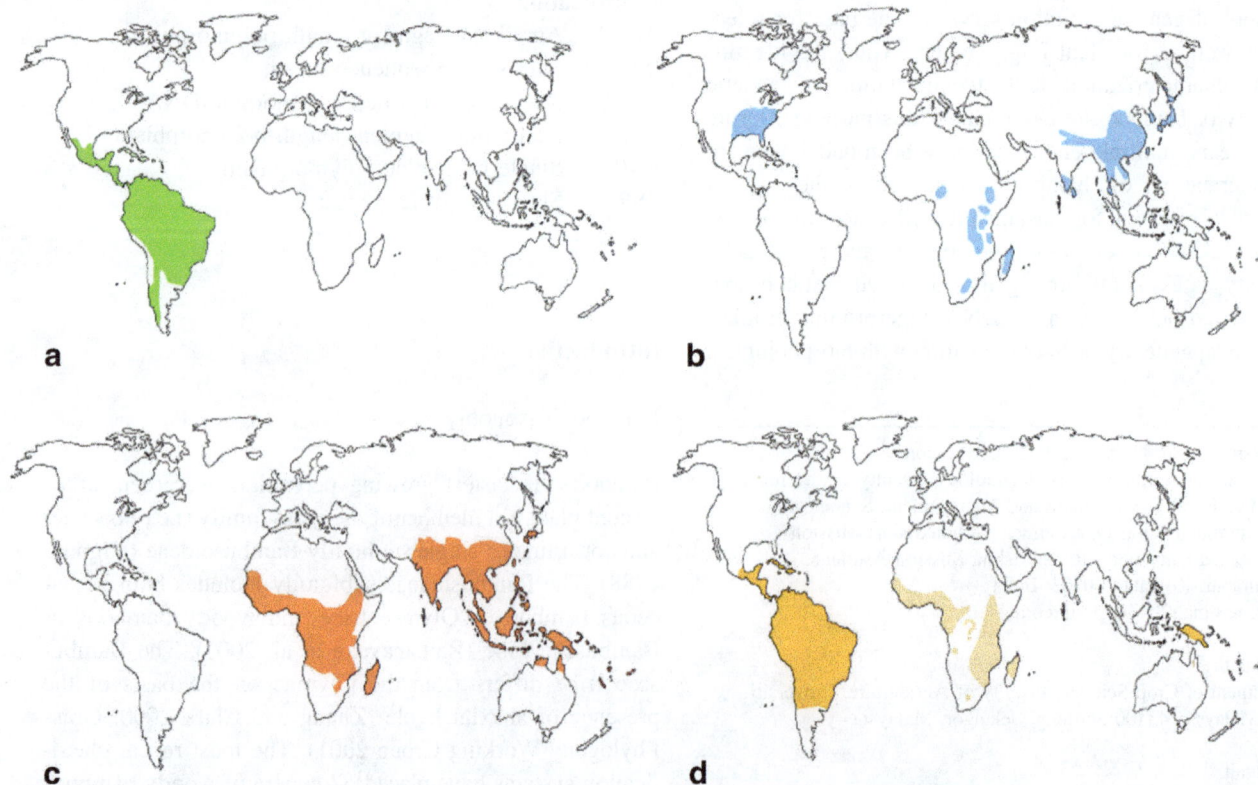

Fig. 1 Worldwide distribution of bamboo. **a** Neotropical woody bamboos, **b** north temperate woody bamboos, **c** paleotropical woody bamboos, and **d** herbaceous bamboos (Source: http://www.eeob.iastate.edu/bamboo/maps.html)

1980). As opined by many scientists, the distribution of bamboo is greatly influenced by human interventions (Boontawee 1988). However, Gamble (1896) has earlier reported that the distribution of bamboo in India is related to rainfall. Varmah and Bahadur (1980) in another report have associated the preferential distribution of different bamboo species with different agroclimatic zones of India. The alpine region comprises *Arundinaria* and *Thamnocalamus*, whereas, these two genera grow in the temperate region along with *Phyllostachys*. *Arundinaria*, *Bambusa* and *Dendrocalamus* grow in the subtropical region, the tropical moist region allows to grow *Bambusa*, *Dendrocalamus*, *Melocanna*, *Ochlandra* and *Oxytenanthera*; on the other hand, in the dry tropical region *Dendrocalamus* and *Bambusa* is predominant (Ahmed 1996) (Fig. 2).

Chromosome and genetic status

Bamboo under the subfamily Bambusoideae is the giant member of grass family (Kigomo 1988). The basic chromosome number of most woody bamboo is 12 ($x = 12$), whereas in herbaceous bamboo it is 11 ($x = 11$) (Grass Phylogeny Working Group 2001). Two different polyploidy groups are present in woody bamboo. The tropical woody bamboo is hexaploid ($2n = 6x = 72$) and the temperate woody bamboo is tetraploid ($2n = 4x = 48$)

Fig. 2 *Dendrocalamus* and *Bambusa*, the two most predominant bamboo species distributed in subtropical, tropical moist and tropical dry agroclimatic zones of India. **a** *Bambusa balcooa*, **b** *Bambusa* bambos, **c** *Bambusa tulda*, **d** *Dendrocalamus asper*, **e** *Dendrocalamus hamiltonii*, **f** *Dendrocalamus strictus* (Source: Authors)

(Clark et al. 1995). After employing an exhaustive chromosome analysis on 185 species from 33 genera and 6 subtribes, Ruiyang (2003) has reported the variation in the chromosome numbers for some species of *Bambusa* and *Dendrocalamus*. The genomic DNA content of tropical woody bamboo is larger than that of the temperate woody bamboo as estimated by flow cytometric analysis (Gielis et al. 1997). The recent flow cytometric analysis of tetraploid bamboo *Phyllostachys heterocycla* var. *pubescens* has estimated its genome size to be 2.075 Gb (Peng et al. 2013). The number of ESTs of bamboo deposited till January 2009 is 3,087 (Peng et al. 2010) and total number of nucleotide sequences deposited till November 2009 is 17,789 (which is <0.1 % of the total sequences from grass family). As reported by Peng et al. (2013), the moso bamboo genome contains 43.9 % GC and 59.0 % transposable elements. In the same article, they have reported 31,987 protein coding genes and the average length of protein coding gene is 3,350 bp.

Bamboo and its usage: the poor man's timber

Bamboo has age-old connection with the material needs of rural people (Mukherjee et al. 2010). Porter-field (1933) suggested that "bamboo is one of those providential developments in nature which, like the horse, the cow, wheat and cotton, have been indirectly responsible for man's own evolution". Bamboo plays manifold role in day-to-day rural life or broadly speaking human life. It plays a crucial role in cultural, artistic, industrial, agricultural, construction and household needs of human beings (McNeely 1995). The utility of bamboo shoots and leaves vary from pickle preparation (Khatta and Katoch 1983) to preparation of traditional medicine. Bamboo has medicinal values too. It has multiple wide uses in ayurveda. Solvent extraction of *P. pubescens* and *P. bambusoideae* showed strong antioxidant activity (Mu et al. 2004). The mature bamboo leaves contain phenolic acids and root contains cyanogenic glycosides (Das et al. 2012). From the adult bamboo culms, high-quality charcoal is produced (Park and Kwon 1998). In some parts of South East Asia, life starts with a knife made of bamboo as the umbilical cord of a new born baby is cut by it (McNeely 1995), and also it is utilized for the procedure of circumcision of a male child (Skeat 1900). Bamboo plays a significant role in paper and pulp industry. As reported by Sharma et al., the national demand of bamboo was 5 million tonnes by 1987 out of which 3.5 million tonnes were required for paper and pulp industry. *B. balcooa* is generally chosen for construction purposes and fiber-based mat board and panel manufacture (Ganapathy 1997), but for its mechanical strength it is also utilized to produce quality paper pulp (Das et al. 2005). Bamboo is considered as "green gold" keeping in mind its

economic importance and multiple end uses in human life (Bhattacharya et al. 2009). Bamboo is widely utilized for making various musical instruments, e.g., flutes are made from hollow bamboo (Kurz 1876). Bamboo is widely utilized for construction purposes. It has manifold application in construction of house such as making pillars, floors, doors and windows, room separator, rafters etc. (Das et al. 2008). It is also utilized for making guard wall of water bodies and river bank. Bamboo is an efficient agent for preventing soil erosion and conserving soil moisture (Christanty et al. 1996, 1997; Mailly et al. 1997; Kleinhenz and Midmore 2001).

Bamboo and biodiversity

In addition to the countless direct human uses, bamboo plays an imperative role in many other ways; as per Kratter's survey, 25 out of 440 bird species living in the Amazon forest are confined to bamboo thickets. Elephants (*Elephas maximus*), wild cattle (*Bos gaurus* and *B. javanicus)* and various species of deer (*Cervidae*) and primates (including macaques Macaca and leaf monkeys *Presbytis),* pigs (Suidae), rats and mice (Muridae), porcupines (Hystricidae) and squirrels (Sciuridae) are subsidiary feeders on Southeast Asian bamboos. More than 15 Asian bird species nests exclusively in bamboo; many of these are rare and threatened using bamboo as the significant proportion of their habitat (Bird Life International 2000). The world's second smallest bat (*Tylonycteris pachypus*, 3.5 cm) nests between nodes of mature bamboo (*Gigantochloa scortechinii*), which it enters through holes created by beetles. The Asian giant panda (*Ailuropoda melanoleuca*), red panda (*Ailurus fulgens*) and the Himalayan black bear (*Selenarctos thibetanus*) are heavily dependent on bamboo for their feed (Bystriakova et al. 2003). The Red Panda (*A. fulgens*) is recently listed as endangered in the Red Data Book of IUCN. Red Panda mainly lives on bamboo leaves and the destruction of bamboo forest is one of the main reasons for its extinction from the wild (Red panda network, www.redpandanetwork.org). Leaves of *Sasa senanesis, S. kurilensis* and *S. nipponica* constitute a major part of the winter diet for Hokkaido voles (*Clethrionomys rufocanus*), it is when most other plants wither.

Importance of characterization

Overexploitation and genetic erosion of bamboo species have made it necessary not only for the collection and conservation of its germplasms (Thomas et al. 1988; Loh et al. 2000) but also to classify and characterize them (Bahadur 1979; Soderstorm and Calderon 1979; Rao and Rao 2005). Characterization of germplasm is an important link between the conservation and utilization of germplasms (Stapleton and Rao 1995; Nayak et al. 2003). To maintain the germplasms and conservation of biodiversity, the investigation of bamboo resources and even study of their local distribution is indispensable (Goyal et al. 2012); which is recorded to be limited till date.

Identification of bamboo

Identification and classification is necessary for collection and conservation of germplasms (Bahadur 1979; Soderstorm and Calderon 1979). Overexploitation and genetic erosion has posed a foremost need for conservation of bamboo germplasm (Thomas et al. 1988; Loh et al. 2000). In case of any plant, the identification keys are mostly based on floral characters. Depending on the flowering cycle, the bamboos are categorized into three major groups, viz. annual flowering bamboos *(Indocalamus wightianus, Ochlandra sp.)*, sporadic or irregular flowering bamboos *(Chimonobambusa sp., D. hamiltonii)* and gregarious flowering bamboos *(B. bambos, B. tulda, D. strictus, T. spathiflora)* (Das et al. 2008; Bhattacharya et al. 2006, 2009). In case of gregarious flowering, all the members of a common cohort (plants from seeds of common origin) go into reproductive phase simultaneously and subsequently die (Bhattacharya et al. 2009). Incidence of flowering of woody bamboo is uncertain (Ramanayake et al. 2007; Mukherjee et al. 2010). As reported earlier, the reproductive cycle of bamboo is too long, from 3 to 120 years (Janzen 1976), hence making the identification depending on reproductive structure difficult (Bhattacharya et al. 2006, 2009). Consequently, the focus on identification of bamboo has shifted from reproductive to vegetative characters (Bhattacharya et al. 2006; Sharma et al. 2008).

Classification of bamboo was traditionally based on morphological characters; however, recently several other useful taxonomic information such as biochemical, anatomical and molecular characters have also been explored (Stapleton 1997). Even though, characterization of bamboos has so far been done based on morphological characters, yet the classification is not reliable since these are often influenced by ecological factors. Das et al. (2007) in their work have shown that only vegetative characters are unable to distinguish closely related species. The clustering pattern they obtained using the key morphological descriptors was not fully in agreement with the classification pattern of Gamble (1896). The reliability of taxonomic groupings based only on the morphological characters has often been questioned due to the involvement of small number of genes for morphological traits that may not truly reflect the entire scenario of the genome (Brown-Guedira et al. 2000 cited in Das et al. 2007).

Nevertheless, DNA-based marker or molecular markers are not influenced by environment (Ram et al. 2008) and it is thus reliable for diversity analysis. Although application of molecular technique for diversity analysis in bamboo was limited till the beginning of twenty-first century (Loh et al. 2000). In recent years, the application of molecular technology for identification and characterization of bamboo species is predominant.

Morphological traits: key to bamboo identification and characterization

As early as in the year of 1896, J.S. Gamble identified old world bamboos based on various vegetative and reproductive characters. Later, botanists discovered different characters of culm-sheath and other vegetative organs as potential descriptors. Chatterjee and Raizada (1963) prepared a key to identification for 22 bamboo taxa based on culm-sheath morphology. According to them, "The general appearance, size, texture and shape of the sheath and their blades afford good characters for distinguishing the different species". Branching pattern is an important characteristic for identification of genus (Bennet and Gaur 1990). Bennet and Gaur (1990) further suggested the study of young vegetative shoots which sprout annually during rainy season revealing that they are of highly distinct character and hence can be utilized as identification of different species.

To assess the relationship and diversity due to ecological and geographical variation within the members of *Chusquea culeou* species complex, 7 vegetative characters and 14 reproductive or floral characters were studied by Triplett and Clark (2003). The Principle Component Analysis based on vegetative and reproductive characters showed that the variation in the characters is continuous and cannot be used to demarcate the species into a morphologically distinct group. Their study emphasized that additional studies are necessary to resolve the classification of *C. culeou* species complex. In the same year, Clark (2003) identified a new species *C. renvoizei* classified within *Chusquea* section *Swallenochloa* based on different morphological characters. The species is endemic to Bolivia. For identification of that particular species viz. *C. renvoizei*, 10 quantitative and qualitative morphological characters were assessed. A number of vegetative and foliage features distinguished the species from the other members of the Swallenochloa section under *Chusquea* species complex. Lately, Bhattacharya et al. (2006) have described 15 culm and 17 culm-sheath characteristics which they have studied for characterization of *B. tulda*, a sporadically flowering bamboo. They surveyed natural bamboo stands from 17 eco-geographical locations in different districts of West Bengal. Their study was in conformity with the prior taxonomic classification given by Gamble (1896), but a detailed description and illustrations are presented in their article. Phylogenetic relationships among 15 bamboo species were evaluated by Das et al. (2007) using 32 key quantitative and qualitative morphological characters (15 culm and 17 culm-sheath characters), which were previously utilized by Bhattacharya et al. (2006). The cluster pattern obtained from key morphological descriptors was not in conforming to classification of Gamble (1896). The comprehensive morphological characterization was done in a gregarious flowering bamboo species, *T. spathiflorus* (Trin.) Munro subsp. *Spathiflorus* (Bhattacharya et al. 2009). An assembly of 28 key vegetative and reproductive characters was studied by them. The vegetative and floral morphology described was in gross agreement with previous reports given by Naithani et al. (2003) and Clayton et al. (2013). Even though, characterization of bamboos has so far been done based on morphological characters, yet the classification is not reliable since these are often influenced by ecological factors.

Limitation of morphological characters

The reproductive cycle of bamboo is too long, from 3 to 120 years (Janzen 1976). So characterization and identification using floral characters is difficult (Bhattacharya et al. 2006, 2009). In case of bamboo, the taxonomical classification is based on mainly vegetative characters (Ohrnberger 2002). The vegetative characters are influenced by environment and therefore, not reliable for taxonomic classification. Triplett and Clark's (2003) study concluded that only morphological descriptors were unable to demarcate the species into morphologically distinct group. As reported by Das et al. (2007), only vegetative characters are unable to distinguish closely related species. The dendrogram pattern of 15 bamboo species is not in agreement with classification given by Gamble (1896). *B. striata* and *B. wamin* of *Bambusa* genus were separated from other species of *Bambusa*. *D. strictus* was grouped with *B. striata*, *B. wamin* and *B. atra*.

Bamboo and molecular descriptors

Molecular methods have become an indispensable part of most of the genetic diversity assay and in the analyses of breeding system, bottlenecks and other key features influencing genetic diversity patterns. The studies may use Restriction Fragment Length Polymorphism (RFLP), Randomly Amplified Polymorphic DNA (RAPD), Amplified Fragment Length polymorphism (AFLP) or Simple Sequence Repeat (SSR). Nevertheless, it is important to understand that different markers have different properties and will reflect different aspects of genetic diversity (Karp and Edwards 1995). Molecular data can provide useful information to deal with various aspects of taxonomic

classification of plants (Das et al. 2008). "Molecular DNA techniques allow researchers to identify genotypes at the taxonomic level, assess the relative diversity within and among the species and locate diverse accessions for breeding purposes" (Nayak et al. 2003). As reported by Loh et al. (2000), the application of molecular techniques for genetic diversity assessment of bamboo was limited till 2000. The study included RFLP in *Phyllostachys* by Friar and Kochert (1994), isozyme analysis of few selection from 5 genera of bamboo by Heng et al. (1996), chloroplast DNA phylogeny of Asian bamboos by Watanabe et al. (1994) and world bamboo by Kobayashi (1997), analysis of *rpl16* intron sequences in determining phylogenetic relationship within the genus *Chusquea* (Loh et al. 2000).

There are many molecular markers available such as hybridization-based marker or RFLP marker, polymerase chain reaction (PCR) based markers such as RAPD, AFLP, SSR, Inter Simple Sequence Repeats (ISSR), Single Nucleotide Polymorphism (SNP) markers etc. RFLP technique was applied to study the genetic variation and evolution of 20 species of *Phyllostachys* by Friar and Kochert (1994). However, RFLP technique requires fine quality DNA and it shows very low polymorphism in comparison to others. RAPD is a low-cost and rapid method and does not require any information regarding the genome of the plant, and it also has been widely used to determine the genetic diversity in several plants (Belaj et al. 2001; Deshwall et al. 2005). As it is a quick and sensitive method, RAPD can be effectively employed to distinguish useful polymorphism (Ko et al. 1998). It requires very small amount of genomic DNA and can produce very high level of polymorphism and can be effective for diversity analysis in plants (Williams et al. 1990). RAPD analysis has proved its significance for diversity study of field crops like rice (Qian et al. 2001; Rabbani et al. 2008; Pervaiz et al. 2010), many horticultural plants such as coffee (Orozco-Castillo et al. 1994), tea (Wachira et al. 1995), almond (Shiran et al. 2007), sesame (Akbar et al. 2011), turmeric (Singh et al. 2012). It has been employed for phylogenetic relationship study and characterization of bamboo by many recent workers (Nayak et al. 2003; Das et al. 2005; Bhattacharya et al. 2006; Ramanayake et al. 2007; Das et al. 2007; Bhattacharya et al. 2009).

A sum of 98 mapped SSR primers from rice and 20 EST-derived SSR primers from sugarcane was utilized for the evaluation of genetic diversity among 23 bamboo species (Sharma et al. 2008). The study showed that 44 rice SSR and 15 SSR of sugarcane primers produced repeatable amplification in at least one species of bamboo. A total number of 42 out of these 59 primers proved to be efficient for species identification. Two species-specific Sequence Characterized Amplified Region (SCAR) markers were developed by Das et al. (2005). They have developed Bb_{836}

for *B. balcooa* and Bt$_{609}$ for *B. tulda*. The species specificity was confirmed by southern hybridization, and validation was done using 80 accessions of *B. balcooa* and *B. tulda* each. Recent genomic studies in bamboo include the study by Peng et al. (2010) in *P. pubescence* var. *heterocycla*. They have reported 10,608 full-length cDNA sequences of bamboo. Approximately 38,000 Expressed Sequence Tags (ESTs) were generated in this study. In the next year, Zhang et al. (2011) reported full genome sequence of six woody bamboo chloroplast genome (cp DNA). A contemporary study conducted by Gui et al. (2010) reports, identification of syntenic genes between bamboo and other grasses, such as rice and sorghum. They found that the content of repetitive elements (36.2 %) in bamboo is similar to that of rice. It was reported that both rice and sorghum express high genomic synteny with bamboo thus suggesting that these could be utilized as model for decoding tropical bamboo genomes.

Recent study includes phylogenetic analysis of Bambusoideae subspecies (Sungkaew et al. 2009), genome-wide full-length cDNA sequencing (Peng et al. 2010), identification of syntenic genes between bamboo and other grasses (Gui et al. 2010), chloroplast genome sequencing (Zhang et al. 2011) and identification of genes involved in fiber development (Rai et al. 2011). For analysis of genetic diversity of 23 bamboo accessions, 59 SSR from rice and sugarcane were utilized (Sharma et al. 2008). Two species-specific SCAR markers were developed and validated for identification of *B. tulda* and *B. balcooa* (Das et al. 2005) and very recently the draft genome of moso bamboo was reported by Peng et al. (2013). Sixteen novel microsatellite markers were developed for *D. sinicus* by Dong et al. (2012) which will be useful for evaluation of genetic diversity of *D. sinicus*. Very recently, the complete genome sequence of moso bamboo (*P. pubescence* var. *heterocycla*) was reported by Peng et al. (2013). The 2.05 Gb assembly covered 95 % of the genomic region and gene prediction modeling identified 31,987 genes.

RAPD and ISSR: two most potential markers
for bamboo genetic diversity study

RAPD is an inexpensive, simple and rapid technology (Belaj et al. 2001; Deshwall et al. 2005) which has been employed in diversity analysis in plants since its development by Williams et al. (1990). It requires small amount of genomic DNA and can produce high level of polymorphism (Williams et al. 1990). As reported by Ko et al. (1998), while studying genetic relationship within *Viola sp.*, RAPD, being a quick and sensitive method, can be utilized to distinguish polymorphism. RAPD analysis has proved its significance for diversity analysis and identification of germplasms of several plants (Kapteyn and

Simon 2002; Welsh and McClelland 1990). RAPD has several limitations including dominance, uncertain locus homology, and especially sensitivity to the reaction conditions as reported by Qian et al. (2001). According to them to solve some of these problems, Inter Simple Sequence Repeat (ISSR) markers can be put into effect.

ISSRs are the regions that lie within the microsatellite repeats (Joshi et al. 2000) and offer great potential to determine intra-genomic and inter-genomic diversity compared to other arbitrary primers, since they reveal variation within unique regions of the genome at several loci simultaneously. The primer is composed of a microsatellite sequence anchored at 3′ or 5′ end by 2–4 arbitrary, often degenerate nucleotides (Qian et al. 2001). Several properties of microsatellites, such as high variability among taxa, ubiquitous occurrence and high copy number in eukaryotic genomes (Weising et al. 1998), make ISSRs extremely useful markers. They exhibit specificity of sequence-tagged-site markers without the requirement of any prior knowledge of genome sequence for primer synthesis (Joshi et al. 2000). ISSR technique has been employed for phylogenetic relationship study in many crops such as rice (Joshi et al. 2000; Qian et al. 2001). ISSR has become a popular technique for genetic relationship study by many scientists working on bamboo (Lin et al. 2010; Mukherjee et al. 2010).

Several studies have been conducted in rice which include work done by Rabbani et al. in the year of 2008. They have employed RAPD analysis for genetic diversity assessment and identification of 10 traditional, 28 improved and 2 Japanese cultivars of Pakistani rice, where 40 genotypes were grouped into 3 main clusters corresponding to aromatic, non- aromatic and japonica group. A number of improved traditional cultivars originating from different sources did not form well-defined groups and interspersed, indicating no association between the RAPD patterns and geographic origin of the cultivars. Another study was conducted by Qian et al. (2001) for assessment of genetic variation within and among the population of *Oryza granulata* from China using 20 RAPD primers and 12 ISSR primers. Their study showed that RAPD markers revealed a high degree (73.85 %) of genetic variation between the populations residing in two regions; whereas genetic diversity between populations within the same regions was recorded in a very low level. The ISSR primers showed great amount of variation (49.26 %) between two regions coupled with a low level of variation within population and between populations within region. Ten RAPD primers and ten ISSR primers were utilized for detecting DNA polymorphism, identification and genetic diversity study in 16 barley cultivars (Fernandez et al. 2002). One RAPD primer and four ISSR primers were able to distinguish all the cultivars and a strong and quite linear relationship was obtained between resolving power (Rp) of a primer and its ability to discriminate genotypes. RAPD analysis was employed for detection of genetic diversity between *Coffee* species and between *Coffea arabica* genotypes (Orozco-Castillo et al. 1994). The dendrograms were consistent with the known history and evolution of the *C. arabica*. Materials originating from Ethiopia and Arabica sub-groups *C. arabica* var. *typica* and *C. arabica* var. *bourbon* were clearly distinguished. RAPD analysis therefore reflects morphological differences between the sub-groups and the geographical origin of the coffee material.

RAPD analysis was used to estimate genetic diversity and taxonomic relationships in 38 clones belonging to three tea varieties (Wachira et al. 1995). Extensive genetic variability was detected between species, which was partitioned into 'between' and 'within' population components. RAPD analysis was able to discriminate all of the 38 commercial clones, even those which cannot be distinguished on the basis of morphological and phenotypic traits.

RAPD marker was utilized by Shiran et al. (2007) for detection of genetic diversity of Iranian almond cultivar and their relationship to important foreign cultivars and their relatives. RAPD proved to be more efficient in discriminating genotypes than the SSR markers for the same set of genotypes. For genetic diversity assessment of 20 accessions of sesame (*Sesamum indicum* L.), the RAPD technology was employed by Akbar et al. (2011). RAPD technique revealed a high level of genetic variation among the sesame accessions collected from diverse ecologies of Pakistan. This high level of genetic diversity among the genotypes recommended that RAPD technique is valuable for taxonomic classification of sesame and can be helpful for the upholding of germplasm banks and the competent choice of parents in breeding programs. Very recently Singh et al. (2012), for evaluation of genetic diversity of 60 accessions of turmeric (*Curcuma longa*) from 10 different agroclimatic zones, utilized both RAPD and ISSR primers. Using both RAPD and ISSR markers for 60 genotypes, 62 % correlation between genetic similarity and geographical location were demonstrated. The highest genetic diversity was observed in western central table land.

Many researchers have employed RAPD and ISSR for genetic diversity analysis and identification of bamboo. Nayak et al. (2003) utilized thirty decamer random primers on 12 bamboo species for their identification and genetic relationship study. Selected primers were used for identification and for establishing a profiling system to estimate genetic diversity. Cluster analysis revealed two main clusters which are again divided into three mini clusters. Das et al. (2005) developed two species-specific SCAR markers. Their work involved thirty random decamer

primers which were initially screened to detect species-specific markers. Two species-specific RAPD marker Bb$_{836}$ for *B. balcooa* was derived from PW-02 and Bt$_{609}$ for *B. tulda* was derived from OPA-08. Bhattacharya et al. (2006) employed RAPD technology for characterization of *B. tulda*, a gregarious flowering bamboo. The study was conducted based on 32 key morphological characters and 30 random decamer primers (RAPD primers). The molecular clustering pattern is in agreement with classification given by Gamble (1896), while the dendrogram generated from morphological characters differ greatly from it. Phylogenetic relationship among 15 bamboo species was evaluated using morphological and molecular markers (Das et al. 2007). The molecular technique involved RAPD markers, and the dendrogram pattern generated, is in conformity with classical taxonomy. Ramanayake et al. (2007) investigated nine species of bamboo, four (genera) of which are from Sri Lanka, using 41 RAPD primers. Among the four *Bambusa* species, the genetic distances between *B. bambos, B. ventricosa* and *B. vulgaris* were smaller, while *B. atra* differed from them for having greater distance. Smaller genetic distance between *G. atroviolacea* and three *Bambusa* species indicates that *G. atroviolacea* has closure affinity with these three species than *B. atra*. *A. hindsii* which shows greatest distance from all others. In the year 2009, Bhattacharya et al. utilized random decamer primers for molecular characterization of a gregarious flowering bamboo, *T. spathiflorus* subsp. *spathiflorus*. DNA fingerprinting using RAPD markers could not detect any polymorphism either 'between populations' or 'within populations'.

Genetic diversity among twelve accessions of *M. baccifera* from Mizoram was evaluated using RAPD and ISSR markers by Lalhruaitluanga and Prasad (2009). Cluster analysis using Dice similarity coefficient by RAPD markers showed two groups. Similar clustering was found, using Dice similarity coefficient, by ISSR markers. ISSR marker was used for genetic diversity study among 10 cultivars of *P. pubescens* (Lin et al. 2010) where 16 ISSR primers were able to distinguish ten cultivars of *P. pubescens*. Genetic distance and cluster analysis showed that genetic similarity existed between all the cultivars of *P. pubescens* under study. In the year 2010, Mukherjee et al. employed 12 ISSR primers and four EST-based random primers for genetic relationship evaluation among 22 taxa of bamboo. The grouping of genotypes based on Jaccard's similarity matrix using Unweighted Pair Group Method Arithmetic Average, and Principle Coordinate Analysis agreed with earlier reported molecular phylogenetic study only with a few deviations.

Lin et al. (2010) performed crossbreeding of two *Phyllostachys* species and for the identification of their hybrid they utilized eight ISSR primers. Using ISSR markers, they identified three hybrids produced by the cross. The fingerprinting pattern and genetic distance measure suggest that two hybrids were authentic, whereas the third one probably an intraspecies offspring. Genetic diversity among 12 natural populations of *D. membranaceus* was assessed as a preliminary analysis for protection of germplasm resources using ISSR markers (Yang et al. 2012). They have reported a large proportion of genetic variation among the members 'within populations', while the lower genetic variation was found 'among populations'. No significant correlation between genetic and geographic distances 'among populations' was found.

Acknowledgments The authors are grateful to Faculty Centre for Integrated Rural Development and Management, School of Agriculture and Rural Development, Ramakrishna Mission Vivekananda University, India, for providing the key research facilities.

Conflict of interest We, the authors of this article, declare that there is no conflict of interest and we do not have any financial gain from it.

References

Ahmed MF (1996) In: Keynote address: proceedings of the National seminar on bamboo, Bangalore, 28–29 Nov, pp 6–8

Akbar F, Rabbani MA, Masood MS, Shinwari ZK (2011) Genetic diversity of sesame (*Sesamum indicum* L.) germplasm from Pakistan using RAPD markers. Pak J Bot 43:2153–2160

Bahadur KN (1979) Taxonomy of bamboos. Ind J For 2:222–241

Bahadur KN, Jain SS (1983) Rare Bamboos of India. In: Jain SK, Rao PR (eds) An assessment of threatened plants of India. Botanical survey of India, Howrah, pp 265–271

Behari B (2006) Status of Bamboo in India. Compilation of papers for preparation of national status report on forests and forestry in India. Survey and Utilization Division, Ministry of Environment and Forest, 109–120

Belaj A, Trujilo I, Rosa R, Rallo L, Gimenez MJ (2001) Polymorphism and discrimination capacity of randomly amplified polymorphic markers in an olive germplasm bank. J Am Soc Hort Sci 126:64–71

Bennet SSR, Gaur RC (1990) Thirty seven bamboos growing in India. Forest Research Institute, Dehradun

Bhattacharya S, Das M, Bar R, Pal A (2006) Morphological and molecular characterization of *Bambusa tulda* with a note on flowering. Ann Bot 98:529–535

Bhattacharya S, Ghosh JS, Das M, Pal A (2009) Morphological and molecular characterization of *Thamnocalamus spathiflorus* subsp. *spathiflorus* at population level. Plant Syst Evol 282:13–20

Bird Life International (2000) Threatened birds of the World. Lynx Edicions and Bird Life International, Barcelona, Cambridge

Boontawee B (1988) Status of bamboo research and development in Thailand. In: proceedings of the International Bamboo workshop held in Cochin, India, 14–18 Nov, Kerala Forest Research

Brown-Guedira GL, Thompson JA, Nelson RL, Warburton ML (2000) Evaluation of genetic diversity of soybean introductions and North American ancestors using RAPD and SSR markers. Crop Sci 40:815–823

Bystriakova N, Kapos V, Lysenko I, Stapleton C (2003) Distribution and conservation status of forest bamboo biodiversity in the Asia-Pacific region. Biodiversity Conserv 12:1833–1841

Chatterjee RN, Raizada MB (1963) Culmsheaths as an aid to identification of Bamboos. Ind For 89:744–756

Christanty L, Mailly D, Kimmins JP (1996) "Without bamboo, the land dies": biomass, litterfall, and soil organic matter dynamics of a Javanese bamboo talun-kebun system. For Ecol Manag 87:75–88

Christanty L, Kimmins JP, Mailly D (1997) "Without bamboo, the land dies": a conceptual model of the biogeochemical role of bamboo in an Indonesian agroforestry system. For Ecol Manag 91:83–91

Clark LG (2003) A new species of *Chusquea* Sect. *Swallenochloa* (Poaceae: Bambusoideae) from Bolivia. Bamboo Science and Culture. J Am Bamboo Soc 17:55–58

Clark LG, Zhang W, Wendel JF (1995) A phylogeny of the grass family (Poaceae) based on ndhF sequence data. Syst Bot 20:436–460

Clayton WD, Harman KT, Williamson H (2013) Grass Base- the online world grass flora. http://www.kew.org/data/grasses-db.html

Das M, Bhattacharya S, Pal A (2005) Generation and characterization of SCARs by cloning and sequencing of RAPD products: a strategy for species- specific marker development in bamboo. Ann Bot 95:835–841

Das M, Bhattacharya S, Basak J, Pal A (2007) Phylogenetic relationships among the bamboo species as revealed by morphological characters and polymorphism analyses. Biol Plant 51:667–672

Das M, Bhattacharya S, Singh P, Filgueiras TS, Pal A (2008) Bamboo taxonomy and diversity in the Era of molecular markers. Adv Bot Res 47:225–268

Das S, Rizvan Md, Basu SP, Das S (2012) Therapeutic potentials of *Bambusa bambos* Druce. Indo Global J Pharma Sci 2:85–87

Deshwall RPS, Singh R, Malik K, Randhawa GJ (2005) Assessment of genetic diversity and genetic relationships among 29 populations of *Azadirachta indica* A. Juss. using RAPD markers. Genet Resour Crop Evol 52:285–292

Dong YR, Zhang ZR, Yang HQ (2012) Sixteen novel microsatellite markers developed for *Dendrocalamus sinicus* (Poaceae), the strongest woody bamboo in the World. Am J Bot 99:347–349

Dransfield S, Widjaja EA (1995) Plant resources of southeast Asia PROSEA No: 7-Bamboos. Backhuys Publishers, Leiden, Holland

Fernandez M, Figueiras A, Benito C (2002) The use of ISSR and RAPD markers for detecting DNA polymorphism, genotype identification and genetic diversity among barley cultivars with known origin. Theor Appl Genet 104:845–851

Filgueiras TS, Goncalves APS (2004) A checklist of the basal grasses and bamboos in Brazil. Bamboo Sci Cult 18:7–18

Friar E, Kochert G (1994) A study of genetic variation and evolution of Phyllostachys (Bambusoideae: Poaceae) using nuclear restriction fragment length polymorphisms. Theor Appl Gen 89:265–270

Gamble JS (1896) The Bambuseae of British India. Ann R Bot Gard Calcutta 7:1–133

Ganapathy PM (1997) Sources of non wood fiber for paper, board and panels production: status, trends and prospects for India. In: Asiapacific forestry sector outlook study working paper series, Working Paper No. APFSOS/WP/10. Forestry Policy and

Planning Division, Rome Regional Office for Asia and the Pacific, Bangkok, 1–59

Gielis J, Everaert I, De Loose M (1997) Genetic variability and relationships in Phyllostachys using random amplified polymorphic DNA. In: Chapman GP (ed) The bamboos, vol 19., Linnaean Society SymposiumAcademic, London, pp 107–124

Goyal AK, Ghosh PK, Dubey PK, Sen A (2012) Inventorying bamboo biodiversity of North Bengal: a case study. Int J Fund Appl Sci 1:5–8

Grass Phylogeny Working Group (2001) Phylogeny and sub-familial classification of the grasses. Ann Mo Bot Gard 88:373–457

Gui YJ et al (2010) Insights into the bamboo genome: syntenic relationships to rice and sorghum. J Integr Plant Biol 52:1008–1015

Heng HP, Yeoh HH, Tan CKC, Rao AN (1996) Leaf isozyme polymorphisms in bamboo species. J Singapore Nat Acad Sci 22:10–14

Janzen DH (1976) Why bamboos wait so long to flower. Ann Rev Ecol Syst 7:347–391

Joshi SP, Gupta VS, Agarwal RK, Ranjekar PK, Brar DS (2000) Genetic diversity and phylogenetic relationship as revealed by inter simple sequence repeat (ISSR) polymorphism in the genus Oryza. Theor Appl Genet 100:1311–1320

Kapteyn J, Simon JE (2002) The use of RAPDs for assessment of identity, diversity, and quality of Echinacea. In: Janick J, Whipkey A (eds) Trends in new crops and new uses. ASHS Press, Alexandria, pp 509–513

Karp A, Edwards KJ (1995) Molecular techniques in the analysis of the extent and distribution of genetic diversity. In: IPGRI workshop on molecular genetic tools in plant genetic resources, 9–11 Oct, Rome, IPGR

Khatta V, Katoch BS (1983) Nutrient composition of some fodder tree leaves available in sub- mountainous region of Himachal Pradesh. Ind For 109:17–24

Kigomo BN (1988) Distribution, cultivation and research status of bamboo in Eastern Africa. KEFRI Ecol Ser Monogr 1:1–19

Kleinhenz V, Midmore DJ (2001) Aspects of bamboo agronomy. Adv Agron 74:99–145

Ko MK, Yang J, Jin YH, Lee CH, Oh BJ (1998) Genetic relationships of Viola species evaluated by random amplified polymorphic DNA analysis. J Hort Sci Biotech 74:601–605

Kobayashi M (1997) Phylogeny of world bamboos analyzed by restriction fragment length polymorphisms of chloroplast DNA. In: Chapman GP (ed) The bamboos, vol 19., Linnean Society SymposiumAcademic, London, pp 227–236

Kurz S (1876) Bamboo and its use. Ind For 1:219–269

Lalhruaitluanga H, Prasad MNV (2009) Comparative results of RAPD and ISSR markers for genetic diversity assessment in Melocanna baccifera Roxb. growing in Mizoram State of India. Afr J Biotechnol 8:6053–6062

Li DZ (1997) The flora of China Bambusoideae project: problems and current understanding of bamboo taxonomy in China. In: Chapman GP (ed) The Bamboos. Academic Press, London, pp 61–81

Liese W, Hamburg FRG (1987) Research on Bamboo. Wood Sci Technol 21:189–209

Lin XC, Lou YF, Liu J, Peng JS, Liao GL, Fang W (2010) Cross breeding of Phyllostachys species (Poaceae) and identification of their hybrids using ISSR markers. Genet Mol Res 9:1398–1404

Loh JP, Kiew R, Set O, Gan LH, Gan YY (2000) A study of genetic variation and relationship within the bamboo subtribe Bambusinae using amplified fragment length polymorphism. Ann Bot 85:607–612

Mailly D, Christanty L, Kimmins JP (1997) "Without bamboo, the land dies": nutrient cycling and biogeochemistry of a Javanese bamboo talun-kebun system. For Ecol Manag 91:155–173

McNeely AJ (1995) Bamboo, Biodiversity and conservation in Asia. Bamboo, people and the environment. In: Proceedings of Vth International bamboo workshop and the IV international bamboo congress, Ubud, Bali, Indonesia

Mu J, Uehara T, Li J, Furuno T (2004) Identification and evaluation of antioxidant activities of bamboo extracts. For Stud China 6:1–5

Mukherjee AK, Ratha S, Dhar S, Debata AK, Acharya PK, Mandal S, Panda PC, Mahapatra AK (2010) Genetic relationships among 22 Taxa of Bamboo revealed by ISSR and EST-Based random primers. Biochem Genet 48:1015–1025

Naithani HB, Pal M, Lepcha STS (2003) Gregarious flowering of Thamnocalamus spathiflorus and T. falconeri, bamboos from Uttaranchal India. Ind For 129:517–526

Nayak S, Rout GR, Das P (2003) Evaluation of genetic variability in bamboo using RAPD Markers. Plant Soil Environ 49:24–28

Ohrnberger D (2002) The bamboos of the World. Second impression. Elsevier, Amsterdam

Orozco-Castillo C, Chalmers KJ, Wauh R, Powell W (1994) Detection of genetic diversity and selective gene introgression in coffee using RAPD markers. Theor Appl Genet 8:934–940

Park SB, Kwon SD (1998) Development of new uses of bamboos (II): development of carbonization kiln and schedule investigation for bamboo charcoal making. FRI J For Sci 59:17–24

Peng Z et al (2010) Genome-wide characterization of the biggest grass, bamboo, based on 10,608 putative full-length cDNA sequences. BMC Plant Biol 10:116

Peng Z et al (2013) The draft genome of the fast- growing non- timber forest species moso bamboo (Phyllostachys heterocycla). Nature Genet 45:456–463

Pervaiz ZH, Rabbani MA, Shinwari ZK, Masood MS, Malik SA (2010) Assessment of genetic variability in rice (Oryza sativa L.) germplasm from Pakistan using RAPD markers. Pak J Bot 42:3369–3376

Porter-field WM (1933) Bamboo, the universal provider. Scientific Mon 36:176–183

Qian W, Ge S, Hong DY (2001) Genetic variation within and among populations of a wild rice Oryza granulata from China detected by RAPD and ISSR markers. Theor Appl Genet 102:440–449

Rabbani MA, Pervaiz ZH, Masood MS (2008) Genetic diversity analysis of traditional and improved cultivars of Pakistani rice (Oryza sativa L.) using RAPD markers. Elect J Biotech 11:1–10

Rai V, Ghosh JS, Pal A, Dey N (2011) Identification of genes involved in Bamboo fiber development. Gene 478:19–27

Ram SG, Parthiban KT, Kumar RS, Thiruvengadam V, Paramathma M (2008) Genetic diversity among Jatropha species as revealed by RAPD markers. Genet Resour Crop Evol 55:803–809

Ramanayake SMSD, Meemaduma VN, Weerawardene TE (2007) Genetic diversity and relationships between nine species of bamboo in Sri Lanka, using random amplified polymorphic DNA. Plant Syst Evol 269:55–61

Ruiyang C (2003) Chromosome Atlas of Major Economic Plants Genome in China. Chromosome Atlas of Various Bamboo Species, 4th edn. Science Press, Beijing, p 646

Sharma YML (1980) Bamboos in the Asia Pacific Region. In: Lessard G, Chorinard A (eds.) Proceedings Workshop on bamboo research in Asia, Singapore, 28–30 May, 1980. International Development Research Centre, Ottawa, Canada, pp 99–120

Sharma RK et al (2008) Evaluation of rice and sugarcane SSR markers for phylogenetic and genetic diversity analyses in bamboo. Genome 51:91–103

Shiran B, Amirbakhtiar N, Kiani S, Mohammadi S, Sayed-Tabatabaei BE, Moradi H (2007) Molecular characterization and genetic relationship among almond cultivars assessed by RAPD and SSR markers. Sci Hort 111:280–292

Singh S, Panda MK, Nayak S (2012) Evaluation of genetic diversity in turmeric (*Curcuma longa* L.) using RAPD and ISSR markers. Ind Crop Prod 37:284–291

Skeat WW (1900) Malay magic: being an introduction to the folklore in popular religion of the Malay Peninsula. MacMillan and Co., Ltd., London

Soderstorm TR, Calderon CE (1979) A commentary on bamboos (Poaceae: Bambusoideae). Biotropica 11:161–172

Stapleton CMA (1997) Morphology of woody bamboos. In: Chapman GP (ed) The Bamboos. Academic Press, London, pp 251–267

Stapleton CM, Rao VR (1995) Progress and Prospects in Genetic Diversity Studies on bamboo and its Conservation. Bamboo, People and the Environment. In the proceedings of Vth International Bamboo Workshop and the IV International Bamboo Congress, Ubud, Bali, Indonesia, June 19–22

Sungkaew S, Stapleton CM, Salamin N, Hodkinson TR (2009) Non-monophyly of the woody bamboos (Bambuseae; Poaceae): a multi-gene region phylogenetic analysis of Bambusoideae s.s. J Plant Res 122:95–108

Tewari DN (1992) A monograph on Bamboo. International Book Distributors, Dehradun

Thomas TA, Arora RK, Singh R (1988) Genetic wealth of bamboos in india and their conservation strategies. Bamboos Current Research. In: Proceedings of International Bamboo workshop, Nov. 14–18, Cochin, India

Triplett J, Clark LG (2003) Ambiguity and an American Bamboo: the *Chusquea culeou Species Complex*. Bamboo Science and Culture. J Am Bamboo Soc 17:21–27

Varmah JC, Bahadur KN (1980) In: Lessard G, Choulnard A (eds.) Country report: India, Bamboo research in Asia, proceedings of a workshop in Singapore, 28–30 May, pp 19–46

Wachira FN, Waugh R, Powell W, Hackett CA (1995) Detection of genetic diversity in tea (*Camellia sinensis*) using RAPD markers. Genome 38:201–210

Watanabe M, Ito M, Kurita S (1994) Chloroplast DNA phylogeny of Asian bamboos (Bambusoideae, Poaceae) and its systematic implication. J Plant Res 107:253–261

Weising K, Winter P, Hutter B, Kahl G (1998) Microsatellite markers for molecular breeding. Crop Sci: Recent Adv 1:113–143

Welsh J, McClelland M (1990) Fingerprinting genomes using PCR with arbitrary primers. Nucleic Acids Res 18:7213–7218

Williams JGK, Kubelik KJ, Livak KJ, Rafalski JA, Tingey SV (1990) DNA polymorphisms amplified by arbitrary primers are useful genetic markers. Nucleic Acids Res 18:6531–6535

Yang HQ, An MY, Gu ZJ, Tian B (2012) Genetic diversity and differentiation of *Dendrocalamus membranaceus* (Poaceae: Bambusoideae), a declining bamboo species in Yunnan, China, as based on inter-simple sequence repeat (ISSR) analysis. Int J Mol Sc 13:4446–4457

Zhang W, Clark LG (2000) Phylogeny and classification of the Bambusoideae (Poaceae). In: Jacobs SWL, Everett JE (eds) Grasses: systematics and evolution. CSIRO Publishing, Collingwood, pp 35–42

Zhang YJ, Ma PF, Li DZ (2011) High-throughput sequencing of six bamboo chloroplast genomes: phylogenetic implications for temperate woody bamboos (Poaceae: Bambusoideae). PLoS One 6:e20596

11

Siderophore biosynthesis genes of *Rhizobium* sp. isolated from *Cicer arietinum* L.

Bejoysekhar Datta · Pran K. Chakrabartty

Abstract *Rhizobium* BICC 651, a fast-growing strain isolated from root nodule of chickpea (*Cicer arietinum* L.), produced a catechol siderophore to acquire iron under iron poor condition. A Tn5-induced mutant (B153) of the strain, BICC 651 impaired in siderophore biosynthesis was isolated and characterized. The mutant failed to grow on medium supplemented with iron chelator and grew less efficiently in deferrated broth indicating its higher iron requirement. The mutant produced less number of nodules than its parent strain. The Tn5 insertion in the mutant strain, B153, was located on a 2.8 kb *Sal*I fragment of the chromosomal DNA. DNA sequence analysis revealed that the Tn5-adjoining genomic DNA region contained a coding sequence homologous to *agbB* gene of *Agrobacterium tumefaciens* MAFF301001. About 5 kb genomic DNA region of the strain BICC 651 was amplified using the primers designed from DNA sequence of agrobactin biosynthesis genes of *A. tumefaciens* MAFF 301001 found in the database. From the PCR product of the strain BICC 651, a 4,921 bp DNA fragment was identified which contained four open reading frames. These genes were designated as *sid*, after siderophore. The genes were identified to be located in the order of *sid*C, *sid*E, *sid*B, and *sid*A. Narrow intergenic spaces between the genes indicated that they constitute an operon. Phylogenetic analyses of deduced *sid* gene products suggested their sequence similarity with the sequences of the enzymes involved in biosynthesis of catechol siderophore in other bacteria.

Keywords Chickpea · Nodulation · *Rhizobium* · Siderophore biosynthesis genes

Introduction

In iron deficient environment, bacteria meet their iron requirement by producing low-molecular-mass iron-chelating compounds called siderophores. The compounds bind available Fe^{3+} with high affinity to form complexes which are internalized by the cells with the help of cognate membrane proteins. The structures of these compounds are quite variable. Many of them are catecholate in nature. Catecholate siderophores were isolated from a variety of bacterial species including *Escherichia coli* which produces enterobactin, the prototype of catecholate siderophore. Enterobactin is biosynthesized from 2,3-dihydroxybenzoic acid (DHBA) which, in turn is synthesized from chorismate via the consecutive actions of three enzymes, viz., isochorismate synthase (EntC), isochorismatase (EntB) and 2,3-dihydro-2,3-dihydroxybenzoate dehydrogenase (EntA) (Earhart 1996). Although, the enzymes involved in the pathway of DHBA biosynthesis are common to all catechol siderophore-producing organisms, the organization of the biosynthetic genes within the operon and the sequences of the genes are different.

The members of the genus *Rhizobium* are soil bacteria that enter into symbiotic relationship with plant hosts. This relationship is iron dependent, since iron is required for nodule formation, and synthesis of leghemoglobin, nitrogenase complex, ferredoxin and other electron transport

B. Datta (✉)
Department of Botany, University of Kalyani, Nadia, Kalyani, West Bengal 741 235, India
e-mail: bejoy.datta@gmail.com; dattabejoysekhar@yahoo.com

P. K. Chakrabartty
Acharya J.C. Bose Biotechnology Innovation Centre, Madhyamgram Experimental Farm, Madhyamgram, Kolkata, West Bengal 700 129, India
e-mail: prankrishna1946@gmail.com

proteins to energise the nitrogenase system during symbiosis (Guerinot 1991). Because a part of life cycle of *Rhizobium* requires invasion, growth and differentiation within its host plant tissue, the role of siderophore for iron acquisition is important for the plant–microbe interaction. Improved iron scavenging properties of *Bradyrhizobium* sp. and *Sinorhizobium meliloti* 1021 positively correlate with rhizospheric growth and nodulation effectiveness in peanut and in alfalfa (O'Hara et al. 1988; Gill et al. 1991). Moreover, siderophore-producing ability helps in the sustenance of rhizobia in iron-deficient soils (Lesueur et al. 1995). Benson et al. (2005) reported that the gene *fegA* required for utilization of ferrichrome of *B. japonicum* 61A152 was also required for symbiosis. When the gene with its native, promoter was cloned and transferred to *Mesorhizobium* sp. GN25, nodulating peanut (Joshi et al. 2008) and *Rhizobium* sp. ST1, nodulating pigeonpea (Joshi et al. 2009), the transconjugants became symbiotically more efficient indicating importance of iron uptake protein in rhizobium-legume symbiosis.

Siderophore biosynthesis genes were studied in *Rhizobium leguminosarum* bv. *viciae* (Carter et al. 2002) and in *Sinorhizobium meliloti* 1021 (Lynch et al. 2001). In both the systems, siderophore biosynthesis genes were located on plasmids and were clustered close to the genes encoding their cognate membrane proteins. Siderophores produced by these strains were not of catechol type. Catecholate siderophores were isolated from *R. leguminosarum* bv. *trifolii* (Skorupska et al. 1989), *R. ciceri* (Roy et al. 1994; Berraho et al. 1997), *Bradyrhizobium* (cowpea) (Modi et al. 1985), *Bradyrhizobium* (peanut) (Nambiar and Sivaramakrishnan 1987), but the biosynthetic genes have not been investigated yet.

A fast-growing *Rhizobium* strain BICC 651 was isolated from a nodule produced on root of a chickpea (*Cicer arietinum* L.) plant grown in the Experimental Farm of Bose Institute at Madhyamgram, West Bengal, India. The strain produced a catechol siderophore and its cognate membrane receptor in response to iron deficiency (Roy et al. 1994). Structurally, the siderophore produced by *Rhizobium* BICC 651 contains 2,3-dihydroxybenzoic acid as core compound with two moles of threonine as the ligand. This suggested that the strain BICC 651 is likely to possess enzymes for siderophore biosynthesis similar to those of other catechol siderophore-producing organisms and the biosynthetic mechanism of the siderophore could be similar to that in other catechol siderophore-producing organisms. In the present study, transposon Tn5-*mob* insertional mutagenesis of the *Rhizobium* strain BICC 651 was performed to generate mutants impaired in siderophore production. One of the mutants was exploited to find out siderophore biosynthetic genes of the *Rhizobium* strain BICC 651.

Materials and methods

Bacterial strains, plasmid and culture media

Bacterial strains and plasmid used in the study are listed in Table 1. *Escherichia coli* and *Rhizobium* strains were cultured in Luria–Bertani (LB) medium and yeast extract mannitol (YEM) medium, respectively. The complete medium described by Modi et al. (1985) [composition (g/l): K_2HPO_4, 0.5; $MgSO_4 \cdot 7H_2O$, 0.4; NaCl, 0.1; mannitol, 10; glutamine, 1; and NH_4NO_3, 1; pH 6.8] was used to study the growth and siderophore production of the strain BICC 651 and its mutants. For the purpose, the medium was deferrated with hydroxyquinoline (Meyer and Abdallah 1978). CAS-agar plate for detection of siderophore production by the organisms was prepared according to Schwyn and Neilands (1987) using the complete medium.

Tn5 mutagenesis

Transconjugates were generated by plate mating method (Mukhopadhyaya et al. 2000). *Escherichia coli* S17.1, harbouring the suicide plasmid pSUP5011::Tn5-*mob* (Simon 1984) was used as the donor and a rifampicin (100 µg/ml) resistant spontaneous mutant of the *Rhizobium* strain BICC 651, designated as *Rhizobium* BICC 651R was used as the recipient. Fresh cultures of both the donor and the recipient were mixed, centrifuged to collect the cells and spread on plates containing tryptone-yeast extract agar medium [composition (g/l): bactotryptone, 8; bactoyeast, 5; NaCl, 5; and agar 20; pH 6.8] (Beringer 1974). The mating mixture was incubated overnight at 30 °C. The cells were then collected, suspended in YEM broth and plated on Rhizobium medium (Himedia) containing neomycin

Table 1 Bacterial strains and plasmid used in the study

Strains and plasmid	Relevant characteristic(s)	References and/or source
Escherichia coli S17.1	Conjugative donor for pSUP5011	Laboratory collection
Rhizobium BICC 651	Wild type, Sid^{+a}	Roy et al. (1994)
Rhizobium BICC 651R	Sid$^+$, Rif^r (spontaneous)	This study
Rhizobium B153	Sid$^-$:: Tn5-*mob*, Rif^r Neo^r	This study
Plasmid pSUP5011	pBR325(Bam$^-$)::Tn5-*mob*, Ap^r Neo^r Cm^r	Simon (1984)

[a] Ability to produce siderophore is denoted by Sid$^+$, inability is denoted by Sid$^-$

(50 µg/ml) and rifampicin (100 µg/ml). The plates were incubated at 28 °C for the development of isolated colonies of transconjugants. The colonies were picked to prepare master plates for post mating selection. Each master plate was replica plated on CAS plate. The mutants which did not produce orange halo were selected as siderophore negative (Sid⁻) mutants.

Bacterial growth and siderophore production

The parent strain BICC 651R, and its Sid⁻ mutant, B153, were examined for their growth on Rhizobium medium in presence of increasing concentrations of bipyridine, a synthetic iron chelator. Subsequently, the parent strain and the mutant were studied for their growth and siderophore production in deferrated complete medium. The cultures were incubated on a shaker at 28 °C and growth was measured by monitoring OD_{590} of the culture. Siderophore in the culture supernatant was assayed using the CAS reagent (Schwyn and Neilands 1987).

Plant inoculation and nodulation assay

For nodulation study, surface-sterilized seeds of chickpea were inoculated with *Rhizobium* strains and planted in sea sand in pots. Sea sand was treated with HCl, washed with distilled water until the pH of the washing was neutral. It was then air dried and fumigated with chloroform overnight under airtight condition, autoclaved and finally kept at 200 °C in a hot air oven for 6 h. Five hundreds gram of sterile sand in an earthen pot was moistened with 100 ml of nitrogen-free plant nutrient medium for sowing the seeds. The composition of the nitrogen-free medium was (g/l): $CaSO_4$. $2H_2O$, 0.34; K_2HPO_4, 0.17; $MgSO_4$. $7H_2O$, 0.25; Fe Citrate, 0.002; KCl, 0.075; trace element solution, 0.5 ml; pH 7.2 (Norris and Date 1976). The composition of the trace element solution was (g/l): $ZnSO_4$. $7H_2O$, 2.25; $CuSO_4$. $5H_2O$, 1.00; $MnSO_4$. $5H_2O$, 0.50; $CaCl_2$. $2H_2O$, 2.00 $Na_2B_4O_7$. $10H_2O$, 0.23; $(NH_4)_6Mo_7O_{24}$, 0.10. Surface sterilization of chickpea seeds was carried out by soaking in concentrated H_2SO_4 for 10 min, then washing the seeds several times with sterile distilled water. The parent strain BICC 651R and its Sid⁻ mutant B153, were grown in YEM broth till late log phase, and the harvested cells were mixed with sand-charcoal (1:3) containing 2 % aqueous sodium carboxymethyl cellulose. The mixture was used to coat the surface-sterilized seeds. The coated seeds ($\sim 10^8$ bacteria/seed) were kept in dark for overnight and the next day, five seeds were transferred to each pot containing the sterile sand soaked in nitrogen-free plant nutrient medium. A control set of seeds which did not receive any inoculum was also included in this study. The pots were kept under well-illuminated condition and watered when necessary to moisten the sand. At 1 week interval, 50 ml of 1/10th dilution of nitrogen-free plant nutrient medium was added to each pot. At 35 days of inoculation, plants were uprooted and observed for development of nodules.

DNA preparations and Southern hybridization

Total DNA of the parent, BICC 651R and the mutant, B153, was isolated, digested with *Sal*I, electrophoresed on 1 % agarose gel according to standard protocol (Sambrook et al. 1989). DNA fragments from agarose gel were blotted on nylon membrane and hybridized with probe prepared from Tn5 fragment containing neomycin resistance (neo^r) gene. The probe was labelled by biotinylation following the manufacturer's instruction (NE Blot Phototope Kit, Biolabs).

DNA ligation and inverse PCR

The *Sal*I digested DNA fragments showing positive signal following hybridization with labelled probe were eluted from the gel and allowed to self-ligate by incubating the DNA at a concentration of 0.3–0.5 µg/ml in presence of 3 U of T4 DNA ligase (Promega)/ml overnight at 4 °C (Huang et al. 2000). The ligation mixture was extracted by phenol: chloroform, the DNA was precipitated with ethanol, and dissolved in sterile distilled water to a concentration of 20 µg/ml. The inverse PCR was carried out using the self-ligated product containing a portion of the Tn5 as template and Tn5Int and NeoF as primers (Table 2). The amplifications were performed using Perkin-Elmer PCR system 2,400 in 50 µl of reaction mixture containing $1 \times$ enzyme buffer, 1.5 mM $MgCl_2$, 200 µM dNTPs (each), 500 nM of each primers and 1.5 U *Taq* DNA polymerase (Fermentas) and 50 ng of purified self-ligated product. The thermal programme used for PCR was as

Table 2 Primers used in the study

Primers	Sequences (5′–3′)	Position of nucleotides	Sources
Tn5Int	CGGGAAAGGTTCCGTTCAGGACGC	21–34 [complementary] or 5775–5798	*E. coli* transposon Tn5 (accession no.
NeoF	CGCATGATTGAACAAGATGG	1548–1567	U00004 L19385).
agbCF	GACAGGATCGACGGACTGAC	1646–1665	Ferric iron uptake gene of *A. tumefaciens*
agbAR	GTAACGAAGGGTGAGGCAAT	6620–6639 [complementary]	MAFF 301001 (accession no. AB083344).

follows: initial denaturation for 10 min at 94 °C; 35 cycles of denaturation for 30 s at 94 °C, annealing for 30 s at 55 °C, and extension at 72 °C for 1 min, followed by a final extension at 72 °C for 10 min. The identity of the PCR product was confirmed by nested PCR and also by ascertaining the profile of bands resulting from its specific endonuclease digestion. The PCR product was purified and sequenced by ABI PRISM 377 automated DNA sequencer (Perkin-Elmer, Applied Biosystem, Inc.).

Amplification of agb homologues from genomic DNA of BICC 651 and site of Tn5 insertion in Sid⁻ mutant

To amplify the homologues of agrobactin biosynthetic genes from the strain BICC 651, PCR amplifications were carried out using the genomic DNA of BICC 651 as template, and agbCF and agbAR as primers (Table 2). A long PCR amplification kit was used for the purpose following the manufacturer's instruction (Genei, India). To identify the position of Tn5 insertion in the genomic DNA of the mutant, B153, two sets of PCR were carried out, one with the primer pair, agbCF and Tn5Int and another with the primer pair agbAR and Tn5Int. The reaction mixture for amplification contained 100 ng of genomic DNA as template and the thermal programme remained the same as described earlier for IPCR. To amplify larger (>2 kb) DNA fragment, extension time was increased by 1 min for each 1 kb of desired amplicon during final extension. The PCR products were analysed by agarose gel electrophoresis, purified and sequenced.

Analysis of the Tn5 adjacent DNA sequence

Genomic DNA sequence was analyzed using BLAST version 2.2.1 of National Center for Biotechnology Information. Nucleotide BLAST and BLASTX were used to search for nucleotide sequences and derivative amino acid sequences, respectively.

Genomic context analysis

Gene context analysis of *sidB* gene of *Rhizobium* strain BICC 651 was carried out using the GeConT programme. The program allowed comparison of adjacent genes of *sidB* and visualization of genomic context of *sidB* homologue in the genomes of other organisms. The program is available at: http://www.ibt.unam.mx/biocomputo/gecont.html (Ciria et al. 2004).

Phylogenetic analysis

Evolutionary analyses were conducted using the software MEGA5 (Tamura et al. 2011). Multiple sequence alignment was carried out using a CLUSTALW and evolutionary history of the sequence was inferred using the neighbor-joining method (Saitou and Nei 1987). The optimal tree with the sum of branch length is equal to 4.31786458 is shown. The tree is drawn to scale, with branch lengths in the same units as those of the evolutionary distances used to infer the phylogenetic tree. The evolutionary distances are computed using the Poisson correction method (Zuckerkandl and Pauling 1965) and are in units of the number of amino acid substitutions per site.

Results and discussion

Transposon mutagenesis, and isolation of Sid⁻ mutants

Transposon insertion mutagenesis in a siderophore producing *Rhizobium* sp. of chickpea was carried out to select the siderophore negative mutants to identify the structural genes required for biosynthesis of siderophore in the organism. Random mutagenesis of the *Rhizobium* strain BICC 651R by mobilization of Tn5-*mob* from the suicide vector pSUP5011 to the recipient cells occurred at a frequency of 10^{-5} per donor as revealed by the number of neomycin-resistant transconjugants. This is about 1,000 times greater than the rate of spontaneous resistance of the recipients to neomycin (10^{-8}). Similar transposition frequency had also been reported for other bacteria (Mukhopadhyaya et al. 2000; Manjanatha et al. 1992). Upon screening of the 1,000 of transconjugants, five Sid⁻ mutants were isolated. Among these, one mutant (B153) was characterized in the present study.

Study of iron requirement

The amount of iron required for optimal growth of many Gram-negative bacteria ranged from 0.36 to 1.8 µM (Lankford 1973). The strain BICC 651R was able to grow on Rhizobium medium containing bipyridine up to a concentration of 400 µM but the Sid⁻ mutant, B153, could not grow even in the presence of 50 µM of bipyridine (Table 3). When the organisms were grown in the complete medium deferrated with hydroxyquinoline, the highest level of growth of the Sid⁻ mutant B153 ($OD_{590} = 1$) was almost half as that of the parent strain ($OD_{590} > 2$). These observations suggested that because of impaired biosynthesis of siderophore, the mutant required more iron for its growth than the parent strain. Siderophore was not detected in the culture filtrate of the mutant at all stages of its growth, whereas the parent strain was observed to produce as much as 70 nmole of siderophore per ml of culture filtrate during late log phase of its growth (at 48 h) in the deferrated complete medium (Fig. 1).

Table 3 Growth of bacterial strains in presence of increasing concentrations of bipyridine

Concentration of bipyridine (μM)	BICC 651R	B153
0	+	+
50	+	−
100	+	−
200	+	−
300	+	−
400	+	−
500	−	−

The organisms were streaked on bipyridine containing Rhizobium medium and growth was scored after three days of incubation; '+' indicates growth, '−' absence of growth

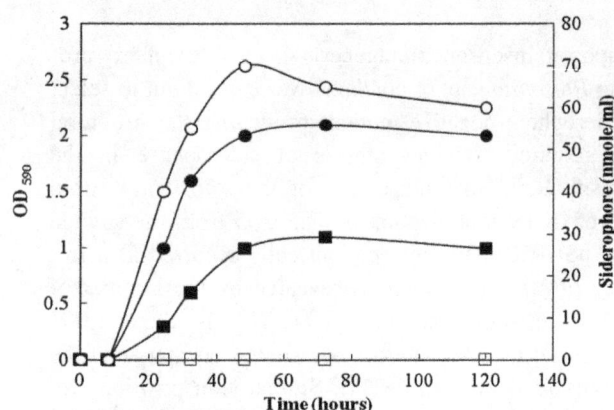

Fig. 1 Growth (*filled circle, filled box*) and siderophore production (*open circle, open box*) of the parent strain BICC 651R and its Sid⁻ mutant B153, respectively, in the deferrated complete medium

Plant nodulation study

Table 4 presents the data of symbiotic performance of the siderophore non producing mutant B153 and its parent strain BICC 651R. The control plants receiving no inoculum produced no nodules. The plants inoculated with B153 produced less number of nodules than those inoculated with the parent strain BICC 651R. The wet weight of nodules produced by the mutant B153 was almost half as compared to those of its parents. Gill et al. (1991) reported that mutants of *Sinorhizobium meliloti* which were unable to produce siderophore were able to nodulate the plants but the efficiency of the nodules in nitrogen fixation was less as

compared to wild type incited nodules indicating the importance of iron in symbiotic N_2 fixation.

Identification of the DNA fragments containing Tn5

*Sal*I cuts the 7.7 kb Tn5-*mob* DNA into two parts: a 2.7 kb part with neo^r gene and a 5 kb part containing the 1.9 kb *mob* region. When the Tn5 probe prepared from Tn5 fragment containing neo^r gene was allowed to hybridize with *Sal*I restricted genomic DNAs of BICC 651R and its Sid⁻ mutant, B153, it showed positive signal only as a single band of 2.8 kb genomic DNA of the mutant indicating insertion of a single copy of Tn5 in the genome of the mutant and no hybridization of the probe was detected with the genomic DNA of the parent, BICC 651R. In the mutant, the 2.8 kb *Sal*I restricted genomic DNA fragment contained the 2.7 kb Tn5 fragment adjoining nearly a 0.1 kb genomic DNA fragment. The 2.8 kb *Sal*I fragment was used to self-ligate more easily than using a larger (>7.7 kb) *Eco*RI restricted fragment. The Tn5 adjoining genomic DNA fragments were amplified by inverse PCR using the Tn5Int and NeoF primer pair (Table 2). A 1.3 kb amplicon containing an almost 0.1 kb genomic DNA was obtained (Fig. 2a). The genomic DNA fragment was sequenced and the sequence showed 94 % identity with the *agbB* gene sequence of *Agrobacterium tumefaciens* MAFF301001 (GenBank accession no. AB083344). The *agbB* gene product, isochorismatase, catalyzes the synthesis of 2,3-dihydro 2,3-dihydroxybenzoic acid from isochorismate (Sonoda et al. 2002). Thus, in the mutant, B153, Tn5 interrupted one of the DHBA biosynthetic genes and resulted in impaired synthesis of siderophore.

Nucleotide sequence analysis of the siderophore biosynthesis region

Using the agbCF and agbAR primer pair, a 5 kb DNA fragment was amplified from the genomic DNA of BICC 651 (Fig. 2b). From the fragment, 4,921 bp was sequenced using appropriate primers and the DNA sequence was submitted to GenBank under accession no. GU251056. Upon nucleotide BLAST search, the 4,921 bp DNA region of BICC 651 was found to match (90 % identity) with the agrobactin biosynthetic (*agb*) genes of *A. tumefaciens* MAFF301001 (accession no. AB083344). Nucleotide

Table 4 Symbiotic performance of *Rhizobium* BICC 651R and its siderophore non-producing mutant

Measurements were made on the 35th day of sowing

Strains	No. of nodules per plant	Wet weight of nodules (mg) per plant	Root length (mm)	Shoot length (mm)
BICC 651R	20	872.5	100	230
B153 (Sid⁻)	14	417.2	100	200
Control	Nil	Nil	80	200

sequence analysis of the 4,921 bp DNA region of BICC 651 revealed the presence of four open reading frames (ORFs) (Fig. 3a). *ORF1* (*sidC*) was 1,029 bp long (positions 213–1,241), potentially coding for a 342-amino-acid-residue-long protein. *ORF2* (*sidE*), identified between positions 1,406–3,034, was 1,629 bp long and coded for a putative protein of 542 amino acid residues. At 97 bp, downstream from the TGA stop codon of *ORF2* was found

the *ORF3* (*sidB*), 870 bp long, potentially coding for a 289-amino-acid-residue protein (positions 3,131–4,000). Finally, at 14 bp from the TGA stop codon of *ORF3*, the ATG start codon of *ORF4* (*sidA*) (positions 4,014–4,772) was located. It was only 759 bp long, with a predicted translation product of 252-amino acid residues. The four ORFs identified in the 5 kb region were closely connected with narrow intergenic spaces indicating their polycistronic organization within an operon. In the sequence of 4,921 bp, a probable ribosome-binding site (AGGAGG) was identified six bp upstream of the ATG start codon of *sidC* of BICC 651 using Promoter prediction search tool (www. softberry.com). A presumable iron box (ACAAAACAT GATTAGC) was also identified 56 bp upstream of the *sidC* start codon similar to the one found in *A. tumefaciens* MAFF301001 (accession no. AB083344) and in *Pseudomonas fluorescens* (accession no. Y09356). Presence of the potential Fur box sequence indicated iron regulation of the functions of *sid* genes in the strain BICC 651.

Site of Tn5 insertion

From the genomic DNA of the mutant, B153, a 3.5 kb product was obtained using agbCF and Tn5Int primers (Fig. 2c) and a 1.5 kb product was amplified using agbAR and Tn5Int primers (Fig. 2d). Matching the sequences, it was found that Tn5 was inserted next to the 210th nucleotide position from the start codon of the *sidB* gene in the mutant B153 (Fig. 3a).

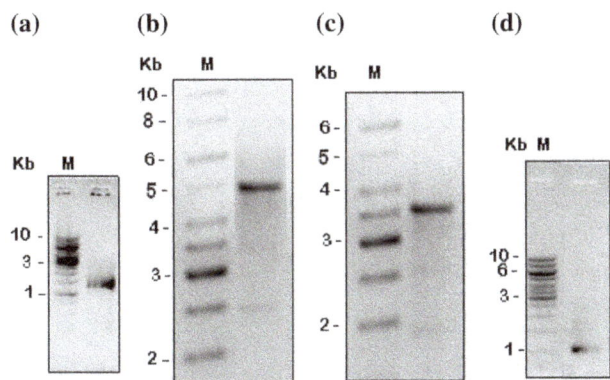

Fig. 2 Ethidium bromide stained agarose gel showing amplified DNA fragments. **a** 1.3 kb fragment containing part of Tn5 and part of *sidB* gene of the mutant, B153. **b** 5 kb fragment containing four *sid* genes of *Rhizobium* BICC 651. **c** 3.5 kb fragment containing whole of the *sidC*, *sidE* genes, and part of the *sidB* gene of B153. **d** 1.5 kb fragment containing part of the *sidB* and whole of the *sidA* gene of B153

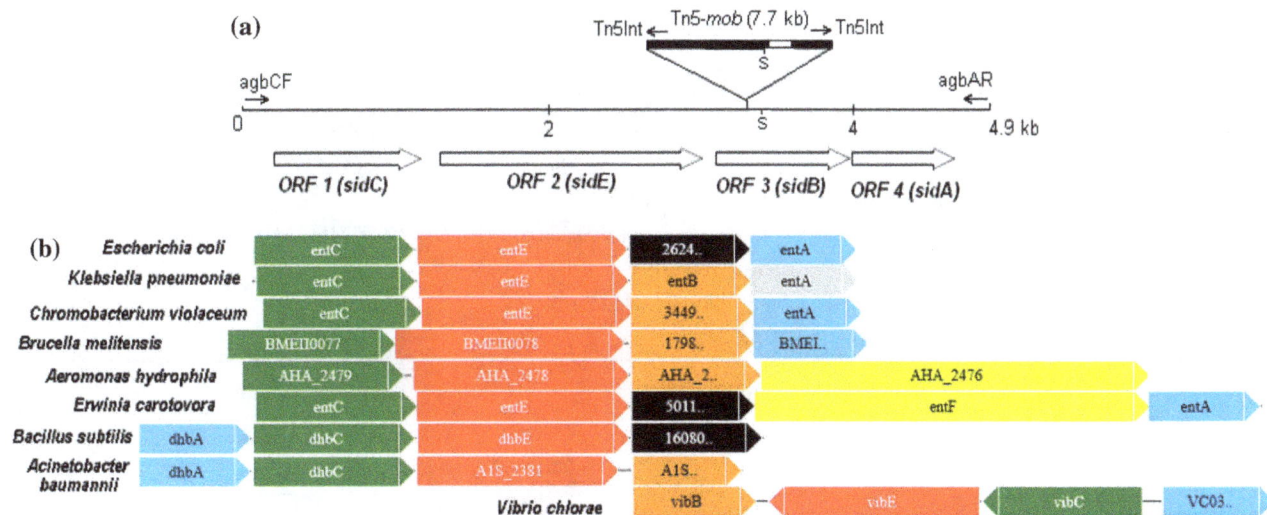

Fig. 3 Genetic organization and genomic context of the siderophore biosynthesis DNA region of *Rhizobium* mutant, B153. **a** The physical map showing organization of *sid* genes, transcriptional directions of four ORFs, site of Tn5 insertion, *Sal*I (S) restriction sites, alignment site of the primers (agbCF, agbAR and Tn5Int) and hybridization site of the Tn5 probe (marked as *white*). **b** Genomic context and organization conservation of the *sidB* gene region in the genome of *Rhizobium* BICC 651 in comparison to the genomes of related catechol-producing organisms. *Arrows* represent genes with their relative orientation in the genomes. Genes are *coloured* according to their functional categories and labelled according to their original gene annotation in the database. *Orange/black* (isochorismatase); *red* (2,3-dihydroxybenzoate-AMP-ligase); *green* (isochorismate synthase); *sky blue/grey* (2,3-dihydro-2,3-dihydroxybenzoate) and *yellow* (entF)

Table 5 Amino acid sequence identities of products of *sid* genes of *Rhizobium* BICC 651 with related catechol-producing organisms

ORFs (accession No.)	Corresponding protein in other catechol-producing organisms (accession No.)	% identity
ORF1/*SidC*	AgbC [*Agrobacterium tumefaciens* MAFF 301001] (BAC16757)	92
(ADB12985)	Isochorismate synthase [*Rhizobium lupini* HPC(L)] (EKJ95799)	85
	Isochorismate synthase [*Brucella melitensis* 16M] (AAL53318)	60
	PmsC [*Pseudomonas fluorescens* WCS374] (CAA70528)	55
	EntC [*Chromobacterium violaceum* ATCC 12472] (AAQ59160)	44
	Isochorismate synthase [*Aeromonas hydrophila* ATCC7966] (ABK37486)	41
	EntC [*Escherichia coli* ATCC 8739] (ACA78675)	39
	EntC [*Klebsiella pneumoniae* subsp. *pneumoniae* MGH 78578] (ABR76061)	39
	VibC [*Vibrio chlorae* Lou15] (AAC45925)	34
	DhbC [*Bacillus subtilis* subsp. *subtilis* 168 (AAC44631)	33
ORF2/SidE	AgbE [*Agrobacterium tumefaciens* MAFF 301001] (BAC16758)	89
(ADB12986)	2,3-dihydroxybenzoate-AMP-ligase [*Rhizobium lupini* HPC(L)] (EKJ95800)	87
	2,3-dihydroxybenzoate-AMP-ligase [*Brucella melitensis* ATCC 23457] (ACO01901)	72
	ATP-dependent activating enzyme/PmsE [*Pseudomonas fluorescens* WCS374] (CAA70529)	64
	EntE [*Chromobacterium violaceum* ATCC 12472] (AAQ59159)	63
	2,3-dihydroxybenzoate-AMP-ligase [*Aeromonas hydrophila* ATCC7966] (ABK38474)	59
	DhbE [*Acinetobacter baumannii* ATCC 17978] (ABO12800)	59
	EntE [*Escherichia coli* ATCC 8739] (ACA78674)	55
	EntE [*Klebsiella pneumoniae* subsp. *pneumoniae* MGH 78578] (ABR76062)	55
ORF3/SidB	AgbB [*Agrobacterium tumefaciens* MAFF 301001] (BAC16759)	97
(ADB12987)	Isochorismatase [*Rhizobium lupini* HPC(L)] (EKJ95801)	95
	Isochorismatase [*Brucella melitensis* 16 M] (AAL53320)	71
	DhbB [*Acinetobacter baumannii* ATCC 17978] (ABO12799)	63
	EntB [*Chromobacterium violaceum* ATCC 12472] (AAQ59158)	62
	Isochorismatse [*Aeromonas hydrophila* ATCC7966] (ABK39064)	60
	DhbB [*Bacillus subtilis* subsp. *subtilis* 168] (AAC44633)	56
	EntB [*Klebsiella pneumoniae* subsp. *pneumoniae* MGH 78578] (ABR76063)	55
	VibB [*Vibrio chlorae* Lou15] (AAC45926)	54
	EntB [*Escherichia coli* ATCC 8739] (ACA78673)	53
ORF4/SidA	AgbA [*Agrobacterium tumefaciens* MAFF 301001] (BAC16760)	90
(ADB12988)	2,3-dihydro-2,3-dihydroxybenzoate dehydrogenase [*Rhizobium lupini* HPC(L)] (EKJ95802)	89
	2,3-dihydro-2,3-dihydroxybenzoate dehydrogenase [*Brucella melitensis* 16 M] (AAL53321)	67
	EntA [*Chromobacterium violaceum* ATCC 12472] (AAQ59157)	58
	2,3-dihydro-2,3-dihydroxybenzoate dehydrogenase [*Aeromonas hydrophila* ATCC7966] (ABK36600)	57
	EntA [*Escherichia coli* ATCC 8739] (ACA78672)	51
	DhbA [*Bacillus subtilis* subsp. *subtilis* 168] (AAC44630)	34
	VibA [*Vibrio chlorae* Lou15] (AAC45924)	34

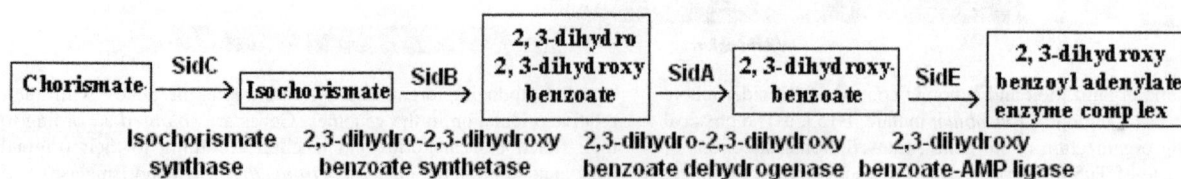

Fig. 4 Deduced functions of the derivative amino acid sequence of *sid* gene products of *Rhizobium* BICC 651

Comparison of genetic organization of *sid* genes of BICC 651 with other species

The genomic context of *sid* DNA region in *Rhizobium* BICC 651, when compared to those in other catechol siderophore-producing members revealed that the organization of the region was not conserved; relative orientation and arrangement of the genes in the region of the genomes of different members varied considerably (Fig. 3b). Genetic organization of *sidC*, *sidE*, *sidB*, and *sidA* of *Rhizobium* BICC 651 was similar to that of *entCEBA* in *E. coli* for enterobactin biosynthesis (Crosa and Walsh 2002), in *A. tumefaciens* MAFF301001 for agrobactin (Sonoda et al. 2002), and siderophore biosynthetic genes in *Brucella melitensis*, *Chromobacterium violaceum*, and *Klebsiella pneumoniae*. In *Aeromonas hydrophila*, *entA* was missing immediately after *entB*, whereas in *Erwinia carotovora*, *entF* was present in between *entB* and *entA*.

The four biosynthesis genes were arranged as *dhbACEB* cluster in *Bacillus subtilis* (Rowland et al. 1996) as well as in *Acinetobactor baumannii* (Dorsey et al. 2003). In *Vibrio cholerae* for vibriobactin biosynthesis, these were organised as *vibACEB* cluster in which *vibC* and *vibE* were co-transcribed, while *vibA* and *vibB* belonged to two independent transcriptional units (Wyckoff et al. 1997). Since organization of catechol siderophore genes of any *Rhizobium* spp. was not found in the database, gene context analysis was carried out from available data of catechol siderophore-producing organisms.

Deduced functions of *sid* gene products

Comparison of the derivative amino acid sequences of the genes with those of related organisms was made by BLASTX analysis (Table 5). The derivative amino acid sequence of *ORF1* (*sidC*) of *Rhizobium* BICC 651 showed

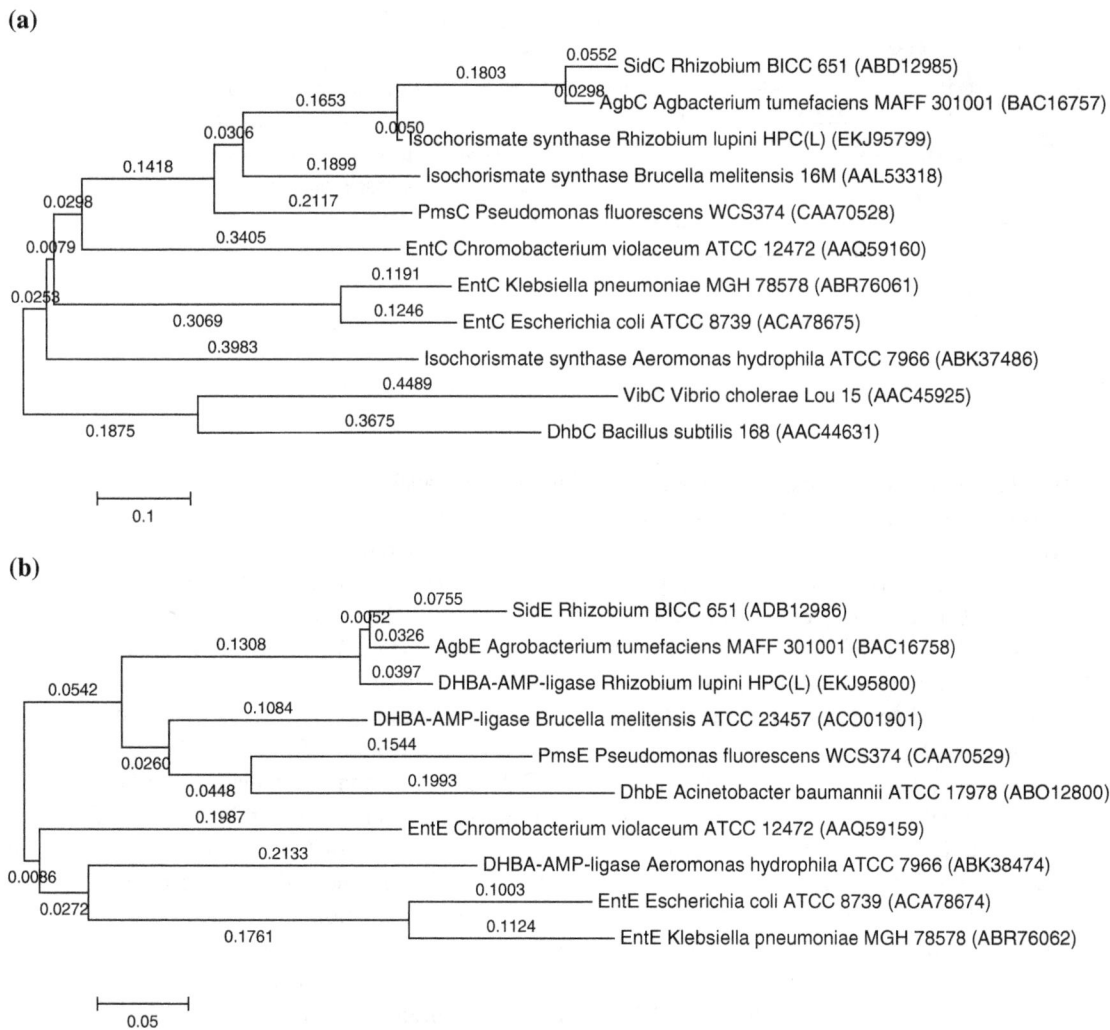

(a)

(b)

Fig. 5 Phylogenetic trees based on comparison of the translational products of the *sid* genes. Trees were constructed by using the neighbour-joining method. Horizontal branch lengths are proportional to the estimated number of nucleotide substitutions. **a** SidC and its homologues, **b** SidE and its homologues, **c** SidB and its homologues, **d** SidA and its homologues

(c)

(d)

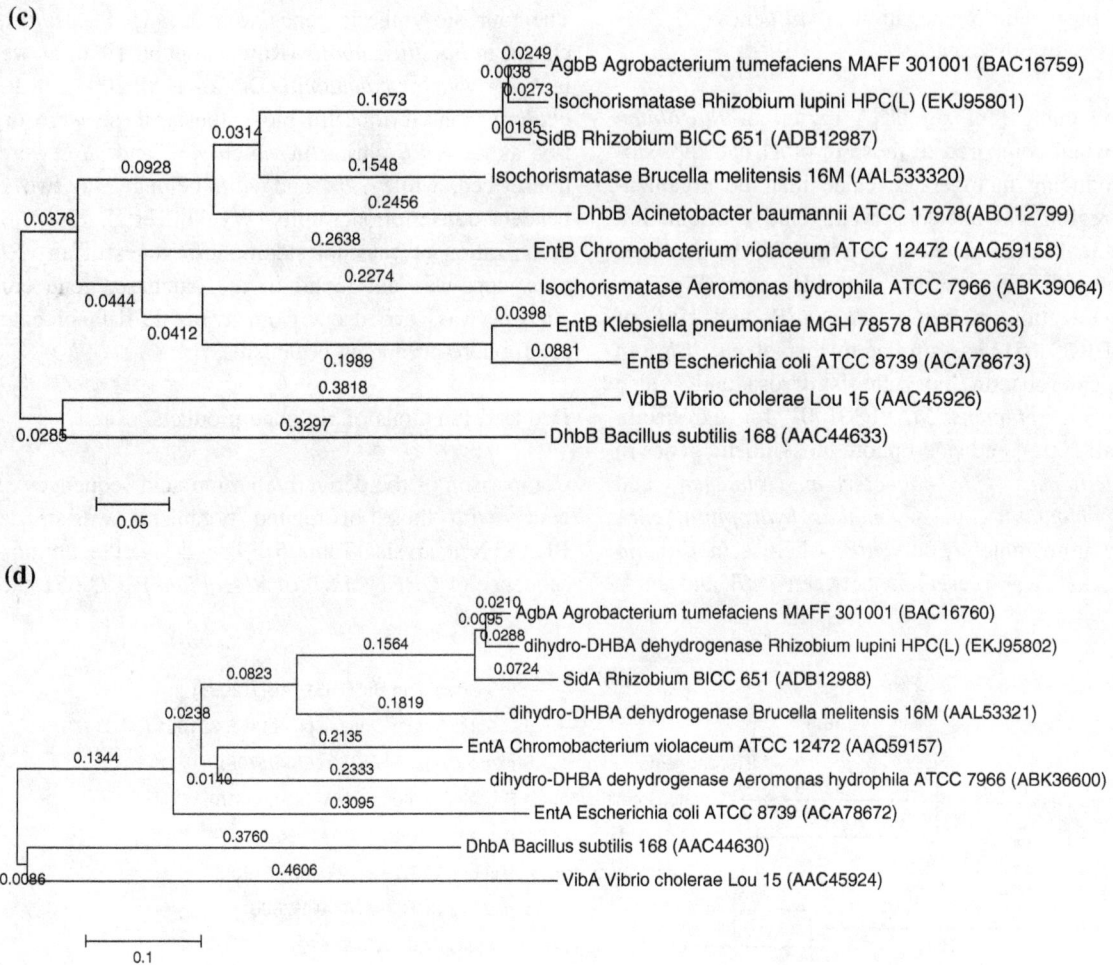

Fig. 5 continued

highest similarity (92 %) to isochorismate synthase of *Agrobacterium tumefaciens* (accession no. BAC16757). The enzyme converts chorismate to isochorismate. The derivative amino acid sequence of *ORF2* (*sidE*) showed highest homology (89 %) to that of 2,3-dihydroxybenzoate-AMP-ligase of *A. tumefaciens* (BAC16758). The enzyme activates 2,3-dihydroxybenzoate by forming a (2,3-dihydroxybenzoyl) adenylate-enzyme complex (Rusnak et al. 1989). Consequently, the deduced gene product of *ORF3* (*sidB*) showed highest similarity (97 %) with 2,3-dihydro-2,3-dihydroxybenzoate synthetase (isochorismatase) of *A. tumefaciens* (BAC16759). The enzyme removes the enolpyruvyl side chain of isochorismate (Rusnak et al. 1990). The *ORF4* (*sidA*) translational product showed 90 % homology with 2,3-dihydro-2,3-dihydroxybenzoate dehydrogenase of *A. tumefaciens* (BAC16760). The enzyme catalyzes the NAD-coupled oxidation of 2,3-dihydro-2,3-dihydroxybenzoate (Liu et al. 1989). The deduced functions of the derivative amino acid sequence of *sid* gene products are presented in the Fig. 4.

Phylogenetic analysis

On the basis of the amino acid sequence identities data given in Table 5, phylogenetic tree of each of the four *sid* translational products was constructed using neighbor-joining method (Fig. 5). In all the trees, the strain BICC 651 was placed in the same evolutionary branch as of *Agrobacterium tumefaciens*. This result along with 16S Rdna sequence analysis of the strain BICC 651 (accession no. DQ839132) demonstrates that the strain BICC 651 has a chromosomal background similar to that of *A. tumefaciens*. Since the strain BICC 651 was isolated from a nodule on the root of a legume and genes responsible for symbiosis including nodulation are located on Sym plasmids of rhizobia it might be possible that the strain BICC 651 had its ancestry in *Agrobacterium* but through horizontal gene transfer acquired the Sym plasmid from some *hitherto* unknown source. A somewhat similar observation was made with a strain T1K, reported to be *Rhizobium trifolii* (*R. leguminosarum* biovar *trifolii*), which nodulated red and white clover and shared many physiological properties with *Agrobacterium tumefaciens*. The strain was more closely related to *Agrobacterium* than to *R. trifolii* (Skotnicki and Rolfe 1978). Based on the gene sequence similarity, it is concluded that the stain BICC 651 might acquire the siderophore gene cluster from other bacteria via horizontal gene transfer during evolutionary process as suggested by Sonoda et al. (2002) who proposed that *Agrobacterium tumefaciens* MAFF301001 had obtained the *agb* gene cluster from *Brucella melitensis*, *Pseudomonas fluorescens*, *Escherichia coli* or *Salmonella typhimurium* in the same process.

Nucleotide sequence analysis revealed that the genes, *sidC, sidE, sidB, sidA* constitute an operon. Disruption of *sidB* by Tn5 insertion results in poor growth under iron-limiting condition, absence of siderophore production and production of less number of nodules by the Sid⁻ mutant, B153 as compared to the wild type strain. The data indicate that the siderophore biosynthetic genes are essential for growth and production of catechol siderophore of the *Rhizobium* BICC 651, which in turn influences its symbiotic efficiency.

Acknowledgments We thank late Dr. Pradosh Roy and Prof. Sujoy Kumar Dasgupta, Department of Microbiology, Bose Institute, Kolkata for their active support. The work was funded by Department of Biotechnology, India and a fellowship (to B. Datta) was provided by the University Grant Commission, India.

References

Benson HP, Boncompagni E, Guerinot ML (2005) An iron uptake operon required for proper nodule development in the *Bradyrhizobium japonicum*-soybean symbiosis. Mol Plant Microbe Interact 18:950–959

Beringer JE (1974) R factor transfer in *Rhizobium leguminosarum*. J Gen Microbiol 84:188–198

Berraho E, Lesuere D, Diem HG, Sasson A (1997) Iron requirement and siderophore production in *Rhizobium ciceri* during growth on an iron-deficient medium. World J Microbiol Biotechnol 13:501–510

Carter RA, Worsley PS, Sawers G, Challis GL, Dilworth MJ, Carson KC et al (2002) The *vbs* genes that direct synthesis of the siderophore vicibactin in *Rhizobium leguminosarum*: their expression in other genera requires ECF sigma factor RpoI. Mol Microbiol 44(5):1153–1166

Ciria R, Abreu-Goodger C, Morett E, Merino E (2004) GeConT: gene context analysis. Bioinformatics 14:2307–2308

Crosa JH, Walsh TC (2002) Genetics and assembly line enzymology of siderophore biosynthesis in bacteria. Microbiol Mol Biol Rev 66:223–249

Dorsey CW, Tolmasky ME, Crosa JH, Actis LA (2003) Genetic organization of an *Acinetobacter baumannii* chromosomal region harbouring genes related to siderophore biosynthesis and transport. Microbiology 149:1227–1238

Earhart CF (1996) Uptake and metabolism of iron and molybdenum. In: Neidhardt FC et al. (eds) *Escherichia coli* and *Salmonella*: Cellular and Molecular Biology. American Society for Microbiology, Washington, DC, pp 1075–1090

Gill PR, Barton LL, Scoble MD, Neilands JB (1991) A high affinity iron transport system of *Rhizobium meliloti* may be required for efficient nitrogen fixation *in planta*. In: Chen Y, Hadar Y (eds) Iron nutrition and interactions in plants. Kluwer Academic Publishers, Netherlands, pp 251–257

Guerinot ML (1991) Iron uptake and metabolism in the rhizobia/legume symbioses. Plant Soil 130:199–209

Huang G, Zhang L, Birch RG (2000) Rapid amplification and cloning of Tn5 flanking fragments by inverse PCR. Lett Appl Microbiol 31:149–153

Joshi FR, Chaudhari A, Joglekar P, Archana G, Desai AJ (2008) Effect of expression of *Bradyrhizobium japonicum* 61A152 *fegA* gene in *Mesorhizobium* sp., on its competitive survival and nodule occupancy on *Arachis hypogea*. Applied Soil Ecology 40:338–347

Joshi FR, Desai DK, Archana G, Desai AJ (2009) Enhanced survival and nodule occupancy of pigeon pea nodulating *Rhizobium* sp. ST1 expressing *fegA* gene of *Bradyrhizobium japonicum* 61A152. J Biol Sci 9:40–51

Lankford CE (1973) Bacterial assimilation of iron. Crit Rev Microbiol 2:273–331

Lesueur D, Carr Del Rio M, Dien HG (1995) Modification of the growth and the competitiveness of *Bradyrhizobium* strain obtained through affecting its siderophore-producing ability. In: Abadia J (ed) Iron nutrition in soil and plants. Kluwer Academic Publishers, PaysBas, pp 59–66

Liu J, Duncan K, Walsh C (1989) Nucleotide sequence of a cluster of *Escherichia coli* enterobactin biosynthesis genes: identification of *entA* and purification of its product 2,3-dihydro-2,3-dihydroxybenzoate dehydrogenase. J Bacteriol 171:791–798

Lynch D, O'Brien J, Welch T, Clarke P, Cuiv P, Crosa JH, O'Connell M (2001) Genetic organization of the region encoding regulation biosynthesis and transport of *rhizobactin* 1021, a siderophore produced by *Sinorhizobium meliloti*. J Bacteriol 183:2576–2585

Manjanatha MG, Loynachan TE, Atherly AG (1992) Tn5 mutagenesis of Chinese *Rhizobium fredii* for siderophore over production. Soil Biol Biochem 24:151–155

Meyer JM, Abdallah MA (1978) The florescent pigment of *Pseudomonas fluorescens* biosynthesis, purification and physical-chemical properties. J Gen Microbiol 107:319–328

Modi M, Shah KS, Modi VV (1985) Isolation and characterization of catechol-like siderophore from cowpea *Rhizobium* RA-1. Arch Microbiol 141:156–158

Mukhopadhyaya P, Deb C, Lahiri C, Roy P (2000) A *soxA* gene, encoding a diheme cytochrome c, and a sox locus, essential for sulfur oxidation in a new sulfur lithotrophic bacterium. J Bacteriol 182:4278–4287

Nambiar PTC, Sivaramakrishnan S (1987) Detection and assay of siderophores in Cowpea Rhizobia (*Bradyrhizobium*) using radioactive Fe (⁵⁹Fe). Appl Microbiol Lett 4:37–40

Norris DO, Date RA (1976) Legume bacteriology. In: Shaw NH, Bryan WW (eds) Tropical Pasture Research – Principles and Methods, Commonwealth Bureau of Pastures and Field Crops Bulletin no. 51. Commonwealth Agricultural Bureau Publishers: Slough, England, pp 134–174

O'Hara GW, Hartzook A, Bell RW, Loneragan JF (1988) Response to *Bradyrhizobium* strain of peanut cultivars grown under iron stress. J Plant Nutr 11:6–11

Rowland BM, Grossman TH, Osburne MS, Taber HW (1996) Sequence and genetic organization of a *Bacillus subtilis* operon encoding 2,3-dihydroxybenzoate biosynthetic enzymes. Gene 178:119–123

Roy N, Bhattacharyya P, Chakrabartty PK (1994) Iron acquisition during growth in an iron-deficient medium by *Rhizobium* sp. isolated from *Cicer arietinum*. Microbiology 140:2811–2820

Rusnak F, Faraci W, Walsh C (1989) Subcloning, expression, and purification of the enterobactin biosynthetic enzyme 2,3-dihydroxybenzoate-AMP ligase: demonstration of enzyme-bound (2,3-dihydroxybenzoyl)adenylate product. Biochemistry 28:6827–6835

Rusnak F, Liu J, Quinn N, Berchtold G, Walsh C (1990) Subcloning of the enterobactin biosynthetic gene entB: expression, purification, characterization, and substrate specificity of isochorismatase. Biochemistry 29:1425–1435

Saitou N, Nei M (1987) The neighbor-joining method: a new method for reconstructing phylogenetic trees. Mol Biol Evol 4:406–425

Sambrook J, Fritsch EF, Maniatis T (1989) Molecular cloning: a laboratory manual, 2nd edn. Cold Spring Harbor Press, New York

Schwyn B, Neiland JB (1987) Universal chemical assay for the detection and determination of siderophores. Anal Biochem 160:47–56

Simon R (1984) High frequency mobilization of gram-negative bacterial replicons by the in vitro constructed Tn5-Mob transposon. Mol General Genet 196:413–420

Skorupska A, Derylo M, Lorkiewiez Z (1989) Siderophore production and utilization by *Rhizobium trifolii*. Biol Met 2:45–49

Skotnicki ML, Rolfe BG (1978) Interaction between the fumarate reductase system of *Escherichia coli* and the nitrogen fixation genes of *Klebsiella pneumoniae*. J Bacteriol 133:518–526

Sonoda H, Suzuki K, Yoshida K (2002) Gene cluster for ferric iron uptake in *Agrobacterium tumefaciens* MAFF301001. Genes Genet Syst 77:137–146

Tamura K, Peterson D, Peterson N, Stecher G, Nei M, Kumar S (2011) MEGA5: molecular evolutionary genetics analysis using maximum likelihood, evolutionary distance, and maximum parsimony methods. Mol Biol Evol 28:2731–2739

Wyckoff EE, Stoebner JA, Reed KE, Payne SM (1997) Cloning of a *Vibrio cholerae* vibriobactin gene cluster: identification of genes required for early steps in siderophore biosynthesis. J Bacteriol 179:7055–7062

Zuckerkandl E, Pauling L (1965) Evolutionary divergence and convergence in proteins. In: Bryson V, Vogel HJ (eds) Evolving genes and proteins. Academic Press, New York, pp 97–166

12

Genetic diversity and chemical profiling of different populations of *Convolvulus pluricaulis* (convolvulaceae): an important herb of ayurvedic medicine

Showkat Hussain Ganie · Zahid Ali ·
Sandip Das · Prem Shankar Srivastava ·
Maheshwar Prasad Sharma

Abstract *Convolvulus pluricaulis* Choisy, commonly known as "Shankhpushpi", is an ayurvedic medicinal plant recommended as a brain tonic to promote intellect and memory, eliminate nervous disorders and to treat hypertension. Because of increasing demand of the drug, this plant species has been over-exploited. As a consequence, many unrelated plants are being sold by the crude drug dealers in India in the name of "Shankhpushpi". Information on its existing gene pool is currently lacking. We developed molecular (Random Amplification of Polymorphic DNA) and chemical (high performance liquid chromatography) markers that could distinguish the genuine plant species from its adulterants. Molecular characterization confirmed higher genetic variation at inter-zonal level as compared to intra-zonal populations. A total of 37 reproducible amplicons were generated of which 22 were polymorphic. The number of amplicons was in the range of 6–11 and genetic distance for the studied primers ranged from 0.07 to 0.34. Fifty nine per cent polymorphism was obtained across different geographical locations. Dendrogram studied through unweighted pair group method of arithmetic analysis differentiated all the genotypes into two major clusters, Cluster I had the single population of Rajasthan and Cluster II was represented by genotypes of Delhi, Haryana, Madhya Pradesh and Rajasthan. The Kaempferol content ranged from 0.07 to 0.49 mg/g and Delhi population was the highest accumulator.

Keywords *Convolvulus pluricaulis* · Genetic diversity · HPLC · Kaempferol · RAPD

Introduction

The indigenous plant-based systems of medicine (Ayurveda, Siddha and Unani) have been in existence for several centuries and continue to serve humanity for infinite time to come. The use of medicinal plants registered a decline with the development of synthetic drugs and antibiotics, but the toxicity and harmful side effects of synthetic drugs have again brought medicinal plants to the forefront of health care system. However, overpopulation and over-exploitation of medicinal plants, particularly in the developing countries, have caused extensive damage to the medicinal plants wealth. Therefore, characterization and conservation of medicinal plants is the need of hour. Herbal medicinal materials are traditionally identified by their organoleptic or microscopic characteristics, including size, shape, colour, odour, flavour, texture and other physical properties. However, these methods are considered to be subjective because the morphological and chemical characters can be influenced by the environment and changes in different developmental stages. Molecular markers are not influenced by environmental factors; tests can be carried out at any time during any stage of plant development; they have the potential of existing in unlimited numbers, covering the entire genome. Moreover, a small amount of sample is sufficient for analysis. A

S. H. Ganie (✉) · M. P. Sharma
Department of Botany, Jamia Hamdard, Hamdard Nagar,
New Delhi 110062, India
e-mail: showkatbotany@gmail.com

Z. Ali · P. S. Srivastava
Department of Biotechnology, Jamia Hamdard, Hamdard Nagar,
New Delhi 110062, India

S. Das
Department of Botany, University of Delhi, New Delhi 110007,
India

number of PCR-based molecular markers have been used for detecting polymorphism at DNA level. Among them, random amplification of polymorphic DNAs (RAPD) gained much popularity because of its simplicity, non-requirement of prior information of nucleotide sequence and can be performed with very small amount of genomic DNA. Random amplification of polymorphic DNA technique has been successfully employed for the estimation of genetic diversity; some of which include *Plantago ovata* (Singh et al. 2009), *Ricinus communis* (Gajera et al. 2010), *Jatropha curcas* (Zhang et al. 2011), *Curcuma longa* (Singh et al. 2012), *Aloe vera* (Nejatzadeh-Barandozi et al. 2012) and *Clitoria ternatea* (Yeotkar et al. 2012). In addition to DNA markers, phytochemical markers high performance liquid chromatography (HPLC) play a role to portray genetic variability and authentication of medicinally important plants (Li et al. 2008). Hence, molecular and chemical characterization can go hand in hand rather than in isolation because such studies would be quite helpful for conservation strategies and the selection of population containing maximum content of active compound.

Convolvulus pluricaulis (Family: Convolvulaceae) is a prostrate spreading wild herb. In India it has a narrow distribution found in plains of Punjab, Uttar Pradesh, Haryana, Rajasthan, Bihar and Chota Nagpur (Sethiya and Mishra 2010). The plant is used in Ayurveda to cure various ailments. It is recommended as a brain tonic to promote intellect and memory, eliminate nervous disorders and to treat hypertension (Bala and Manyam 1999); is anti-helmintic, good in dysentery, hair tonic, cures skin ailments and reduces high blood pressure (Rai 1987). The leaves are recommended for depression and mental disturbance (Singh and Mehta 1977). The herb has been widely used to treat nervous disorders, similar to the use of kava kava (*Piper methysticum*) and valerian (*Valeriana officinalis*) prescribed by American herbalists (Husain et al. 2007).

The plant shows the presence of alkaloids, glycosides, coumarins and flavonoides (Bhowmik et al. 2012). One of the flavonoid in *C. pluricaulis* is Kaempferol (Andrade et al. 2012); it is reported to have significant biological activity being used for chemo-preventive purposes like inhibition of cell growth (Jin et al. 1995), induction of apoptosis (Jin et al. 1995; Braig et al. 2005; Chen and Kong 2005; Niering et al. 2005) and inhibition of proteasome activity (Chen and Kong 2005).

Being found in open waste lands, the natural populations of this herb has depleted to a great extent due to over-exploitation and habitat degradation, especially in urban areas, and therefore, need protection. This species, therefore, appears well suited for study using molecular and chemical markers to determine its genetic diversity. The

Fig. 1 *Convolvulus pluricaulis*

present study is based on RAPD and HPLC analyses, of the materials collected from different locations in India.

Materials and methods

The samples of C. *pluricaulis* were collected from Aravali foothills (Delhi and Kurukshetra, Haryana), Gangetic Plains (Lucknow, Uttar Pradesh), Arid zone (Jodhpur, jaipur, Udaipur, Rajasthan) and Vindhyachal (Bhopal, Madhya Pradesh). Whole herb was collected, transferred to sealable polythene bags and transported to the laboratory within 10–24 h depending upon the distance from the collection site. The samples were subjected to stringent method of botanical identification (Fig. 1); voucher specimens of the same were prepared and are kept in the Herbarium, Department of Botany, Hamdard University, New Delhi, 110062. The identified specimens were compared with authenticated voucher specimens preserved in the herbarium of National Institute of Science Communication and Information Resources (NISCAIR). The lyophilized leaves were used for DNA isolation. A part of the plant material was dried at 40 °C for HPLC analysis.

DNA isolation and RAPD assay

The modified CTAB protocol of Doyle and Doyle (1990) and purification kit (HiPurA) were used to isolate DNA from the leaves. The leaves (1 g) were pulverized to fine powder using liquid nitrogen in a chilled mortar and pestle followed by the addition of 100 mg of PVP (insoluble) and 10 ml pre-heated CTAB buffer (CTAB 2 %, 2 M Sodium Chloride, 100 mM Tris HCl- pH 8, 20 mM EDTA). The slurry was transferred into autoclaved 50 ml centrifuge tube and incubated at 60 °C for 1 h. After incubation, the tubes were kept at room temperature for 20 min and 10 ml of chloroform, isoamyl alcohol (CHCl$_3$: IAA, 24:1) was added and mixed carefully for 15 min. The content was centrifuged at 8,000 rpm for 15 min at 15 °C. The upper phase was collected in fresh autoclaved centrifuge tubes to which 10 µg/ml of RNAase was added and the tubes were incubated at 37 °C for 30 min. To inactivate RNAase A, 10 ml CHCl$_3$: IAA (24:1) was added and the content was centrifuged at 8,000 rpm for 15 min at 15 °C. The upper phase was transferred again into autoclaved centrifuge tube and 0.5 vol of 3 M Potassium acetate (pH 5.2) was added. To precipitate the DNA, two volumes of chilled absolute ethanol was used and the tubes were kept at −20 °C for 2 h. It was recentrifuged at 8,000 rpm for 15 min at 4 °C. The supernatant was discarded and the pellet was washed with 70 % ethanol, air dried and dissolved in 250 µl of sterile water. The DNA obtained was not ideal for PCR analysis and therefore was purified by DNA purification kit (HiPurA, India) according to manufacturer's instructions.

The polymerase chain reaction was carried out in 15 µl reaction volume containing 50 ng DNA, 0.5 units Taq DNA polymerase, 1.66 mM MgCl$_2$, 30 pmol 11-mer primers, 200 µM of each dNTPs, 1 × Taq polymerase buffer. The final volume was made up with sterile MilliQ water. The amplifications were carried out in DNA thermal cycler (Eppendorf, Germany). The PCR amplification conditions for RAPD consisted of initial step of denaturation at 94 °C for 4 min, 35 cycles of denaturation at 94 °C for 1 min, annealing at 35 °C for 1 min, extension at 72 °C for 2 min, followed by final extension at 72 °C for 10 min. The amplified DNA was loaded on 1.2 % agarose gel in 0.5 × TBE buffer containing 10 µl of EtBr (10 mg/ml) and photographed using gel documentation system (UVP, Germany). Lambda DNA EcoR 1- Hind 111 double digest was used as molecular marker (Bangalore Genei, Bangalore, India) to know the size of the fragments. Twenty-five 11-mer RAPD primers of OPN and G series, purchased from Eurofins MWG Operon, Germany), were screened. The data analysis was carried out by band scoring of well-marked amplified fragments.

Data and statistical analysis

PCR products were scored for the presence (1) or absence (0) irrespective of band intensity since each product of identical molecular weight was supposed to represent a single locus. Genetic analysis was carried out using Nei genetic similarity index (Nei and Li 1979) using the formula $F_{xy} = 2n_{xy}/(n_x + n_y)$, where n_{xy} is the number of common RAPD fragments shared by two samples and n_x and n_y are the total number of bands scored in each sample. The genetic distance was calculated using Hillis and Mortiz equation (1990), $D = 1 − F$, where "F" is species similarity. The dendrogram was constructed using the NTSYS-pc software (Numerical taxonomy and multivariate system) (Rohlf 1993).

HPLC analysis

The leaf powder (400 mg) was first defatted by pre-extraction with 20 ml chloroform and refluxed for 1 h with 30 ml 95 % aqueous methyl alcohol (MeOH) and 9 ml of 25 % hydrochloric acid. After filtration with Whatman paper, the samples were extracted twice with 20 ml of MeOH for 10 min. The combined hydrolysates were diluted with MeOH to 100 ml and filtered through syringe filter (0.45 µm). Each extract was injected in triplicate. The kaempferol reference standard (Sigma) in different concentrations was prepared in parallel to generate standard curve for quantification. Kaempferol quantification was made by comparing its retention time of the peak area with the Kaempferol peak area from the standard. The retention time of standard Kaempferol was 2.13. HPLC analysis conditions were: Waters 600E HPLC system equipped with 125 × 4 mm C18 column with PDA detector 996 and auto sampler 2701. The mobile phase was 2 % acetic acid and methanol: acetic acid: water (18:18:1 v/v) at 1 ml/min constant flow rate, 35 °C column temperature and 370 nm wavelength for detection. The injection volume was 20 µl.

Results and discussion

In order to study the diversity at DNA level in C. pluricaulis, 25 primers were used for RAPD analysis. Only five primers generated clear and reproducible bands, with the size range of 0.5–2.0 kb. A total of 37 bands (Table 1), with an average of 7.4 bands per primer, were produced.

Table 1 RAPD data and percentage of polymorphism in C. pluricaulis

S. no	Primer code	Nucleotide sequence (5´-3´)	Total no. of bands	Polymorphic bands	% Monomorphism	% Polymorphism
1	OPN-01	CCTCAGCTTGG	11	09	18.82	81.18
2	OPN-02	AACCAGGGGCA	07	05	28.58	71.42
3	OPN-04	GGACCGACCCA	06	06	0.00	100.00
4	OPN-09	TTGCCGGCTTG	07	01	85.72	14.28
5	G-01	ATGCTCTGCCC	06	01	83.34	16.66
	Average		37	22	40.55	59.45

Number of polymorphic bands per primer ranged from 01 to 09 with an average of 4.8 polymorphic bands per primer. The mean percentage of polymorphic bands was 59.45. Such average polymorphism might be due to non-effective gene flow, low fecundity, low pollen flow, local selection procedure (environment and struggle for existence), inbreeding systems (Loveless and Hamrick 1984), biotic factors like human interference, habitat destruction and commercial exploitation (Vijay et al. 2009). The earlier work carried out on similar aspects lend support to our study that also recorded average polymorphism at individual provenance/genotype level in *Lycoris longituba* (65.96 %, Deng et al. 2006), *Asparagus racemosus* (54.92 %, Vijay et al. 2009), *Typha angustifolia* (71 %, Na et al. 2010).

The primers OPN-04, OPN-01 and OPN-02 generated highest percentage polymorphisms (100, 81 and 71.42, respectively), and the lower polymorphisms (16.66 and 14.28 %) were obtained with G-01 and OPN-09 (Figs. 2, 3). Of the total bands generated, 15 were monomorphic across all the genotypes. The genetic similarity and distance was in the range of 0.66–0.93 and 0.07–0.34, respectively (Table 2).

Population-specific bands generated through different primers represented the identification marks for various genotypes. The unique bands of 0.8 and 0.6 kb amplified by primer OPN-01, one band of 0.35 kb amplified by OPN-02 and bands of 2.1 and 1.0 kb developed by primer OPN-09 are specific to Rajasthan genotypes. The drug Shankhpushpi is being equated with three different plant species (*Clitoria ternatea*, *Convolvulus pluricaulis* and *Evolvulus alsinoides*); studies based on relative efficacy and usage suggest that *C. pluricaulis* can be considered as the actual source, while *E. alsinoides* and *C. ternatea* as the alternative sources of Shankhpushpi (Nair et al. 1997). In addition to these three plants, different unrelated drugs are being sold in crude drug markets of India in the name of Shankhpushpi (Singh and Viswanathan 2001). The specific bands obtained, therefore, will help distinguish the authentic drug from its substitutes and adulteraterants. Molecular markers have been used in the identification of herbal drugs (Rout 2006; Rivera-Arce et al. 2007; Irshad et al. 2009; Heubl 2010; Ganie et al. 2012).

The similarity coefficients were used to generate a tree for cluster analysis using the UPGMA method. The resulting dendrogram differentiated two major clusters. First represented a solitary genotype of Jodhpur (Rajasthan) and the second, the genotypes of Kurukshetra (Haryana), Delhi, Bhopal (Madhya Pradesh), Udaipur, Jaipur (Rajasthan) and Lucknow (Uttar Pradesh) (Fig. 4). The genotypes of Delhi and Bhopal shared highest level of similarity coefficient (0.87). The sample collected from

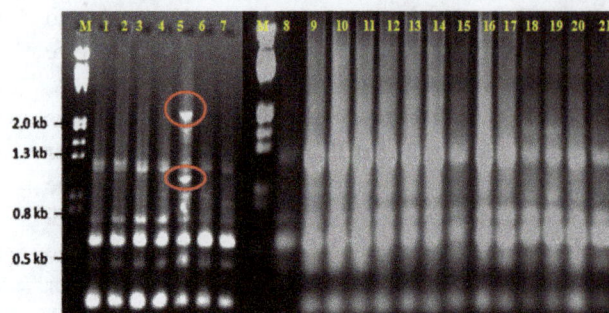

Fig. 2 RAPD fingerprint obtained with OPN-09 primer in different accessions of *C. pluricaulis: M* Marker (λ DNA digested with *Hind* III and *Eco*R I); *1, 2* Kurukshetra Campus (Haryana); *3, 4* Arjun Herbal Park Kurukshetra (Haryana); *5* Jodhpur (Rajasthan); *6, 7* Lucknow (Uttar Pradesh); *8–11* Hamdard Campus (Delhi); *12–15* Bhopal (Madhya Pradesh); *16–18* Udaipur (Rajasthan); *19–21* Jaipur (Rajasthan). *Circles* represent region-specific bands

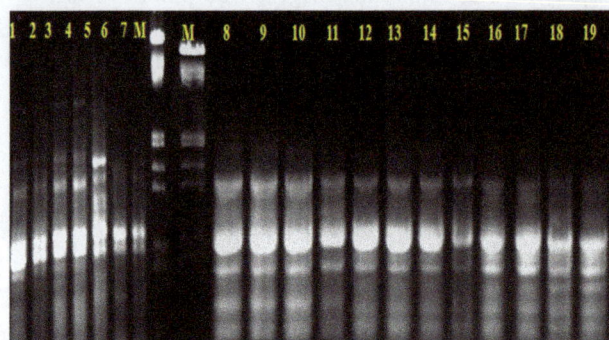

Fig. 3 RAPD fingerprint obtained with G-01 primer in different accessions of *C. pluricaulis: M* Marker (λ DNA digested with *Hind* III and *Eco*R I); *1, 2* Kurukshetra Campus (Haryana); *3, 4* Arjun Herbal Park Kurukhshetra (Haryana); *5* Jodhpur (Rajasthan); *6, 7* Lucknow (Uttar Pradesh); *8–11* Hamdard Campus (Delhi); *12–15* Bhopal (Madhya Pradesh); *16–17* Udaipur (Rajasthan); *18–19* Jaipur (Rajasthan)

Table 2 Average genetic distance in different genotypes of *C. pluricaulis*

Genotype	01	02	03	04	05	06	07	08
01	–	0.07	0.27	0.14	0.09	0.08	0.23	0.18
02	0.07	–	0.25	0.15	0.11	0.16	0.20	0.26
03	0.27	0.25	–	0.30	0.34	0.29	0.31	0.30
04	0.14	0.15	0.30	–	0.17	0.16	0.14	0.20
05	0.09	0.11	0.34	0.17	–	0.07	0.11	0.17
06	0.08	0.16	0.29	0.16	0.07	–	0.11	0.13
07	0.23	0.20	0.31	0.14	0.11	0.11	–	0.14
08	0.18	0.26	0.30	0.20	0.17	0.13	0.14	–

01 Haryana (Kurukshetra Campus), 02 Haryana (Arjun Herbal Park-Kurukshetra), 03 Rajasthan (Jodhpur), 04 Uttar Pradesh (Lucknow), 05 Delhi (Hamdard Campus), 06 Madhya Pradesh (Bhopal), 07 Rajasthan (Udaipur), 08 Rajasthan (Jaipur)

Fig. 4 Dendrogram showing the similarity coefficients in different accessions of *C. pluricaulis*. *Har. 1* (Haryana, Kurukshetra Campus), *Har. 2* (Haryana, Arjun Herbal Park, Kurukshetra), *Del* (Delhi, Hamdard Campus), *M. P.* (Madhya Pradesh, Bhopal), *Raj.* (Rajasthan—Udaipur, Jaipur and Jodhpur), *U. P.* (Uttar pradesh, Lucknow)

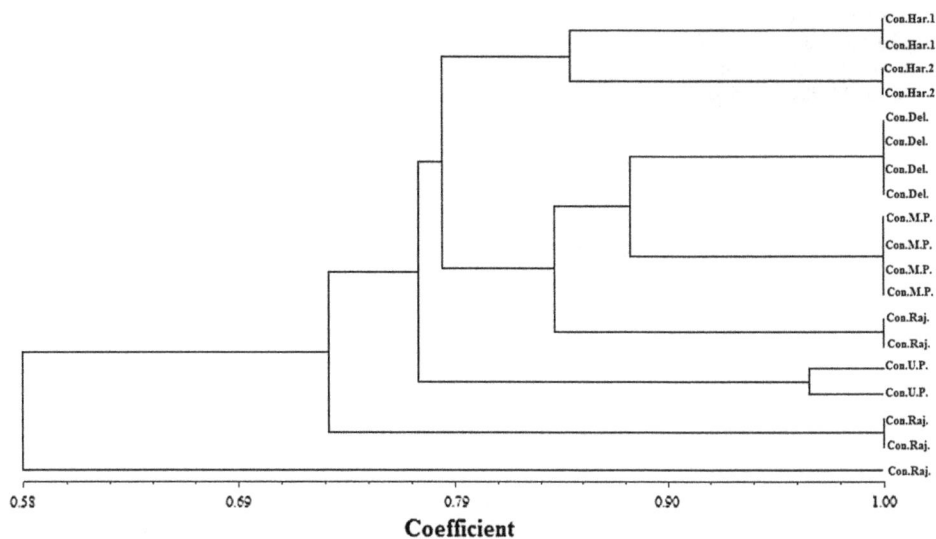

Jodhpur (Rajasthan) had the highest degree of divergence and shared a similarity coefficient of only 0.58 with the rest of the genotypes. The intra-zonal diversity was maximum among Rajasthan populations followed by Haryana populations whereas in other samples, intra-zonal diversity was negligible. The clustering obtained was not region specific, thus showing the lack of any defined population structure in this species. Similar finding have been reported for *Prunus cerasifera* (Ayanoglu et al. 2007) and *Punica granatum* (Jbir et al. 2008). However, Ali et al. (2013) reported well-structured clustering in *Clitoria ternatea* and Belaj et al. (2010) for Andalusian and most Catalonian olive cultivars. Other than Rajasthan genotypes that remained scattered all over the dendrogram, all other genotypes formed perfect clusters by remaining confined within their respective clades.

The HPLC analysis also showed considerable variation in kaempferol content in different genotypes investigated. The results are presented in Table 3. Delhi population was found to be the highest accumulator (0.495 ± 0.014 mg/g dry wt) of kaempferol, followed by populations of Haryana (0.375 ± 0.002 mg/g dry wt), and the least kaempferol content was observed in Madhya Pradesh (0.076 ± 0.002 mg/g dry wt) populations (Fig. 5). The plant collection was made in different geographical locations; variation found, therefore, might be due to different environmental conditions. However, variation in kaempferol content could also be due to intra-specific nucleotide sequence differences. Hence, both genetic as well as environmental factors may be responsible for variation in kaempferol. The variation could be correlated by Sabu et al. (2001) equation: VK = VG + VE, where VK is the variation in kaempferol content, VG is the genetic variance and VE is the environmental variance. The average climate of a particular geographical region remains almost

Table 3 Kaempferol content in different samples of *C. pluricaulis*

S. no	Plant samples	Amount of kaempferol (mg/g dry wt) ± SE
1	Delhi, Hamdard Campus	0.495 ± 0.014
2	Haryana, Kurukshetra	0.375 ± 0.002
3	Uttar Pradesh, Lucknow	0.184 ± 0.007
4	Rajasthan, Jodhpur	0.279 ± 0.003
5	Rajasthan, Jaipur	0.23 ± 0.002
6	Rajasthan, Udaipur	0.19 ± 0.001
7	Madhya Pradesh, Bhopal	0.076 ± 0.002

constant; therefore, our aim was to screen the different populations of *C. pluricaulis* on the basis of kaempferol content. Furthermore, secondary metabolite content generally increases with the maturity and hence, the genotypes were collected at the flowering stages.

Though simultaneous molecular and chemical characterization is rare in medicinal plants, there are some reports where both the techniques have been taken into consideration. Boszormenyi et al. (2009) correlated RAPD profile with the essential oil composition among sage cultivars. Han et al. (2008) studied the genetic and chemical profile in different populations of *Fructus xanthii* in China. Similarly genetic and the terpenoid profile of *Zataria multiflora* collected from different locations in Iran have been analyzed (Hadian et al. 2011).

On the basis of differences in kaempferol content, the populations of *C. pluricaulis* could be divided into three groups. The first group comprised the accessions of Delhi and Haryana (0.495, 0.375), the group II, genotypes of Rajasthan (Jodhpur, Jaipur and Udaipur) and Uttar Pradesh (0.279, 0.23, 0.19 and 0.184), and the group III represented accession from Madhya Pradesh (0.076). Other than kaempferol, a number of medicinally important secondary

Fig. 5 Chromatograms showing kaempferol content of *C. pluricaulis* collected from (**a**) Delhi, (**b**) Madhya Pradesh

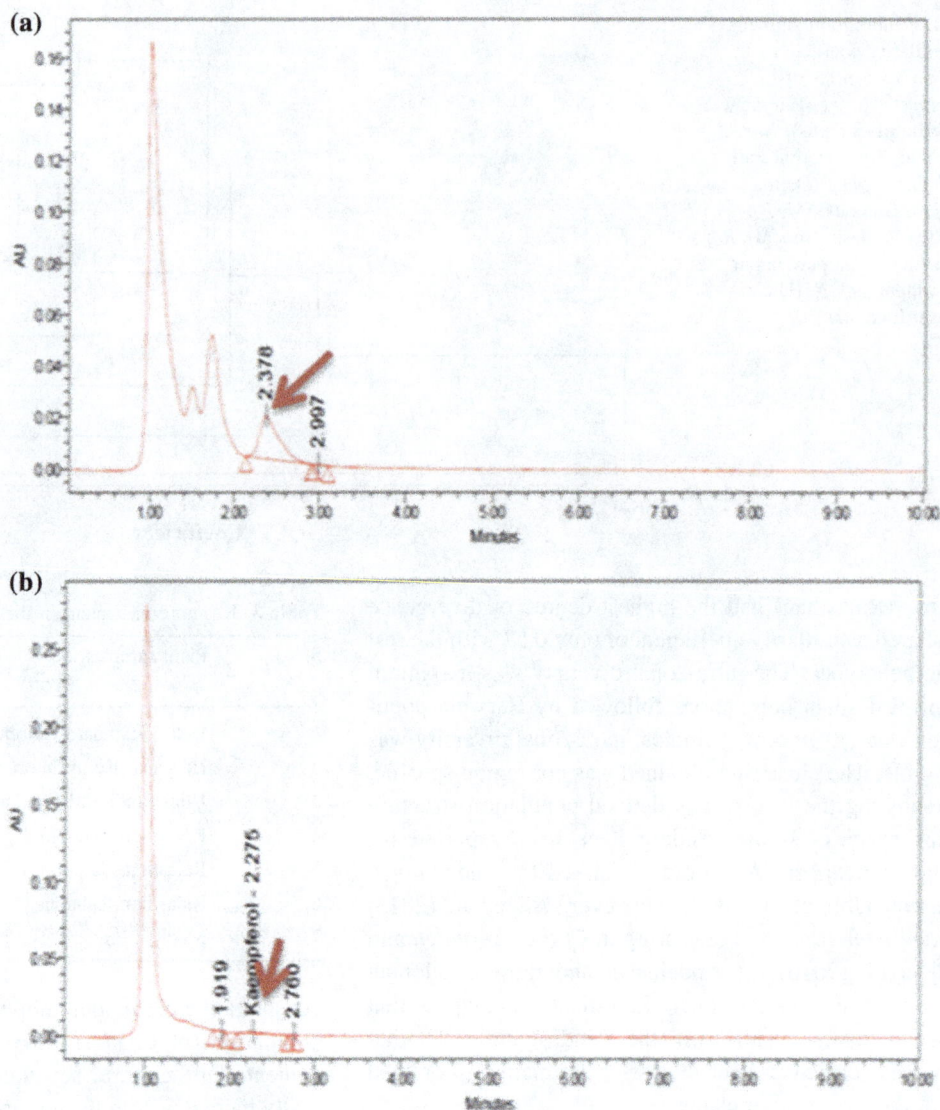

metabolites of *C. pluricaulis* have also been isolated of which scopoletin has been quantified by HPLC using different solvent systems (Kapadia et al. 2006; Upadhyay et al. 2013). The scopoletin content obtained by Upadhyay et al. (2013) was highest in hydro-alcoholic extract (0.1738 %) followed by methanolic extract (0.0932 %) and aqueous extract (0.0435 %). Our results declared that the herb contained less kaempferol in comparison to scopoletin. Delhi populations being rich in kaempferol may potentially be multiplied and used on a large scale for commercial cultivation. The kaempferol content could further be increased by tissue culture and transformation studies. Our next focus is to isolate the kaempferol from *C. pluricaulis* and determine its anti-cancer activity using rat models.

A comparative molecular and chemical analysis of different population of *C. pluricaulis* was achieved in this work.

RAPD technology could be efficiently used to demonstrate genetic relationships among different populations of *C. pluricaulis*. By analysing the genetic and chemical profiling, it is possible to identify the elite population. This information could be employed for devising strategies for authentication and conservation of *C. pluricaulis*.

Acknowledgments This work was financed by the National Medicinal Plants Board, Ministry of Health and Family Welfare, Government of India. The authors are thankful to unknown reviewers for their constructive suggestions that helped us to reshape the manuscript.

Conflict of interest There is no conflict of interest.

References

Ali Z, Ganie SH, Narula A, Sharma MP, Srivastava PS (2013) Intra-specific genetic diversity and chemical profiling of different accessions of *Clitoria ternatea* L. Ind Crops Prod 43:768–773

Andrade C, Monteiro I, Hegde RV, Chandra JS (2012) Investigation of the possible role of Shankapushpi in the attenuation of ECT induced amnestic deficits. Indian J Psychiatry 54:166–171

Ayanoglu H, Bayazit S, Inan G, Bakır M, Akpınar AE, Kazan K, Ergu lA (2007) AFLP analysis of genetic diversity in Turkish green plum accessions (*Prunus cerasifera* L.) adapted to the Mediterranean region. Sci Hortic 114:263–267

Bala V, Manyam MD (1999) Dementia in Ayurveda. J Altern Complement Med 5:81–88

Belaj A, Munoz-Diez C, Baldoni L, Satovic Z, Barranco D (2010) Genetic diversity and relationships of wild and cultivated olives at regional level in Spain. Sci Hortic 124:323–330

Bhowmik D, Kumar KPS, Paswan S, Srivatava S, Yadav A, Dutta A (2012) Traditional Indian herbs *Convolvulus Pluricaulis* and its medicinal importance. J Pharmacogn Phytochem 1:44–51

Boszormenyi A, Hethelyi E, Farkas A, Horvath G, Papp N, Lemberkovics E, Szoke E (2009) Chemical and genetic relationships among sage (Salvia officinalis L.) cultivars and judean sage (*Salvia judaica* Boiss.). J Agric Food Chem 57:4663–4667

Braig M, Lee S, Loddenkemper C, Rudolph C, Peters AH, Schlegelberger B, Stein H, Dorken B, Jenuwein T, Schmitt CA (2005) Oncogene-induced senescence as an initial barrier in lymphoma development. Nature 436:660–665

Chen C, Kong AN (2005) Dietary cancer-chemopreventive compounds: from signalling and gene expression to pharmacological effects. Trends Pharmacol Sci 26:318–326

Deng CL, Zhou J, Gao WJ, Sun FC, Qin RY, Lu LD (2006) Assessment of genetic diversity of *Lycoris longituba* (Amaryllidaceae) detected by RAPDs. J Genet 85:05–207

Doyle JJ, Doyle JL (1990) Isolation of plant DNA from fresh tissue. Focus 12:13–15

Gajera BB, Kumar N, Singh AS, Punvar BS, Ravikiran R, Subhash N, Jadeja GC (2010) Assessment of genetic diversity in castor (*Ricinus communis* L.) using RAPD and ISSR markers. Ind Crops Prod 32:491–498

Ganie SH, Srivastava PS, Narula A, Ali Z (2012) Authentication of shankhpushpi by RAPD markers. Eurasia J Biosci 6:39–46

Hadian J, Ebrahimib SN, Mirjalilia MH, Azizi A, Ranjbard H, Friedtc W (2011) Chemical and genetic diversity of *Zataria multiflora* Boiss. accessions growing wild in Iran. Chem Biodiv 8:176–188

Han T, Hu Y, Zhou SY, Li HL, Zhang QY, Zhang H, Huang BK, Rahman K, Zheng HC, Qin LP (2008) Correlation between the genetic diversity and variation of total phenolic acids contents in *Fructus Xanthii* from different populations in China. Biomed Chromatogr 5:478–486

Heubl G (2010) New aspects of DNA-based authentication of Chinese medicinal plants by molecular biological techniques. Planta Med 76:1963–1974

Hillis DM, Moritz C (1990) Molecular systematics. Sinauer Associates, Sunderland

Husain GM, Mishra D, Singh PN, Rao CV, Kumar V (2007) Ethnopharmacological review of native traditional medicinal plants for brain disorders. Phcog Rev 1:20–29

Irshad S, Singh J, Kakkar P, Mehrotra S (2009) Molecular characterization of *Desmodium* species- an important ingredient of 'Dashmoola' by RAPD analysis. Fitoterapia 80:115–118

Jbir R, Hasnaoui N, Mars M, Marrakchi M, Trifi M (2008) Characterization of Tunisian pomegranate (*Punica granatum* L.)

cultivars using amplified fragment length polymorphism analysis. Sci Hortic 115:231–237

Jin X, Nguyen D, Zhang WW, Kyritsis AP, Roth JA (1995) Cell cycle arrest and inhibition of tumor cell proliferation by the p16INK4 gene mediated by an adenovirus vector. Cancer Res 55:3250–3253

Kapadia NS, Acharya NS, Acharya SA, Shah MB (2006) Use of HPTLC to establish a distinct chemical profile for Shankhpushpi and for quantification of scopoletin in *Convolvulus pluricaulis* and in commercial formulations of Shankhpushpi. J Planar Chromatogr 19:195–199

Li WF, Jiang JG, Chen J (2008) Chinese medicine and its modernization demands. Arch Med Res 39:246–251

Loveless MD, Hamrick JL (1984) Ecological determinants of genetic structure in plant populations. Ann Rev Ecol Syst 27:237–277

Na HR, Kim C, Choi HK (2010) Genetic relationship and genetic diversity among *Typha taxa* from East Asia based on AFLP markers. Aquat Bot 92:207–213

Nair KV, Nair AR, Nair CPR (1997) Standardization of ayurvedic drugs. Bull Med Ethnobot Res 18:151–156

Nei M, Li WH (1979) Mathematical model for studying genetic variation in terms of restriction endonucleases. Proc Natl Acad Sci 76:5269–5273

Nejatzadeh-Barandozi F, Naghavi MR, Enferadi ST, Mousavi A, Mostofi Y, Hassani ME (2012) Genetic diversity of accessions of Iranian Aloe vera based on horticultural traits and RAPD markers. Ind Crops Prod 37:347–351

Niering P, Michels G, Watjen W, Ohler S, Steffan B, Chovolou Y, Kampkotter A, Proksch P, Kahl R (2005) Protective and detrimental effects of kaempferol in rat H4IIE cells: implication of oxidative stress and apoptosis. Toxicol Appl Pharmacol 209:114–122

Rai MK (1987) Ethnomedicinal Studies of Patalkot and Tamiya (Chhindwara): plants used as tonic. Anc Sci Life 3:119–121

Rivera-Arce E, Gattuso M, Alvarado R, Zarate E, Aguero J, Feria I, Lozoya X (2007) Pharmacognostical studies of the plant drug *Mimosae tenuiflorae* cortex. J Ethnopharmacol 113:400–408

Rohlf EJ (1993) NTSYS-pc: numerical taxonomy and multivariate analysis system. Version 1.80. Applied Biostatistics Inc., Setauket

Rout GR (2006) Identification of *Tinospora cordifolia* (Willd.) Miers ex Hook F & Thomas using RAPD markers. Z Naturforsch C 6:118–122

Sabu KK, Padmesh P, Seeni S (2001) Intraspecific variation in active principle content and isozymes of *Andrographis paniculata* Nees (Kalmegh): a traditional hepatoprotective medicinal herb of India. J Med Aromat Plant Sci 23:637–647

Sethiya NK, Mishra SH (2010) Review on ethnomedicinal uses and phytopharmacology of memory boosting herb *Convolvulus pluricaulis* Choisy. Aust J Med Herbal 22:19–25

Singh RH, Mehta AK (1977) Studies on the psychotropic effect of the Medhya Rasayana drug 'Shankhpushpi' (*Convolvulus pluricaulis*) part 1 (Clinical Studies). J Res Indian Med Yog Homeo 12:3–18

Singh HB, Viswanathan MV (2001) Need for authentication of market samples of crude drug Shankhpushpi *Convolvulus microphyllus*. J Med Aromat Plant Sci 22:612–618

Singh N, Lal RK, Shasany AK (2009) Phenotypic and RAPD diversity among 80 germplasm accessions of the medicinal plant isabgol (*Plantago ovata*, Plantaginaceae). Genet Mol Res 8:1273–1284

Singh S, Panda MK, Nayak S (2012) Evaluation of genetic diversity in turmeric (*Curcuma longa* L.) using RAPD and ISSR markers. Ind Crops Prod 37:284–291

Upadhyay V, Sharma Tiwari K A, Joshi HM, Malik A, Singh B, Kalakoti SB (2013) Standardization of HPLC method of Scopoletin in different extracts of *Convolvulus pluricaulis*. Int J Pharm Sci Drug Res 5:28–31

Vijay N, Sairkar P, Silawat N, Garg RK, Mehrotra NN (2009) Genetic variability in *Asparagus racemosus*(Willd.) from Madhya Pradesh, India by random amplified polymorphic DNA. Afr J Biotechnol 8:3135–3140

Yeotkar SD, Malode SN, Waghmare VN, Thakre P (2012) Genetic relationship and diversity analysis of *Clitoria ternatea* variants and *Clitoria biflora* using random amplified polymorphic DNA (RAPD) markers. Afr J Biotechnol 10:18065–18070

Zhang Z, Guo X, Liu B, Tang L, Chen F (2011) Genetic diversity and genetic relationship of *Jatropha curcas* between China and Southeast Asian revealed by amplified fragment length polymorphisms. Afr J Biotechnol 10:2825–2832

13

Exploring specific primers targeted against different genes for a multiplex PCR for detection of *Listeria monocytogenes*

Ashwani Kumar · Sunita Grover ·
Virender Kumar Batish

Abstract The efficacy of six different sets of primers targeted against *16S rRNA* and virulence genes such as '*iap*', '*hly*' and '*prf*' was evaluated in separate PCR assays. The primer pairs targeted against *16S rRNA* resulted into amplification of 1.2 kb PCR product. However, sets of primers targeted against different regions of '*iap*' produced 371 and 660 bp PCR products, respectively. The primer pair targeted against '*prf*' gene could produce 508 bp product. Three primer pairs targeted against different regions of '*hly*', i.e., '*hly*', '*hly A*' and '*hly K9*' were able to amplify 713, 276 and 384 bp products, respectively. The PCR conditions were also optimized in respect of two internal sets of primers falling within '*iap*' and '*hly*' genes that amplified 119 and 188 bp products to verify the PCR results obtained with respective external sets of primers. Three different combinations involving four sets of primers based on *16S rRNA*, '*iap*', '*hly*' and '*prf*' were explored in respective multiplex PCR assays in order to select a suitable combination. Combination 1 and 3 worked successfully as revealed by amplification of all the four bands of expected sizes on agarose gel. However, while optimizing the different parameters for developing a functional multiplex PCR, it was observed that in both these combinations, only two of the amplified products, i.e., 1.2 kb and 713 bp could be invariably detected. Hence, these two primers were combined in the multiplex PCR and the conditions were optimized for application in dairy foods for detection of *Listeria monocytogenes*.

Keywords Primers · Evaluation · Multiplex PCR · *Listeria monocytogenes* · Detection · Dairy foods

Introduction

Listeria monocytogenes, a high-risk emerging food pathogen, has recently assumed lot of interest as a result of its association with several outbreaks of listeriosis across the world through implication with wide variety of foods, both raw and processed (Dalton et al. 1997; CDC 2000; Kumar et al. 2012, 2014). The reports available in India regarding the incidence of *L. monocytogenes* have been analyzed by many workers (Khan et al. 2011; Vinothkumar et al. 2013; Trimulai 2013). Moreover, the occurrence of *L. monocytogenes* in India has been underreported in many cases because of the inefficient surveillance and monitoring system. The ability of this emerging food pathogen to survive and grow in many foods during processing and storage has been attributed to its ubiquitous nature, resistance to diverse environmental conditions such as low pH and high salt concentrations and its microaerobic and psychrotrophic nature. The psychrotrophic nature to grow and survive at a wide range of temperature (2–40 °C) in or on foods for prolonged periods under adverse conditions has made *L. monocytogenes* a major concern for the agri-food industry during the last decade. Mandatory compliance issued by food and drug administration for zero tolerance ruling for this organism in processed/ready-to-eat foods has emphasized the need for development of molecular-based rapid methods for detection of *L. monocytogenes*.

A. Kumar · S. Grover · V. K. Batish (✉)
Molecular Biology Unit, Dairy Microbiology Division, National Dairy Research Institute, Karnal 132001, Haryana, India
e-mail: vkbatish@gmail.com

A. Kumar
Department of Biotechnology, Seth Jai Parkash Mukand Lal Institute of Engineering and Technology, Radaur, Yamuna Nagar 135133, Haryana, India

Keeping in view the limitations associated with conventional, immunological and nucleic acid probe assays, several PCR-based formats have been evolved for detection of *L. monocytogenes* (Kumar et al. 2012). In this context, the gene cassette along with '*iap*' gene involved in pathogenicity of *L. monocytogenes* has been made real targets for its rapid detection by means of PCR-based assays. The expression of three different *L. monocytogenes* virulence genes (*iap*, *hly* and *prf* A) was examined by several investigators to determine the suitable target for specific DNA amplification using gene-specific primers. The '*iap*' gene encodes for a 60 kDa basic extracellular protein (p60) which acts as murein hydrolase involved in septum formation (Wuenscher et al. 1993). The exact role of p60 in invasion is, however, still not clear. Another virulence gene hemolysin encodes two virulence factors involved in the lysis of vacuole namely listerolysin O (LLO) and phosphatidylinositol-specific phospholipase C (PI-PLC) by means of pore-forming activity and hydrolysis of glycophosphatidyl inositol anchors, thereby, leading to escape of organisms from the vacuole in primary macrophages (Portnoy et al. 1992; Sheehan et al. 1994). The use of *16S rRNA* gene as a distinct signature for a bacterial species has become the method of choice for identifying and differentiating microorganisms because of multiple copies (10^4) of rRNA present in cell (Wang et al. 1992). Based on *16S rRNA* gene sequences, several PCR assays have been developed for detection of *Listeria monocytogenes* at both genus and species level (Wiedmann et al. 1993; Czajka et al. 1993). The virulence gene '*hly A*' has also been targeted by different investigators for the development of PCR-based assays intended for detection of *L. monocytogenes* (Deneer and Boychuk 1991; Fluit et al. 1993; Norton and Batt 1999). The '*iap*' gene common to all members of the genus Listeria had also been chosen as a suitable target after finding that there were conserved gene portions at 5' and 3' ends, while internal portions are highly specific (Bubert et al. 1992). In this investigation, we evaluated the efficacy of a few selected pair of primers individually and in different combinations with the objective of developing a reliable Multiplex PCR for detection and identification of *L. monocytogenes*.

Materials and methods

Bacterial cultures and their maintenance

The bacterial cultures used in this investigation included pathogenic strains of *L. monocytogenes* along with other cultures. *Listeria monocytogenes* ATCC 7644 was purchased from Thermo Scientific, UK and *L. monocytogenes* Scott A was procured from DM Division, NDRI, Karnal. The cultures used in this study were propagated in BHI (brain heart infusion)/TSB (Trypticase soya broth) at 37 °C for 18 h. The cultures were preserved on BHI/Trypticase soya agar slants and stored in refrigerator or as glycerol stocks stored at −70 °C ultra low deep freezer (New Brunswick Scientific, USA) until further use. The cultures were activated in BHI broth prior to use by sub-culturing at biweekly intervals.

Preparation of template DNA

Broth cultures

The template/genomic DNA was prepared from broth cultures of *Listeria monocytogenes* by following boiled lysate method (Witham et al. 1996) as well as the method of Pospiech and Neikmann (1995). The boiled lysate was prepared by harvesting the overnight grown culture of the test organism followed by heating the bacterial suspension in 50 MilliQ water for 5 min in a boiling water bath and then centrifuging for 5 min at 10,000 rpm to separate the supernatant containing DNA. For Pospeich and Neikmann's method, the cells were harvested from one ml of overnight grown cultures of *L. monocytogenes* and resuspended in 0.5 ml of SET buffer (75 mM NaCl, 25 mM EDTA, 20 mM Tris) and lysozyme was added at a concentration of 1 mg/ml (25 mM Tris, lysozyme, 10 mg, 5 M NaCl) followed by incubation at 37 °C for 1 h. The subsequent step was the addition of 1/10th volume of 10 % SDS and 0.5 mg/ml of proteinase K and incubation further continued for 2 h at 55 °C. One-third volume of 5 M NaCl and one volume of chloroform were added and incubated at room temperature for 30 min with frequent inversions. The samples were centrifuged and upper aqueous phase transferred to a new tube and the DNA was precipitated by adding one volume of isopropanol or two volumes of ethanol. The DNA was pelleted, dried and dissolved in TER buffer containing 10 µg/ml of RNase A.

PCR assay

The PCR amplification for detection of *Listeria monocytogenes* was performed using Eppendorf master cycler gradient, 5331, Germany. The selected oligonucleotide primers for detection of *Listeria monocytogenes* were got custom synthesized (Bangalore Genei, India). The description of the primer pairs used in this study is given in Table 1. The PCR assay was performed in 25 µl reaction mixture comprising of 100 ng of template DNA, 10× PCR buffer (containing $MgCl_2$), 0.2 mM (each of primers), 0.2 mM (each) dNTPs and 1 unit of Taq polymerase (Boehronger Mannheim). Appropriate positive and negative controls with each reaction were also set up. The PCR cycling parameters used for each set of primers are as per

Table 1 Description of the primers used in the present investigation

S. no.	Target gene	Primers	Primer sequence	Size of amplified product (bp)	References
1	*16S rRNA*	Lm3	5'-ggA CCg ggg CTA ATA CCg AAT gAT AA-3'	1,200	Wiedmann et al. (1993)
		Lm5	5'-TTC ATg TAg gCg AgT TgC AgC CTA-3'		
2	*iap*	ELMIAPF	5'-CAA ACT gCT AAC ACA gCT ACT-3'	371	Klein and Juneja (1997)
		ELMIAPR	5'-gCA CTT gAA TTg CTC TTA TTg-3'		
		Mono A	5'-CAA ACT gCT AAC ACA gCT ACT-3'	660	Bubert et al. (1999)
		Lis 1B	5'-TAA TAC gCg ACC gAA gCC AAC-3'		
3	*Hemolysin*	Hly 1	5'-ATT TTC CCT TCA CTg ATT gC-3'	276	Cooray et al. (1994)
		Hly 2	5'-CAC TCA gCA TTg ATT TgC CA-3'		
		ELMHLYF	5'-TCC gCC TgC AAg TCC TAA gA-3'	713	Klein and Juneja (1997)
		ELMHLYR	5'-gCg CTT gCA ACT gCT CTT TA-3'		
		ILMHLYF	5'-gCA ATT TCg AgC CTA ACC TA-3'	188	Klein and Juneja (1997)
		ILMHLYR	5'-ACT gCg TTg TTA ACg TTT gA-3'		
		HF9	5'-gTT Tgg TTA ATg TCC ATg TT-3'	384	Wagner et al. (2000)
		HR9	5'-TAT TCT AgT CCT gCT gTC CC-3'		
4	*prf A*	ELMPRFF	5'-Cgg gAT AAA ACC AAA ACA ATT T-3'	508	Klein and Juneja (1997)
		ELMPRFR	5'-TgA gCT ATg TgC gAT gCC ACTT-3'		

the published literature and will be described in the "Results and Discussion".

Multiplex PCR assay

For developing a multiplex PCR assay for detection of *L. monocytogenes* in foods, four sets of different primers evaluated previously were tried in three different combinations simultaneously in one assay.

Optimization of multiplex PCR amplification conditions

Amplification conditions were optimized with respect to annealing temperature (Gradient PCR using 60 °C with a gradient of 2 °C), Taq polymerase concentration (0.5–3.0 units), MgCl₂ concentration (1 mM–3.0 mM), Primer concentration (25 ng–100 ng), annealing time (30, 45 and 60 s), extension time (30 s and 1 min) and number of cycles (25, 30, 35 and 40) for the four sets of primers used in the multiplex PCR assay.

Analysis of PCR products

The PCR amplified products were electrophoresed on 2 % agarose gel containing 0.5 μg/ml of ethidium bromide. The gel was visualized under UV transilluminator and photographed using Polaroid 667 packfilm with MP4 system polaroid camera (Photodyne, USA). The molecular size marker consisted of 100 bp DNA ladder comprising of 100–1,000 bp bands (Bangalore Genei, India).

Results and discussion

Evaluation of primers for *Listeria monocytogenes*

During the initial part of this study, we tested the efficacy of six different sets of primers targeted against *16S rRNA* (genus specific) and virulence genes such as 'iap', 'hly' and 'prf' (*L. monocytogenes* specific) in their respective PCR assays using common PCR cycling parameters as their annealing temperatures were pretty close. These include initial denaturation at 95 °C for 4 min followed by 35 cycles each of denaturation at 94 °C for 30 s, annealing at 60 °C for 1 min and extension at 72 °C for 1 min and the final extension of 72 °C for 5 min. The results pertaining to the amplification of the *Listeria monocytogenes*-specific template DNA extracted by Pospiech and Neikmann's method/boiled lysate with individual primers pairs have been presented in Figs. 1 and 2.

Genus-specific primers for *Listeria* spp.

16S rRNA-based primers Lm3/Lm5

16S rRNA has been targeted in the identification of a number of bacteria both at genus and species level by exploring the conserved and variable regions of the gene. (Gopo et al. 1988; Maureau et al. 1989). The choice for targeting *16S rRNA* gene has been dictated by the presence in microorganisms of multiple copies (10⁴) of rRNA, thereby, increasing the ease of signal generation of the assays. Specific DNA probes or PCR primers have been

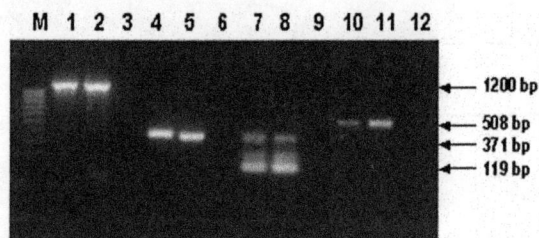

Lanes: M; 100bp Marker; 1-3 (16s rRNA primers): 1, LmATCC 7644 2, Lm Scott A; 3, Negative control; 4-6 ('iap' primers):4, Lm ATCC 7644 5, Lm Scott A; 6, Negative Control; 7-9 (Internal 'iap' primers): 7, Lm ATCC 7644; 8 Lm Scott A; 9, Negative Control; 10-12 ('prf' primers) : 10, Lm ATCC 7644; 11, Lm Scott A: 12, Negative control.

Fig. 1 Evaluation of different sets of primers targeted for detection of *Listeria* and *Listeria monocytogenes* by PCR amplification

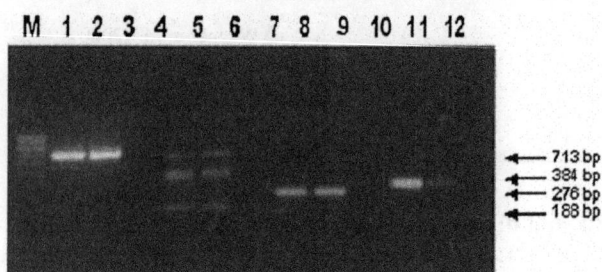

Lanes: M; 100bp Marker; 1-3 ('Hly' primers): 1, Lm ATCC 7644 ; 2, Lm Scott A; 3, Negative control; 4-6 (Internal 'Hly' primers): 4, Lm ATCC 7644 ; 5, Lm Scott A; 6, Negative Control; 7-9 ('Hly A' primers): 7, Lm ATCC 7644 ; 8, Lm Scott A; 9, Negative Control; 10-12 ('Hly K9' primers) : 10, Lm ATCC 7644 ; 11, Lm Scott A: 12, Negative control

Fig. 2 Evaluation of primers using different regions of hemolysin gene of *Listeria monocytogenes*

designed from variable ribosomal RNA regions and used for the detection of specific target cells, e.g., the detection of *L. monocytogenes* (Wang et al. 1991), Aeromonas (Barry et al. 1990), Lactic acid bacteria (Klijn et al. 1991) and Salmonella species (Lin and Tsen 1996). Prompted by the advantages offered by 16S rRNA, we also targeted this gene in our study for exploring a set of primers based on the *16S rRNA* for determining its suitability in a PCR assay for detection of *Listeria*.

The first primer targeted against *16S rRNA* included in this study was intended for detection of all types of Listeria at genus level. A PCR assay was standardized using the primer pair in combination with template DNA from two of the *Listeria monocytogenes* strains ATCC and Scott A. An amplified PCR product of 1,200 bp was detected on the agarose gel with template from both the strains (Fig. 1, Lanes 1 and 2). The PCR amplification conditions include initial denaturation at 95 °C for 4 min followed by 35 cycles each of denaturation at 94 °C for 30 s, annealing at 60 °C for 1 min and extension at 72 °C for 1 min and the final extension of 72 °C for 5 min. The primer pair appears

to be highly specific for *Listeria* only at these amplications parameters, as no specific bands could not be observed.

Our results in this regard are consistent with the earlier findings of Wiedmann et al. (1993) who also achieved the amplification of 1.2 Kb fragment with the help of primers Lm3 and Lm5 in their PCR assay used in conjunction with LCR with all *Listeria* spp. except *L. grayi*. This assay was based on a single base pair difference in the V9 region of the sequence of the genes coding for ribosomal RNA which distinguished *L. monocytogenes* from other closely related *Listeria* spp. The results from our study clearly demonstrate that the two PCR primers used in the PCR assay were genus specific since the 1.2 Kb amplicon could not be detected in any other organism other than *Listeria*.

16S rRNA was also explored previously (Wang et al. 1991, 1992) for detection of *Listeria monocytogenes* in foods spiked with the target organism. They used pair of primers based on a unique region in the *16S rRNA* sequence in *L. monocytogenes* to yield a specific nucleic acid probe. The method was found to be extremely sensitive as it could detect as low as 2–20 cfu/ml of *L. monocytogenes* in pure cultures and as few as 4–40 cfu in inoculated diluted food samples.

Species-specific primers targeted against virulence genes of *L. monocytogenes*

The pathogenicity of *L. monocytogenes* is associated with a number of virulence factors which are encoded on a multigene family common to all *Listeria monocytogenes* strains. Some of these virulence genes could also be very attractive candidates for targeting in the development of PCR-based assays for detection of *L. monocytogenes*. In this study, we evaluated five sets of primers based on 'iap', 'hly' and 'prf' genes in their respective PCR assays. The results pertaining to the suitability of these primers are discussed below.

'iap'-based primers ELMIAPF/R

For this investigation, we had specifically chosen 'iap' gene common to all members of the genus *Listeria* as target because the comparison of all 'iap' genes has indicated that there were conserved regions at 5′ and 3′ ends, while the internal portions are highly specific (Bubert et al. 1992). The 'iap' gene of *L. monocytogenes* encodes the major extracellular protein (P60) (Kuhn and Goebel 1989), which has been shown to be basically an essential murein hydrolase required for adherence/invasion of the organism to the targeted eucaryotic cell. It has been recently shown that the corresponding iap gene portion is also hypervariable in length in different isolates belonging to the same serotypes, thereby, can help in identification of different strains of *L. monocytogenes.*

In order to delineate the species identity of *Listeria* spp., some *Listeria monocytogenes*-specific primers targeted against selected virulence genes were initially explored individually in the study for PCR assays. The first primer pair selected for the purpose was targeted against '*iap*' gene as used previously by Klein and Juneja (1997). The PCR amplification conditions were exactly the same as indicated above for *16S rRNA*-based PCR assay. Agarose gel picture as shown in Fig. 1 (Lanes 4 and 5) revealed a PCR amplified product of 371 bp size with template DNA from both the strains of *Listeria monocytogenes* used in the study. In order to explore the possibility of confirming the authenticity of PCR products (371 bp) of above '*iap*'-based primers (external) specific for *Listeria monocytogenes*, a pair of primers targeted against internal region of '*iap*' gene with the amplified PCR product was also tested in this study. The template used in the PCR assay based on internal '*iap*' primers was the amplified 371 bp product obtained from the previous PCR assay. The PCR assay using internal '*iap*' primers and the amplified product of the external '*iap*'-based PCR assay resulted into the amplification of 119 bp product as can be evidenced from Fig. 1 (Lanes 7 and 8). However, two additional non-specific bands were also detected on the gel albeit at a relatively low intensity. One such band corresponds with the 371 bp product of external '*iap*'-based PCR indicating the possible carry over of the template DNA. The nested PCR conditions used in the study were optimized that resulted into 119 bp product only using 0.5 μl of 1:10 diluted PCR amplified product from external *iap* primers (data not shown). Our results pertaining to amplification of the targeted DNA with '*iap*'-based primers are in close agreement with those of Klein and Juneja (1997) who had previously used there primers for RT-PCR instead of direct PCR with the sole objective of detecting viable cells of *L. monocytogenes.* The RT-PCR assay developed by these investigators could amplify a 371 bp product with ELM*IAP*F/R from only the viable cells when cDNA synthesized from mRNA of *L. monocytogenes* was used as the template. However, in our study these primers were intended to amplify *L. monocytogenes* template DNA through direct PCR for subsequent application of the assay in detection of the targeted organism in raw milk and paneer which do not require any harsh processing treatments. Our results with regard to the use of internal primers-based '*iap*' gene are also comparable to those of Klein and Juneja (1997), although the purpose of using these primers in the two studies was different. Klein and Juneja (1997) had used this internal set of primers just to produce a probe for confirmation of their RT-PCR results for detection of viable *L. monocytogenes.* However, we have used these primers for confirming the specificity of 371 bp product.

'iap'-based primers (Mono A and Lis 1B)

Another set of primers targeted against a different region of '*iap*' was also included in our study for determining their possible application in detection of *Listeria monocytogenes* by PCR assay. The PCR assay set up with this primer pair Mono A and Lis 1B could amplify a 660 bp product with template from *Listeria monocytogenes.* Lis 1 set of primers in the assay worked reasonably well with the PCR amplification parameters as used for other '*iap*'/'*PrfA*'-based PCR assays (data not shown). Our PCR results obtained with Mono A and Lis1B set of primers are consistent with the observations recorded by Bubert et al. (1999) who could also get the amplification of the 660 bp product with this set of primers and exploited it for development of a multiplex PCR assay for *L. monocytogenes.*

'prfA'-based primers ELMPRFF/R

Since the virulence genes in *Listeria monocytogenes* are coregulated by '*prfA*' gene which codes for a 27.1 kDa protein (Chakraborty et al. 1992) which positively regulate all the virulence genes, this gene can also be targeted for detection of *L. monocytogenes* by PCR. We also explored '*prfA*' in the PCR assay using a pair of primers which resulted into the amplification of 508 bp product as seen in Fig. 1 (Lanes 10 and 11). Our results in this regard are again exactly compatible with those of Klein and Juneja (1997) who could also obtained 508 bp products in RT-PCR-based assay used for detection of *L. monocytogenes.*

Primers targeted against '*hly*' gene

Listeria monocytogenes is also capable of producing hemolysins which are involved in the lysis of vacuole and the erythrocytes. The main factor involved in the lysis of the vacuole is the protein that has pore forming activity. 'Listeriolysin O', a 58 kDa protein, is encoded on '*hly A*' gene. '*Hly A*' gene has also been targeted for development of PCR-based assays intended for detection of *L. monocytogenes.* In this study, we used three sets of primers targeted against different regions of '*hly*' gene of *Listeria monocytogenes* as used previously by other workers and evaluated their efficacy in their respective PCR assays. The results pertaining to the same have been presented in Fig. 2.

'hly'-based primers ELMHLYF/R

The primer pair namely ELM*HLY*F/R targeted against '*hly*' gene when used in the PCR assay produced an amplified product of 713 bp with template DNA from *Listeria monocytogenes* (Fig. 2, Lanes 1 and 2). The PCR amplification conditions were exactly the same as those used for

16S rRNA-, 'iap'- and 'PrfA'-based PCR assays. Here also to further check the authenticity of the 713 bp product obtained with *hly*-based primer set, another pair of primers namely ILM*HLY*F/R representing the internal region of 713 bp amplified product was also tested separately in a different PCR assay using the 713 bp amplified product as template and the same amplification parameters as described for *hly*-based PCR assay. This resulted into the amplification of 188 bp product falling within the internal region of 713 bp product as shown in Fig. 2 (Lanes 4 and 5). However, two more additional non-specific bands one corresponding with 713 bp product along with a smaller band could also be detected on the agarose gel. The non-specific bands could be eliminated by following the steps as described previously. Our findings in this regard are further supported by similar observations made by Klein and Juneja (1997) who could also achieve a 713 bp product with their RT-PCR assay using ELM*HLY*F/R set of primers and 188 bp product with ILM*HLY*F/R primers. In this particular case also, these investigators used the internal primers to develop a *L. monocytogenes*-specific probe. On the other hand, the use of these internal primers in our study was intended to confirm the identity of 713 bp product from *L. monocytogenes* template.

'hlyA'-based primers Hly1/2

Another primer pair targeted against '*hly*A' gene when subjected to PCR with the template from *Listeria monocytogenes* yielded an amplified DNA band of 276 bp as can be revealed from Fig. 2 (Lanes 7 and 8). The PCR conditions used in the assay were more or less similar to those used for the other above-mentioned PCR assays except that an annealing temperature of 55 °C was used in place of 60 °C. The results pertaining to efficacy of '*hly A*' gene-based primers with template from *L. monocytogenes* in the PCR assay used in this study are in close agreement with those of Cooray et al. (1994) who could also detect a 276 bp amplified product in their multiplex PCR assay where they had combined the primers targeted against '*hly A*' gene with those from '*prf*A' and 'plcB' genes. However, the PCR cycling conditions used in the two studies were slightly different.

'hly K9'-based primers HF9/HR9

Since K-9 region derived from the non-coding region of '*hly*' gene has been found to be highly polymorphic, this particular region can also be explored for developing a PCR assay for distinguishing different strains of *L. monocytogenes*. This type of assay can be extensively valuable for epidemiological typing. The PCR assay based on *hly K9* set of primers produced an amplified PCR product of 384 bp with *Listeria monocytogenes* template. The PCR

results pertaining to the same have been recorded in Fig. 2 (Lanes 10 and 11). Our results in this regard can be substantiated by similar observations made by Wagner et al. (2000) who also recorded the amplification of 384 bp product in their PCR-based study only when *L. monocytogenes* template was used in the assay.

A critical appraisal of overall PCR results obtained with the respective sets of primers in this study clearly indicates that all the primers worked quite reasonably even under same PCR cycling conditions and hence, there is a possibility of using them simultaneously in conjunction for possible development of a multiplex PCR assay which could authentically detect *L. monocytogenes* in dairy foods rapidly.

Development of multiplex PCR using different combinations of primers for detection of *Listeria monocytogenes*

After evaluation of individual sets of primers in their respective PCR assays, an attempt was then made to explore different combination of these primers to develop a reliable multiplex PCR assay for detection of *Listeria* as such or more specifically *Listeria monocytogenes*. Initially, we tried three combinations involving four different sets of primers in three different multiplex PCR assays using identical PCR parameters. Combination 1 included primer sets targeted against *16S rRNA*, '*hly*', '*prf*A', '*iap*' genes, Combination 2 comprised of primers targeted against *16S rRNA*, '*Lis 1B*', '*prf*A', '*hly K9*' and Combination 3 used primer pairs targeted against *16S rRNA*, '*Lis 1B*', '*prf*A' and '*hly*A'. The PCR parameters used included initial denaturation at 95 °C for 4 min followed by 35 cycles each of denaturation at 94 °C for 30 s, annealing at 58 °C for 1 min and extension at 72 °C for 1 min and an additional step of extended extension at 72 °C for 10 min. The results pertaining to the same have been presented in Fig. 3. As is

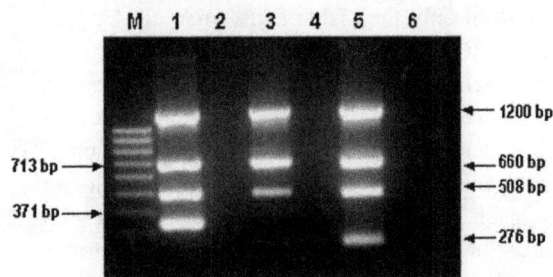

Lanes: M, 100 bp Marker; 1-2, (16S rRNA, Hly, prf, iap primers) :1, Lm ATCC 7644; 2 Negative control, 3-4 (16SrRNA, Lis 1B, prf, Hly K-9): 3, Lm ATCC 7644; 4, Negative Control; 5-6 (16S rRNA, Lis 1B, prf, Hly A): 5, Lm ATCC 7644 : 6, Negative control.

Fig. 3 Evaluation of different combinations of primers targeted against different genes for detection of *Listeria* spp. and *Listeria monocytogenes* by respective multiplex PCR assays

quite evident from the banding patterns revealed on agarose gel, the multiplex PCR assay set up with primers based on *16S rRNA*, *hly*, *prfA* and *iap* (combination 1) amplified the targeted DNA specifically producing four distinct bands of 1,200, 713, 508, 371 bp, respectively (Fig. 3, Lane 1), which matched exactly the same with the PCR products obtained with individual primer pairs as described previously. The multiplex PCR with combination 2-based primers could amplify *L. monocytogenes* template resulting into formation of 1,200, 660 and 508 bp bands, respectively, representing *16S rRNA, Lis 1B and prfA genes* (Fig. 3, Lane 3). However, the fourth expected amplified PCR product of size 384 bp representing *hly K9* could not be detected on the agarose gel. This discrepancy could possibly be attributed to the use of high annealing temperature in the multiplex PCR which may not be suitable for amplification of K-9 fragment. The combination 3 on the other hand again resulted into the amplification of all the four targeted genes in their respective multiplex PCR assay as indicated by the formation of 1,200, 660, 508 and 276 bp bands on the gel (Fig. 3, Lane 5). A comparative evaluation of the three combinations of the primers in their respective multiplex PCR assays clearly indicates that combination 1 and combination 3 performed reasonably well in their respective multiplex PCR assays and the performance of combination 2 was comparatively lower due to non amplification of 384 bp product. Hence, the latter was not considered for further study. Out of the above combinations, 1 and 3 were selected for further improvements in the study.

The main purpose of developing a multiplex PCR in this study was to minimize the possibility of missing *L. monocytogenes* from detection. With this objective in mind, we had combined four sets of primers including one targeted against genus-specific *16S rRNA* and the other three targeted against different regions of virulence genes 'iap', 'hly' and 'prfA' for simultaneous amplification of template DNA from *L. monocytogenes*. Our results in this regard are consistent with similar findings made by Cooray et al. (1994) who also combined three sets of primers targeted against *hly A, prf A and plc B* genes in their multiplex PCR assay which could amplify specifically 795, 571 and 276 bp products with *L. monocytogenes* only. It is quite plausible that some *L. monocytogenes* may lack one or more virulence determinants either because of some mutations or inability of some genes to express under certain conditions. Though the frequency of spontaneous mutations or deletions in these virulence associated genes is not precisely known as yet, we cannot deny the possibility that some mutants like those produced by transposon insertions may exist in nature (Kathariou et al. 1987; Sun et al. 1990) as has been proposed by Cooray et al. (1994). Therefore, it seemed logical and relevant to develop a

procedure in which various virulence associated genes could be detected simultaneously in a single step. Working on similar lines, Klein and Juneja (1997) made an attempt to use three pairs of primers targeted against 'hly', 'iap 'and 'prfA' genes simultaneously in their RT-PCR assay to rule out the possibility of false positive results in PCR-based assays for detection of viable *L. monocytogenes*. However, they were not able to get amplification of all the targeted genes consistently with stronger signal achieved only with primer pair targeted against 'iap' gene, thereby, greatly limiting the applicability of such multiplex RT-PCR assay.

Optimization of multiplex PCR based on combination 3

While further improving the multiplex PCR in terms of using different concentrations of Taq polymerase, dNTPs, primers, manganese chloride, annealing time and number of amplification cycles, it was experienced that two of the amplified products viz. 508 and 276 bp were quite inconsistent in the multiplex PCR and many times either could not be detected on the agarose gel or produced diminished bands (data not shown) that are unable to be visualized. However, the two products namely 1,200 bp based on *16S rRNA*, 713 bp with *ELMHLYF/R* and 660 bp based on *Lis 1B* could always be detected in the multiplex PCR. This inconsistency in the behavior of some primers when used in combination in the multiplex PCR is not an unusual phenomenon due to possible structural interactions between different pieces of oligomers. Although this contention cannot be substantiated as yet, the unequal level of amplifications observed by Klein and Juneja (1997) in their RT-PCR assay using primers targeted against 'iap', 'prf A' and 'hly' genes may indirectly explain the reason for this discrepancy.

In the light of these observations, we finally resorted to develop a multiplex PCR based on two sets of primers one specific for *Listeria* (*16S rRNA*) and another for *Listeria monocytogenes* ('hly') for eventual application in dairy foods.

Multiplex PCR based on two sets of primers
for detection of *Listeria monocytogenes*

Since difficulties were encountered in terms of amplification of all the expected products during optimization of a multiplex PCR based on simultaneous use of four sets of primers, efforts were then directed to develop a multiplex PCR assay exploring only two sets of primers i.e. one genus specific and second *Listeria monocytogenes* specific to get consistent and reproducible results. Two such selected primer pairs which were targeted against *16S rRNA* (genus-specific, 1,200 bp PCR product) and 'hly' (*Listeria monocytogenes*-specific,

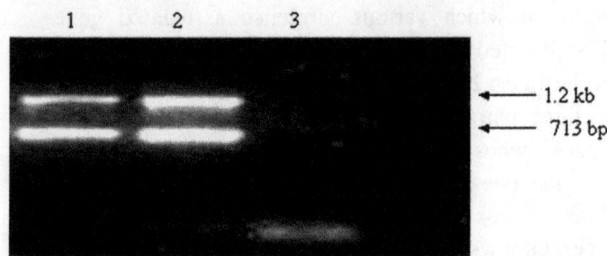

Lanes : 1, Lm ATCC 7644; 2, Lm Scott A; 3, Negative Control

Fig. 4 Multiplex PCR assay using two pairs of primers targeted against *16S rRNA* and *Hly*

713 bp PCR product) since these two primer pairs consistently amplified the specific product. Our previous trials during standardization using the PCR amplification conditions of initial denaturation at 95 °C/4 min followed by 35 cycles each of denaturation at 95 °C/30 s, annealing at 60 °C/1 min and extension at 72 °C/1 min followed by final extension at 72 °C for 5 min led to the amplification of 1,200 bp product specific for *16S rRNA* gene of *Listeria* and 713 bp product specific for *hly* gene of *Listeria monocytogenes* (Fig. 4, Lanes 1 and 2). The multiplex PCR based on *16S rRNA* and '*hly*' genes was successfully applied to selected spiked and natural market dairy food samples for detection of *Listeria monocytogenes* that forms the subject of a separate communication.

From the foregoing presentation, it can be concluded that the multiplex PCR assay developed with two sets of primers targeted against *16S rRNA* and '*hly*' genes could be extremely valuable and of considerable practical utility in dairy industry for monitoring dairy foods for *L. monocytogenes* and hence could help in protecting the health of the consumers against this high-risk food pathogen.

Acknowledgments The work reported here is part of the DBT sponsored project on "PCR Kits" vide Sanction No. BT/PR/958/PID/23/028/98. The authors duly acknowledge the financial support received from Department of Biotechnology, Govt. of India and also to The Director, NDRI, Karnal for providing the necessary infrastructure. The technical assistance provided by Mr. Inder Kumar is also acknowledged.

Conflict of interest The authors report no conflicts of interest.

References

Barry T, Powel R, Gannon F (1990) A general method to generate probes for microorganisms. Biotechniques 8:233–236

Bubert A, Kohler S, Goebel W (1992) The homologous and heterologous regions with in the '*iap*' gene allow genus and

species specific identification of *Listeria* spp. by polymerase chain reaction. Appl Environ Microbiol 58:2625–2632

Bubert A, Hein I, Rauch M, Lehner A, Yoon B, Goebel W, Wagner M (1999) Detection and differentiation of *Listeria* spp. by a single reaction based on multiplex PCR. Appl Environ Microbiol 65:4688–4692

CDC (2000) Food borne outbreak of Listeriosis associated with pork tongue jelly. Morb Mortal Weekly Rep 49:221–222

Chakraborty T, Domann E, Hartl M, Goebel W, Notermans S (1992) Coordinate regulation of virulence genes in *Listeria monocytogenes* requires the product of the *prf* A gene. J Bacteriol 174:568–574

Cooray KJ, Nishibori T, Xiong H, Matsuyama TC, Fujita M, Mitsuyama M (1994) Detection of multiple virulence-associated genes of *Listeria monocytogenes* by PCR in artificially contaminated milk samples. Appl Environ Microbiol 60:3023–3026

Czajka J, Bsat N, Piani M, Russ W, Sultana K, Weidman M, Whitaker R, Batt CA (1993) Differentiation of *Listeria monocytogenes* and *Listeria innocua* by *16S rRNA* genes and intraspecies discrimination of *Listeria monocytogenes* by Random Amplified Polymorphic DNA Polymorphism. Appl Environ Microbiol 59:304–308

Dalton CB, Austin CC, Sobel J (1997) An outbreak of gastroenteritis and fever due to *Listeria monocytogenes* in milk. N Engl J Med 336:100–105

Deneer HG, Boychuk I (1991) Species-specific detection of *Listeria monocytogenes* by DNA amplification. Appl Environ Microbiol 57:606–609

Fluit AC, Torensma R, Visser MJC, Aarsman CJM, Poppelier MJJG, Keller BHI, Klapwijk P, Verhoef J (1993) Detection of *Listeria monocytogenes* in cheese with the magnetic immuno-Polymerase Chain Reaction assay. Appl Environ Microbiol 59:1289–1293

Gopo JM, Melis R, Filipska E, Filipski J (1988) Development of a specific biotinylated DNA probe for rapid identification of Salmonella. Mol Cell Probes 2:271–280

Kathariou S, Metz P, Hof H, Goebel W (1987) Tn916-induced mutations in the hemolysin determinant affecting virulence of *Listeria monocytogenes*. J Bacteriol 169:1291–1297

Khan S, Sujath S, Harish BN, Praharaj I, Parija SC (2011) Neonatal Menigitis due to *Listeria monocytogenes*: a case report from Southern India. J Clin Diagn Res 5(3):608–609

Klein PG, Juneja VK (1997) Sensitive detection of viable *Listeria monocytogenes* by reverse transcription-PCR. Appl Environ Microbiol 63:4441–4448

Klijn N, Weerkamp AH, De Vos NM (1991) Identification of mesophilic lactic acid bacteria by using PCR aplified variable regions of *16S rRNA* and specific DNA probes. Appl Environ Microbiol 57:3390–3393

Kuhn M, Goebel W (1989) Identification of extracellular protein of *Listeria monocytogenes* possibly involved in intracellular uptake by mammalian cells. Infect Immun 57:55–61

Kumar A, Grover S, Batish VK (2012) Monitoring paneer for *Listeria monocytogenes*—a high risk food pathogens by multiplex PCR. African J Biotech 11(39):9452–9456

Kumar A, Grover S, Batish VK (2014) A multiplex PCR assay based on *16S rRNA* and *hly* for Rapid Detection of *L. monocytogenes* in Milk. J of Food measurement and characterization.

Lin CK, Tsen HY (1996) use of two 16S rDNA targeted oligonucleotides as PCR primers for the specific detection of Salmonella in foods. J Appl Bacteriol 80:659–666

Maureau P, Derclaye I, Gregorie D, Janssen M, Cornelis GR (1989) *Campylobacter* species identified on polymorphism of DNA encoding rRNA. J Clin Microbiol 27:1514–1517

Norton DM, Batt CA (1999) Detection of viable *Listeria monocytogenes* with a 5′-Nuclease PCR assay. Appl Environ Microbiol 65:2122–2127

Portnoy DA, Chakraborty T, Goebel W, Cossart P (1992) Molecular determinants of *Listeria monocytogenes* pathogenesis. Infect Immun 60:1263–1267

Pospiech A, Neikmann B (1995) A versatile quick preparation of genomic DNA from Gram-positive bacteria. TIG 11:217–218

Sheehan B, Kocks S, Dramsi S, Gaulin E, Klarsfeld A, Mengaud J, Cossart P (1994) Molecular and genetic determinants of the *Listeria monocytogenes* infectious process. Curr Top Microbiol Immunol 192:187–216

Sun AN, Camilli A, Portnoy DA (1990) Isolation of *Listeria monocytogenes* small plaque mutants defective for intracellular growth and cell-to-cell spread. Infect Immun 58:3770–3778

Tirumalai PS (2013) Listeriosis and Listeria monocytogenes in India. Wudpecker J Food Technol 1(6):98–103

Vinothkumar R, Arunagiri K, Sivakumar T (2013) Studies on pathogenic *Listeria monocytogenes* from marine food resources. Int J Curr Microbiol App Sci 1(1):86–93

Wagner M, Lehner A, Klein D, Bubert A (2000) Single strand confirmation polymorphisms in the *hly* gene and polymerase chain reaction analysis of a repeat regain in the *iap* gene to identify and type *L. monocytogenes*. J Food Prot 63:332–336

Wang RF, Cao WW, Johnson MG (1991) Development of *16S rRNA* based oligomer probe specific for *Listeria monocytogenes*. Appl Environ Microbiol 57:3666–3670

Wang RF, Cao WW, Johnson MG (1992) *16S rRNA* based probes and polymerase chain reaction method to detect *Listeria monocytogenes* cells added to foods. Appl Environ Microbiol 58:2819–2831

Wiedmann M, Barany F, Batt CA (1993) Detection of *Listeria monocytogenes* with a nonisotopic polymerase chain reaction assay. Appl Environ Microbiol 59:2743–2745

Witham PK, Yamashiro CT, Livak KJ, Batt CA (1996) A PCR based assay for detection of *Escherichia coli* shiga like toxin genes in ground beef. Appl Environ Microbiol 62:1347–1353

Wuenscher M, Kohler S, Bubert A, Gerike U, Goebel W (1993) The *iap* gene of *Listeria monocytogenes* is essential for cell viability and its gene product, p^{60}, has bacteriolytic activity. J Bacteriol 175:3491–3501

14

Multiplex PCR for simultaneous identification of *Ralstonia solanacearum* and *Xanthomonas perforans*

S. Umesha · P. Avinash

Abstract   *Ralstonia solanacearum* is a causative agent of bacterial wilt in many economically important crops, and *Xanthomonas perforans* is the causal organism of bacterial spot, one of the most important diseases of vegetables. A multiplex PCR protocol has been developed for the simultaneous, specific and rapid identification of *R. solanacearum* and *X. perforans* in plant materials. Species-specific primers RS-F-759 and RS-R-760 for *R. solanacearum*, RST2 and RST3 for *X. perforans* were used for identification of both pathogens at primer concentrations of 1:4 by optimization of multiplex PCR at annealing temperature of about 61 ± 1 °C. With these primer sets, specific amplification of 281- and 840-bp PCR products was obtained for *R. solanacearum* and *X. perforans*, respectively. The multiplex PCR assay was validated with susceptible plants mechanically inoculated with both the pathogens; specific PCR products confirmed the presence of *R. solanacearum* and *X. perforans*. The multiplex PCR is valuable in identification as well as primary screening of cultivars of both pathogens. The present study is a rapid and easy method for early identification of pathogens from asymptomatic and symptomatic plant materials.

Keywords   Multiplex PCR · *Ralstonia solanacearum* · *Xanthomonas perforans* · Bacterial wilt · Bacterial spot-tomato

S. Umesha (✉) · P. Avinash
Department of Studies in Biotechnology, University of Mysore, Manasagangotri, Mysore 570006, Karnataka, India
e-mail: umeshgroup@yahoo.co.in; pmumesh@gmail.com

Introduction

Globalization of world agriculture increases easy movement of infectious plant material across the countries, which leads to severe problems and difficulty in controlling the spread of plant pathogens, and set back in world agricultural economy by severe yield losses. Robust and inexpensive diagnostic tools are not available for easy identification and classification of many plant pathogens. The primary hurdle in developing highly specific, easily usable diagnostic tools for any pathogen has been difficult in finding unique features, whether they are cell surface antigens or DNA sequences. Therefore, there is a principal need to develop reliable, fast and specific identification methods to prevent spread of diseases caused by several phytopathogenic bacteria.

Ralstonia solanacearum causes bacterial wilt, which is one of the most important and widely spread bacterial diseases of Solanaceous crops in tropics, subtropics and warm temperate regions of the world. This disease has also been recorded in more than 200 plant species, representing over 50 families (Hayward 1994). It exhibits a strong tissue-specific tropism within the host, specifically invading and highly multiplying in the xylem vessels. In addition, it causes vascular browning, stunting, wilting and often leading to rapid death (Remenant et al. 2010). *R. solanacearum* is metabolically versatile and survives not only in soil but also in latently infected plants and water. The pathogen enters the plant through the roots (Xue et al. 2011). Thus, reliable methods to detect the pathogen in tubers as also in soil and soil-related habitats are required. *Xanthomonas perforans* is an important bacterial pathogen of vegetables particularly in tomato (*Lycopersicon esculentum* Mill.), causing serious economic losses worldwide (Kuflom et al. 1997). Bacterial spot disease of tomato occurs in warm, moist regions throughout the world. *X. perforans* is a rod-shaped Gram-negative aerobic bacterial plant pathogen. This spot pathogen is seed-borne, persist as epiphytic populations in asymptomatic seedling and mature plants. The bacterium possesses a polar flagellum that propels them in water, which facilitates the infection of wet leaves. The primary symptoms are necrotic lesions that occur on leaves, stem and fruits. In warm and rainy weather, bacterial spot may cause severe defoliation of plants that result in reduced yield, and the diseased fruits are not suitable for fresh-market sale. Blister-like fruit lesions and stem lesions are observed in late summer. *X. perforans* can survive in dry seeds for 10 years and on plants debris redundancy up to 1 year (Isabelle et al. 2006). The impact of both the pathogens in India and world over is very high.

The most commonly used methods are not accurate for identification of disease on visual symptoms. Isolation of pathogens from plants as well as seeds by conventional techniques is often difficult. Additionally, biochemical tests, pathogenicity test and serological tests take several weeks for the final confirmation of pathogens. The sensitivity of ELISA is limited only for the detection of symptomatic plants; enrichment of target bacteria on semi-selective media before ELISA improves specificity, but it is time consuming. Therefore, there is a need for rapid and specific method for routine indexing of asymptomatic and symptomatic plant materials (Lang et al. 2010).

Diagnostic tests based on the characteristics of the genomics of the bacterial pathogens can provide more reliable results, and these are not dependent on symptom expression and environmental conditions. An early DNA-based approach to distinguishing *R. solanacearum* and *X. perforans* involved amplification of 16S rRNA; the approach could be useful only if supported by other sequence information such as 16S–23S rRNA internal transcribed spacers, which is able to distinguish both pathogens (Schonfeld et al. 2003; Santana et al. 2012). Species-specific PCR-based methods are commonly used in the detection and identification of pathogens with the help of species-specific primers (Adachi and Oku 2000). A multiplex PCR amplification provides reliable pathogen detection in routine testing and allows for the simultaneous amplification of more than one DNA region of interest, which is possible in a single PCR reaction mixture.

Multiplex PCR is widely applied in simultaneous, rapid, accurate detection and identification of major pathogens having different DNA or RNA targets in a single reaction. Co-amplifications of *X. axonopodis* pv. *vesicatoria* and *Clavibacter michiganensis* sub spp. *michiganensis* were performed in tomato plant from a single PCR tube (Ozdemir 2005). A multiplex PCR assay targeting *avrBsT* and *xopL* for the molecular identification of *Xanthomonas*

axonopodis pv. *phaseoli* was validated by comparison with other molecular identification assays aimed at *X. axonopodis* pv. *phaseoli*, on a wide collection of reference strains. This multiplex was further validated on a blind collection of *Xanthomonas* isolates for which pathogenicity was assayed by stem wounding and by dipping leaves in calibrated inoculation (Boureau et al. 2013).

Identification of *R. solanacearum* and *X. perforans* is an important step, as both pathogens have been considered as potential disease threat agents all over the world. Therefore, the primary objective of the present study was to develop a multiplex PCR, specific to *R. solanacearum* and *X. perforans* for identification of pathogen in symptomatic and asymptomatic samples.

Materials and methods

Bacterial isolation

Different cultivars of tomato seed samples were collected from the public and private seed traders, Mysore, India and agriculture fields Mysore, India. Suspected plant material viz, bacterial wilt and bacterial spot symptoms along with suspected soil samples from agriculture fields was collected, brought to the laboratory and subjected to laboratory assays viz, direct plating and liquid assay methods (ISTA 2005). The suspected plant material was cut into pieces (5 mm); plant materials as well as seeds were surface disinfected with sodium hypochlorite (3 %) followed by repeated washing with sterile distilled water and plated on semi-selective medium [Kelman's triphenyl tetrazolium chloride (TZC; dextrose; 10 g, tryptone; 1 g, peptone; 10 g, agar; 18 g, dis H_2O; 1,000 ml: triphenyl tetrazolium chloride; 0.075 g in 7.5 ml)] [Tween B media (peptone; 10 g, KBr; 10 g, $CaCl_2$; 0. 25 g, boric acid; 0. 30 g, agar; 15 g, Tween 80; 10 ml, dis H_2O; 1,000 ml)]. Following surface-sterilization, liquid assay of the collected seeds was carried out in plant material and seed samples were macerated using sterile mortar and pestle in 10 ml sterile distilled water. The supernatant (1 ml) was mixed with 9 ml of sterile distilled water to obtain a dilution of 10^{-1}, and further serial dilutions were prepared up to 10^{-5}. Fifty microliters of each dilution was placed on TZC and Tween B semi-selective media. In addition, the suspected soil samples from different fields were subjected to serial dilution up to 10^{-5} dilution, and aliquots of fifty microliters of each dilution were spread on TZC and Tween B semi-selective media using Drigalski's spreaders in triplicates. Plates were incubated at 28 ± 2 °C for 24–48 h. The yellow colonies with hydrolytic zones were observed for *X. perforans* around the pieces of plant material and seeds, whereas typical mucoid creamy white colonies with pink centers were indicative for the presence of *R. solanacearum* (Hayward 1994). For positive control of *R. solanacearum* (DOBCPR 12), a pure culture used was kindly provided by Prof. Ashok Gadewar, Central Potato Research Institute, Shimla, India (Vanitha et al. 2009). These bacteria were isolated in their pure form and subjected to biochemical/physiological, hypersensitive and pathogenicity tests for confirmation of pathogens.

Biochemical characterization of bacterial isolates

Biochemical characterization of *R. solanacearum* and *X. perforans* was performed based on biochemical tests: Gram staining, KOH solubility, starch hydrolysis test (Fahy and Persley 1983), lipase activity, Kovacs' oxidase test (Kovac's 1956), gelatin hydrolysis and oxidative/fermentative metabolism of glucose.

Pathogenicity assay and hypersensitivity tests

Pathogenicity of bacterial isolates was assessed using two distinct tests. First, the virulence of *R. solanacearum* was tested with a pathogenicity assay on susceptible plants of tomato under the screen house conditions at 28 ± 2 °C. Susceptible cultivars of tomato (Cv. PKM-I) were sown in earthen pots, plants were allowed to grow for 5 weeks and used for bacterial inoculation. *R. solanacearum,* DOB-R strain was cultured in nutrient broth for 24–36 h. The density of cell suspension was adjusted to 1×10^8 cfu/ml using spectrophotometer (Beckman Coulter, California, USA) (Lelliot and Stead 1987). This bacterial suspension was inoculated to the roots of tomato plants just below the soil surface.

Xanthomonas perforans, DOB-X strain suspension (1×10^8 cfu/ml) was sprayed to the aerial parts of the tomato plants to completely run-off level, and plants were covered with polythene bags sprayed with water to maintain a high-humid condition. Pathogen-inoculated plants were closely monitored for the typical symptoms of both bacterial wilt and spot disease. Experiments were conducted in three replicates and repeated twice with appropriate controls.

Two milliliter of each bacterial suspension (1×10^8 cfu/ml) was infiltrated to the leaves of 1 month old tobacco plant (*Nicotiana tabaccum* L.) for hypersensitivity tests, with water as a controls (Carlton et al. 1998). Infiltrated tobacco plants were maintained under green house conditions with 25–30 °C in day and 15–18 °C at night and were maintained until the symptoms appeared.

Specific PCR assay and sequence analysis

Bacterial isolates from soil, seed and plant material were used for the extraction of genomic DNA from *R.*

solanacearum and *X. perforans.* Bacterial DNA was isolated using bacterial genomic DNA isolation kit from BangaloreGeNei (Bangalore, India) according to the manufacturer's instruction. For identification of *R. solanacearum,* uniplex PCR was performed to amplify *speI* region of *R. solanacearum* using RS-F-759 and RS-R-760 primers as described by Opina et al. (1997). The specific primers RS-F-759 and RS-R-760 were custom synthesized from Chromous Biotech, Bangalore, India, RS-F (5′-GTCGCCGTCAACTCACTTTCC-3′) and RS-R (5′-GTCGCCGTCAGCAATGCGGAATCG-3′). DNA was amplified in 25 µl of reaction mixture prepared in 0.2-ml PCR tubes by adding PCR reaction mixture containing 1 µl of 100 mM dNTPs, 2.5 µl of 10× buffer, 2.0 µl of 25 mM MgCl$_2$, 1 U of *Taq* DNA polymerase (Chromous Biotech, Bangalore, India), 1 µl of 10–100 ng genomic DNA and 2.0 µl of each primers of 25 pmol of *R. solanacearum* in a total volume of 25 µl. The PCR tubes were placed in a PCR thermocycler (Labnet, Multigene gradient, California, USA) and programmed thermal cycle as initial denaturation at 94 °C for 3 min, annealing at 53 °C for 1 min and extension at 72 °C for 1 min 30 s, followed by 30 cycles of denaturation at 94 °C for 15 s, annealing at 60 °C for 15 s, elongation step at 72 °C for 15 s and final extension at 72 °C for 5 min.

The isolates of *X. perforans* were subjected to PCR using specific primer which was selected at *hrp B* region of *Xanthomonas campestris* pv. *vesicatoria* strain 75-3 as stated by Leite et al. (1994) and custom synthesized as follows (RST2-5′AGGCCCTGGAAGGTGCCCTGGA3′) and (RST3-5′-ATCGCACTGCGTACCGCGCGCGA 3′). The PCR reaction mixture contained 1 µl of 100 mM dNTPs, 2.5 µl of 10× buffer, 2.0 µl of 25 mM MgCl$_2$, 1 U of *Taq* polymerase (Chromous Biotech, Bangalore, India), 1 µl of 10–100 ng genomic DNA of *X. perforans,* 2.0 µl of each primers 25 pmol in a total volume of 25 µl. The PCR parameters were initial denaturing at 94 °C for 3 min, followed by 30 cycles at 94 °C for 30 s, 62 °C for 30 s, and 72 °C for 1 min 30 s, and a final extension at 72 °C for 7 min.

Amplified PCR products were purified using QIAquick gel extraction kit (Qiagen, Hilden, Germany) following the manufacturer's instructions and sequenced commercially (Eurofins, Bangalore, India), and sequences were compared with other *R. solanacearum and X. perforans* sequence from data base using multiple sequence alignment software. Sequences of both pathogens were deposited in the GenBank database.

Optimization of multiplex PCR

Multiplex PCR was carried out for bacterial DNA templates isolated from mixed culture of both pathogens. A primer mix was used with final concentration of 100 pmol of each primer in a ratio of 1:4, which was standardized among 1:1, 1:2, 1:3 and 1:4 for *R. solanacearum* and *X. perforans*, respectively. Reactions in a total volume of 25 µl were performed with 1 U of *Taq* DNA polymerase (Chromus Biotech, Bangalore, India). DNA was amplified in 25 µl of reaction mixture containing $10\times$ thermo pol buffer, 2 mM $MgCl_2$, and 25 mM of dATP, dCTP, dGTP and dTTP. The cycling was performed with a Master cycler Gradient (Labnet, Multigene gradient, California, USA). Gradient PCR was performed to optimize the multiplex PCR annealing temperature from 55 to 65 °C. The PCR parameters were followed an initial denaturing step at 95 °C for 3 min, followed by 30 cycles of 95 °C for 30 s, 61 ± 1 °C for 1 min and 72 °C for 1 min. The amplified PCR product was mixed with 2 µl of loading dye and separated in 1.2 % agarose gel electrophoresis by TAE buffer using 75 V. Further, gels were documented using Geldoc 1000 System-PC (Bio-Rad, Gurgaon, India). *C. michiganensis* subsp. *michiganensis* and *Pseudomonas fluorescens* were used as negative controls (Table 1).

Validation of multiplex PCR

Developed multiplex PCR was validated by artificially infecting susceptible tomato plants with DOB-R and DOB-X strains, respectively, with 20 different cultivars (five plants with four replicates) (Table 2), along with infected plants collected from agricultural fields. The seed samples were macerated using sterile mortar and pestle in 1 ml sterile distilled water, and soil samples were serially diluted up to 10^{-3}, and 10 µl of macerated seed sample and diluted soil samples were used as template to amplify multiplex PCR. The sensitivity of the primers was checked by diluting single-colony bacterial cultures from 10^{-1} up to 10^{-5} in sterile distilled water (100 µl in 900 µl of sterile distilled water) without isolating DNA from the bacterial cultures in this multiplex PCR, and gel was documented using Geldoc 1000 System-PC (Bio-Rad, Gurgaon, India).

Results and discussion

The bacterial wilt-causing pathogen *R. solanacearum* and bacterial spot-causing pathogen *X. perforans* in tomato plants were isolated and characterized. Plant material, seed samples and soil samples were subjected to laboratory assays such as direct plating and liquid assay. The samples showed the presence of both the pathogens. Isolates of *R. solanacearum* from soil, plant material and seeds were cultured using semi-selective media, and typical mucoid creamy white colonies with pink centers were observed. *X. perforans* colonies on Tween B media exhibited typical

Table 1 Mixed culture, isolation of *R. solanacearum* and *X. perforans* from different sources and their reaction to multiplex PCR

Source	Isolates of *R. solanacearum*	Isolates of *X. perforans*	Multiplex PCR	
			R. solanacearum	*X. perforans*
Soil	RS 1	XP 1	+	+
Soil	RS 2	XP 2	+	+
Soil	RS 3	XP 3	+	+
Soil	RS 4	XP 4	+	+
Soil	RS 5	XP 5	+	+
Plant material	RS 6	XP 6	+	+
Plant material	RS 7	XP 7	+	+
Plant material	RS 8	XP 8	+	+
Plant material	RS 9	XP 9	+	+
Plant material	RS 10	XP 10	+	+
Seed	RS 11	XP 11	+	+
Seed	RS 12	XP 12	+	+
Seed	RS 13	XP 13	+	+
Seed	RS 14	XP 14	+	+
Seed	RS 15	XP 15	+	+
Seed	RS 16	XP 16	+	+
Seed	RS 17	XP 17	+	+
Seed	RS 18	XP 18	+	+
Seed	RS 19	XP 19	+	+
Seed	RS 20	XP 20	+	+
Seed	Cmm	Cmm	–	–
Seed	PS	PS	–	–

RS 1–RS 5. Bacterial isolates isolated from serial dilution method, RS 6–RS 10. Bacterial isolates isolated from direct plating method of plant materials (stem), RS 11–Rs 15. Bacterial isolates isolated from direct plating method of seeds, RS 16–RS 20. Bacterial isolates isolated from liquid assay method for seeds

XP 1–XP 5. Bacterial isolates isolated from serial dilution method, XP 6–XP 10. Bacterial isolates isolated from direct plating method of plant materials (leaves), XP 11–XP 15. Bacterial isolates isolated from direct plating method of seeds, XP16–XP 20. Bacterial isolates isolated from liquid assay method for seeds

Cmm. *Clavibacter michiganensis* subsp. *michiganensis* and PS. *Pseudomonas fluorescens* bacterial isolate isolated from direct plating method of seed and soil, respectively

morphological characteristics such as yellow colonies with hydrolytic zones. Further, different isolates of *R. solanacearum* and *X. perforans* were subjected to biochemical/physiological assays along with hypersensitivity and pathogenicity tests. Phytobacterial pathogens, *C. michiganensis* subsp. *michiganensis* and *P. fluorescens* were also isolated from the collected seed and soil samples, respectively (Table 1).

Table 2 Mixed infection of pathogens in susceptible plants compared by pathogenicity test and multiplex PCR

Bacterial isolates	Tomato cultivars	Response to pathogenicity test		Response to multiplex PCR	
		R. solanacearum	*X. perforans*	*R. solanacearum*	*X. perforans*
1	Ark-Abha	+	+	+	+
2	Ashwini-FI	+	+	+	+
3	Arunodaya	+	+	+	+
4	Indosem	+	+	+	+
5	Indam	+	+	+	+
6	Madanapalli	+	+	+	+
7	Malini	+	+	+	+
8	MPH-I	+	+	+	+
9	OK Seed	+	+	+	+
10	PKM-I	+	+	+	+
11	Sarapana	+	+	+	+
12	HCl-IV	+	−	+	−
13	Local-I	+	−	+	+
14	PHS	+	−	+	+
15	Ashoka	−	−	+	+
16	Alrounder	−	−	+	+
17	Rasi	−	+	−	−
18	Quality	−	+	−	−
19	Vignesh	−	−	+	−
20	Mrytunjaya	−	−	−	+

"+" indicates the positive reaction; "−" indicates negative reaction of phytopathogenic bacteria

R. solanacearum and *X. perforans* were subjected to biochemical characterization, both pathogens stained pink red for Gram's reaction and thin viscid mucoid strand for KOH solubility indicating Gram's negative in nature. *X. perforans* liquefied gelatin media when compared control and show a clear zone of hydrolysis around the bacterial colonies when flooded with Lugol's iodine on starch hydrolysis' test. Lypolytic activity was confirmed positive by the presence of a white precipitate around the colonies of *X. perforans* in Tween 80 agar plates. *R. solanacearum* designate negative for gelatin hydrolysis, starch hydrolysis test as well as lipase activity tests. Kovac's oxidase tests show a positive result by immediate change in the color to blue in *R. solanacearum*, but it was negative in *X. perforans*. *R. solanacearum* changed the color of media from green to yellow, indicating positive results for oxidation test, whereas in the fermentation test, both these pathogens did not show any reaction.

Susceptible tomato cultivars when inoculated with *R. solanacearum* isolated from the soil, seed and plant material showed bacterial wilt symptoms such as vascular browning, stunting, wilting and often with rapid death of tomato plants. Control plants did not show any disease symptoms. *X. perforans* isolates showed bacterial spot symptoms, whereas control plants with nutrient broth did not show any symptoms. Yellowing of leaf followed by necrosis was evident in tobacco plants within 48 h of infiltration with both bacterial isolates, whereas control

leaves did not show any change in leaf morphology. Biochemical/physiological tests, hypersensitivity and pathogenicity tests were used in the identification and confirmation of the isolated pathogens as *R. solanacearum* and *X. perforans*, but these tests are time consuming. Our main objective was to develop a reliable multiplex PCR assay for the identification of these pathogens. This new PCR assay combines two different tests first, high specificity in the identification of the pathogens. The results of present study are in confirmation with the studies reported, earlier (Chandrashekar et al. 2012; Avinash and Umesha 2014).

Polymerase chain reaction technique has found wide application in detecting plant pathogenic bacteria (Adachi and Oku 2000). Targeting 16S–23S spacer region is not specific, but it is also fast and easy. A multiplex "Taq man" PCR method was described by Weller et al. (2000) for specific identification of *R. solanacearum* in potato tubers extract, the developed multiplex PCR described here is a simple, with lower cost and an alternative to the "Taq man" method. A multiplex PCR method for detection and differentiation of *R. solanacearum* strains compared with 16S–23S rRNA primers with the use of sequence analysis which differentiate *R. solanacearum* subclasses which is also cost effective for large-scale screening. The 16S–23S spacer region has large copy number (more than 10,000 per cell) and high degree of sequence conservation, which are the drawbacks of 16S–23S primers Pastrik et al. (2002).

Fig. 1 Optimized multiplex PCR for both *R. solanacearum* and *X. Perforans*. Lanes *1* and *2* amplification of *R. solanacearum* annealing temperature at 57 and 61 °C. Lanes *3* and *4* amplification of *X. perforans* annealing temperature at 58 and 64 °C. Lanes *5–7* multiplex PCR of *R. solanacearum* and *X. perforans* 61 ± 1 °C. 50 bp (*M*) Gene ladder

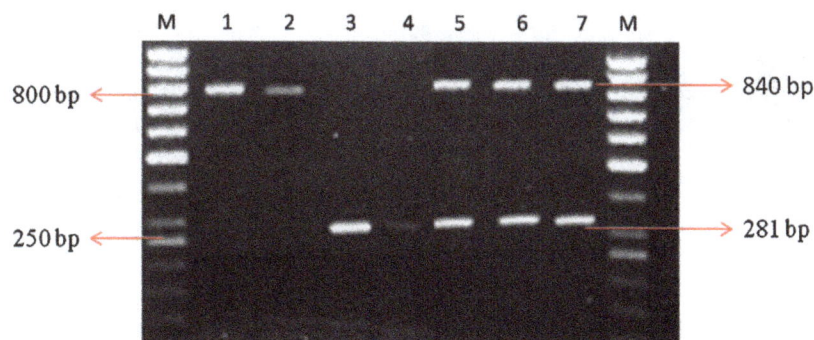

Fig. 2 Multiplex PCR for both *R. solanacearum* and *X. perforans*. Lanes *1–20* mixed culture of *R. solanacearum* and *X. Perforans* amplified at 281 and 840 bp. (N) Negative control *Clavibacter michiganensis* subsp. *michiganensis* and *Pseudomonas fluorescens*. 100-bp DNA marker (*M*)

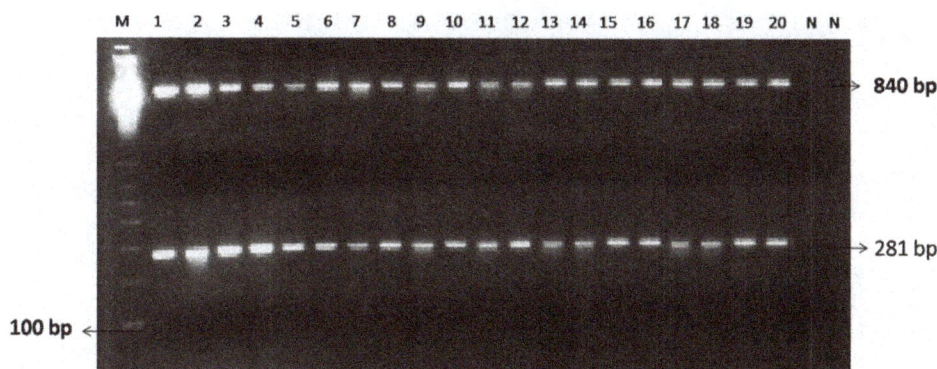

R. solanacearum subjected to specific PCR assay using specific primers viz, RS-F-759 and RS-R-760 showed a specific amplification at 281 bp, confirmed as *R. solanacearum*. Primers RST2 and RST3 showed an amplification of 840-bp amplicon in genome confirmed as *X. perforans*. To confirm primer specificity, homology test for primers was carried out using BLAST search considering query coverage and percentage identity for both pathogens, in which RS-F-759 and RS-R-760 primers showed 100, 95 and 75 % homology to *R. solanacearum, R. syzygii* and *Oryza minuta*, respectively, whereas same primers did not show homology with *Xanthomonas*. The *hrp B* primers did not show 100 % identity with *X. oryzae* pv. *oryzae*, *X. oryzae* pv. *oryzicola* and *X. fuscans* subsp. *fuscans*, whereas *hrp B* primers did not show homology with *R. solanacearum*. Identities of pathogens were further confirmed by sequencing of amplified products of *R. solanacearum* and *X. perforans*, and sequences were deposited in the GenBank database (accession numbers JX628912 and JX628913). To examine further the extant of sequence variation present in *R. solanacearum* and *X. perforans*, nucleotide sequence of other strains of both pathogens from GenBank database was compared. The result revealed that 100 % homology with other *R. solanacearum* strains, whereas *X. perforans* sequence showed partial similarity with *X. oryzae* pv. *oryzae*, but it showed a non-specific amplification when cross-tested with primers RST2 and RST3.

A single gradient multiplex PCR was used to detect *R. solanacearum* and *X. perforans* by determining the range of annealing temperature. The gradient multiplex PCR was carried out using annealing temperature from 55 to 65 °C, PCR amplified an amplicon of 281 bp for *R. solanacearum* in gradient PCR, whereas the intensity of bands was decreased continuously at increasing the annealing temperature from 57 to 61 °C and amplifications were not observed above 61 °C for *R. solanacearum*. *X. perforans* showed decreased intensity of the band for annealing temperature range from 57 to 61 °C, whereas amplifications were observed at 840-bp amplicon at 61 °C, thereby standardizing the concentration of DNA and concentration of primers efficacy with respect to *X. perforans*. In primer concentration of 1:4, *R. solanacearum* and *X. perforans*, respectively, showed optimum amplifications at the annealing temperature 61 ± 1 °C (Fig. 1). *C. michiganensis* subsp. *michiganensis* and *P. fluorescens* did not show any amplification in multiplex PCR. All isolates collected from field conditions viz, seed samples, plant material along with soil samples (Table 1) amplified in a multiplex PCR simultaneously which confirms both pathogens as *R. solanacearum* and *X. perforans* (Fig. 2). Earlier, a multiplex PCR assay was developed for simultaneous detection of *C. michiganensis* subsp. *michiganensis*, *Pseudomonas syringae* pv. *tomato* and *X. axonopodis* pv. *vesicatoria* (Ozdemir 2009). The present multiplex PCR technique developed could be a very useful approach for early

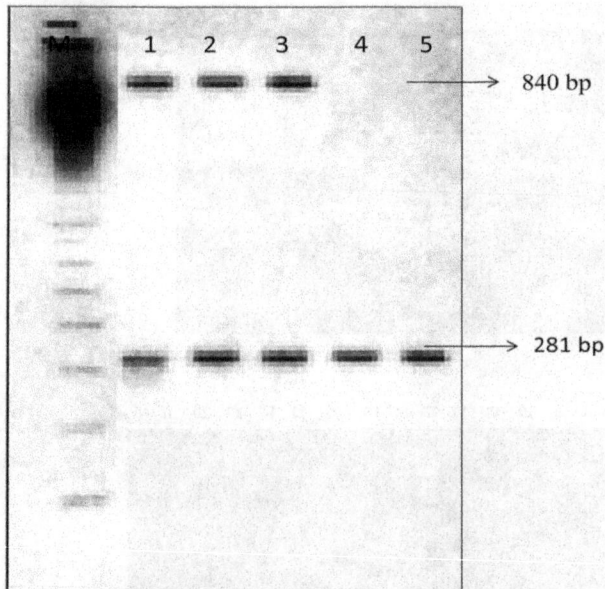

Fig. 3 Multiplex PCR amplification of *R. solanacearum* and *X. perforans* from different soil samples. Amplification at 840 bp indicates *X. perforans*, and amplification at 281 bp indicates *R. solanacearum* 100-bp DNA (*M*)

identification of *R. solanacearum* and *X. perforans* in the current agriculture. On the basis of 23S rRNA gene sequences, one universal forward and four taxon-specific reverse primers were designed for multiplex PCR to aid in identification and differentiation of *Agrobacterium rubi, A. vitis* and *A.* biovars 1 and 2 (Puławska et al. 2006). The multiplex PCR can be a better tool for rapid classification of similar species and taxa of phytopathogenic bacteria which can be used to identify different biovars among *R. solanacearum* and *X. perforans*. The developed multiplex PCR is more efficient than a single PCR reaction in saving time and reduces reaction cost by simultaneous amplification of pathogens.

Twenty different tomato cultivars (five plants in four replicates) were artificially inoculated with the isolates of *R. solanacearum* and *X. perforans* to validate multiplex PCR (Table 2). Leaves were collected from inoculated plants after 8–10 days, and pathogens were isolated by plating the leaves on nutrient agar media. In addition, multiplex PCR performed to identify *R. solanacearum* and *X. perforans* from single-colony PCR to increase sensitivity of amplification. Both the pathogens amplified up to 10^{-3} dilution, indicating that the developed method is useful for identification of both the phytobacterial pathogens without DNA isolation (Fig. 2), correspondingly seed and soil samples which also showed amplification in 10^{-2} dilution which validate the developed multiplex is extremely valuable in early diagnosis of pathogen from seed, soil (Fig. 3) as well as plant materials. The number of contaminations within a system may be low, thereby making it difficult to detect

infested soil, plant material and seeds. To tackle these issues, a multiplex PCR technique was developed in the present study is of great value to distinguish *R. solanacearum* and *X. perforans*. The multiplex PCR was found to be a better technique as it could detect more positive-infected seed samples when compared to the conventional method. The pathogenicity test conducted in susceptible plants of tomato isolates 12–16 exhibited negative results, whereas it showed positive results in multiplex PCR. Interestingly, the isolate 17 (Rasi) gave positive for pathogenicity, but negative in multiplex PCR which overcomes the false positive results (Table 2). Multiplex PCR protocols have been developed to detect several pathogens or genetically heterogeneous strains of a single pathovar simultaneously. Even detection of one bacterium and four viruses was reported by multiplex PCR in olive plants (Bertolini et al. 2003). The early diagnosis of bacterial wilt and spot diseases in this crop is very essential in current scenario to develop suitable management strategies which lead to improvement in the yield of agricultural products.

Conclusion

It can be concluded that the multiplex PCR technique described in the present study is a reliable, sensitive and cost effective procedure for identifying *R. solanacearum* and *X. perforans*. The present technique developed is useful for international sanitary surveillance of planting material exchanges. Additionally, this molecular tool can be useful to determine the relationship between soil/seed contamination and disease incidence of *R. solanacearum* and *X. perforans* in tomato. Furthermore, this method can assess the relative importance of different stages of dissemination of these pathogens.

Acknowledgments This work was supported by the grants from major research project on "Development of PCR-SSCP technique for specific diagnosis of bacterial spot and bacterial wilt pathogens in tomato" awarded by University Grants Commission, Government of India, New Delhi, India, under grant 11th Plan [36-281/2008 (SR)].

Conflict of interest There is no conflict of interest.

References

Adachi N, Oku T (2000) PCR-mediated detection of *Xanthomonas oryzae* pv. *oryzae* by amplification of the 16S–23S rDNA spacer region sequence. J Gen Plant Pathol 66:303–309

Avinash P, Umesha S (2014) Identification and genetic diversity of bacterial wilt pathogen in brinjal. Arch Phytopathol Plant Prot 47(4):398–406

Bertolini E, Olmos A, Lopez MM, Cambra M (2003) Multiplex nested reverse-transcription polymerase chain reaction in a single tube for sensitive detection of four RNA viruses and *Pseudomonas savastanoi* pv. *savastanoi* in olive trees. Phytopathology 93:286–292

Boureau T, Kerkoud M, Chhel F, Hunault G, Darrasse A, Brin C, Durand K, Hajri A, Poussier S, Manceau C, Lardeux F, Saubion F, Jacques MA (2013) A multiplex-PCR assay for identification of the quarantine plant pathogen *Xanthomonas axonopodis* pv. *phaseoli*. J Microbiol Methods 92:42–50

Carlton WM, Braun EJ, Gleason ML (1998) Ingress of *Clavibacter michiganensis* ssp. *michiganensis* into tomato leaves through hydathodes. Phytopathology 88:525–529

Chandrashekar S, Umesha S, Chandan S (2012) Molecular detection of phytopathogenic bacteria using polymerase chain reaction signal-strand confirmation polymorphism. Acta Biochim Biophys Sin 44:217–223

Fahy PC, Persley GJ (1983) Plant bacterial diseases, a diagnostic guide. Academic Press, USA 393

Hayward AC (1994) Characteristics of *Pseudomonas solanacearum*. J Appl Bacteriol 27:265–277

Isabelle RS, Philippe L, Lionel G, Emmanuel J, Olivier P (2006) Specific detection of *Xanthomonas axonopodis* pv. *dieffenbachiae* in Anthurium (*Anthurium andreanum*) tissues by nested PCR. Appl Environ Microbiol 72:1072–1078

ISTA (2005) International rules for seed testing. In: Draper SR ed. Rules. Switzerland: International Seed Testing Association, Zurich, 1: 520

Kovac's N (1956) Identification of *Pseudomonas pyocyanea* by the oxidase reaction. Nature 178:703

Kuflom M, Kuflu DA, Cuppels (1997) Development of a diagnostic DNA Probe for Xanthomonads causing bacterial spot of peppers and tomatoes. Appl Environ Microbiol 63:4462–4470

Lang JM, Hamilton JP, Diaz MG, Van SMA, Burgos MRG, Vera CM (2010) Genomics-based diagnostic marker development for *Xanthomonas oryzae* pv. *oryzae* and *X. oryzae* pv. *oryzicola*. Plant Dis 94:311–319

Leite RP, Minsavage GV, Bonas U, Stall RE (1994) Detection and identification of phytopathogenic Xanthomonas strains by amplification of DNA sequence related to the *hrp* genes of *Xanthomonas campestris* pv. *vesicatoria*. Appl Environ Microbiol 60:1069–1077

Lelliot RA, DE Stead (1987) Methods for the diagnosis of bacterial diseases of plants. In: Preece TF (ed) Methods in plant pathology. Blackwell Scientific Publications, New Jersey, p 216

Opina N, Tavner F, Hollway G, Wang JF, Li TH, Maghirang R, Fegan M, Hayward AC, Krishnapillai V, Hong WF, Holloway BW, Timmis J (1997) A novel method for development of species and strain-specific DNA probes and PCR primers for identifying *Burkholderia solanacearum* (formerly *Pseudomonas solanacearum*). Asia Pac J Mol Biol Biotechnol 5:19–30

Ozdemir Z (2005) Development of a multiplex PCR assay for concurrent detection of *clavibacter michiganensis* subsp. *michiganensis* and *xanthomonas axonopodis* pv. *vesicatoria*. Plant Pathol J 4:133–137

Ozdemir Z (2009) Development of a multiplex PCR assay for the simultaneous detection of *clavibacter michiganensis* subsp. *michiganensis, pseudomonas syringae* pv. *tomato* and *Xanthomonas axonopodis* pv. *vesicatoria* using pure cultures. J Plant Pathol 91:495–497

Pastrik KH, Elphinstone GH, Pukall R (2002) Sequence analysis and detection of *Ralstonia solanacearum* by multiplex PCR amplification of 16S–23S ribosomal intergenic spacer region with internal positive control. Eur J Plant Pathol 108:831–842

Puławska J, Willems A, Sobiczewski P (2006) Rapid and specific identification of four Agrobacterium species and biovars using multiplex PCR. Syst Appl Microbiol 29:470–479

Remenant B, Goutaland BC, Guidot A, Cellier G, Wicker E, Allen C, Fegan M, Pruvost O, Elbaz M, Calteau A, Salvignol G, Mornico D, Mangenot S, Barbe V, Médigue C, Prior P (2010) Genomes of three tomato pathogens within the *Ralstonia solanacearum* species complex reveal significant evolutionary divergence. BMC Genom 11:379

Santana GB, Lopes AC, Elba A, Cristine CB, Allen C, Quirino BF (2012) Diversity of Brazilian biovar 2 strains of *Ralstonia solanacearum*. J Gen Plant Pathol 78:190–200

Schonfeld J, Heuer H, van-Elsas JD, Smalla K (2003) Specific detection of *Ralstonia solanacearum* in soil on the basis of PCR amplification of fliC fragments. Appl Environ Microbiol 69:7248–7256

Vanitha SC, Niranjana SR, Umesha S (2009) Role of phenylalanine ammonia lyase and polyphenol oxidase in host resistance to bacterial wilt of tomato. J Phytopathol 157:552–557

Weller SA, Elphinstone JG, Smith NC, Boonham N, Stead DE (2000) Detection of *Ralstonia solanacearum* strains with a quantitative multiplex, real-time, fluorogenic PCR (*Taq*Man) assay. Appl Environ Microbiol 66:2853–2858

Xue QY, Ni YY, Wei Y, Holger H, Philippe P, Hua GJ, Kornelia S (2011) Genetic diversity of *Ralstonia solanacearum* strains from China assessed by PCR-based fingerprints to unravel host plant and site-dependent distribution patterns. FEMS Microbiol Ecol 75:507–519

15

RNAi-mediated down-regulation of *SHATTERPROOF* gene in transgenic oilseed rape

Hadis Kord · Ali Mohammad Shakib · Mohammad Hossein Daneshvar ·
Pejman Azadi · Vahid Bayat · Mohsen Mashayekhi · Mahboobeh Zarea ·
Alireza Seifi · Mana Ahmad-Raji

Abstract Oilseed rape is one of the important oil plants. Pod shattering is one of the problems in oilseed rape production especially in regions with dry conditions. One of the important genes in Brassica pod opening is *SHATTERPROOF1* (*SHP1*). Down-regulation of *BnSHP1* expression by RNAi can increase resistance to pod shattering. A 470 bp of the *BnSHP1* cDNA sequence constructed in an RNAi-silencing vector was transferred to oilseed rape cv. SLM046. Molecular analysis of T2 transgenic plants by RT-PCR and Real-time PCR showed that expression of the *BnSHP* alleles was highly decreased in comparison with control plants. Morphologically, transgenic plants were normal and produced seeds at greenhouse conditions. At ripening, stage pods failed to shatter, and a finger pressure was needed for pod opening.

Keywords *BnSHP* gene · Gene silencing · Oilseed rape · RNAi · Pod shattering

Introduction

Oilseed rape (*Brassica napus* L.) is the third most important oilseed crop in the world (Basalma 2008). Seeds have about 40–48 % oil with a high amount oleic acid and low linolenic acid suitable for frying applications and cooking. Dehiscence of pods causes significant yield loss (Raman et al. 2011). Ordinary yield losses are in the range of 10–25 % (Price et al. 1996). Seed losses have been reported as much as 50 % of the expected yield when adverse climatic conditions delayed harvesting (Macleod 1981; Child and Evans 1989). The process of pod shatter begins with degradation and separation of cell walls along a layer of few cells, termed the dehiscence zone (Meakin and Roberts 1990). Resistance to shattering is an important and necessary trait for oilseed rape improvement (Kadkol 2009). Attempts to solve this problem by interspecific hybridization using related species such as *B. nigra*, *B. juncea* and *B. rapa* have been faced with some difficulties as other undesirable traits will be integrated too (Prakash and Chopra 1990; Kadkol 2009). In Arabidopsis, which is in the same family of brassicaceae, several genes including the ALCATRAZ (ALC), INDEHISCENT (IND), *SHATTERPROOF1* (*SHP1*) and *SHATTERPROOF2* (*SHP2*) and *FRUITFUL* (*FUL*) have been shown to be involved in pod dehiscence (Raman et al. 2011). Genes for a number of hydrolytic enzymes, such as endopolygalacturonases, have also roles in dehiscence (Petersen et al. 1996). In Arabidopsis, *SHP* genes are specifically expressed in flowers with strong expression in the outer replum (Savidge et al. 1995; Flanagan et al. 1996). *SHP* gene also has mainly effect in the ripening of strawberries (Daminato et al. 2013). In *B. napus*, three *BnSHP* alleles (*BnSHP1*, *BnSHP2a* and *BnSHP2b*) have been identified (Tan et al. 2009). *BnSHP1* and *BnSHP2* show 80 % identity at nucleotide sequence. The expression of *BnSHP2a* and *BnSHP2b* (Two alleles of *BnSHP* gene which differ only in downstream sequences) are mainly in root, floral buds and pods, and most strongly in floral buds (Tan et al. 2009). It

H. Kord · A. M. Shakib (✉) · P. Azadi · V. Bayat ·
M. Mashayekhi · M. Zarea · A. Seifi · M. Ahmad-Raji
Department of Tissue culture and Genetic Engineering,
Agricultural Biotechnology Research Institute of Iran (ABRII),
Karaj, Iran
e-mail: a_shakib@abrii.ac.ir

P. Azadi
e-mail: azadip22@gmail.com

H. Kord · M. H. Daneshvar
Ramin University of Agricultural and Natural Resources,
Ahvaz, Iran

is suggested that less severe phenotype of indehiscence will be better, and the *SHP*, *IND* and *ALC* genes are ideal candidates for research and application in breeding new lines suitable for mechanized harvest (Liljegren et al. 2000; Tan et al. 2009). Recent advances about the role of MADS-box genes in dehiscence zone development have been reviewed (Ferrándiz and Fourquin 2014). In this study, we report the effect of the silencing cassette on expression of *BnSHP* alleles in transgenic oilseed rape plants using RNAi approach.

Materials and methods

Nucleic acid isolation

DNA was isolated from 100 mg leaf tissues using the procedure of Dellaporta et al. (1983). For RNA isolation, total RNA was extracted from floral buds using RNeasy Mini Kit (Qiagene Co.). The quantity and quality of RNA samples were checked using nano spectrophotometry and agarose gel electrophoresis. First-strand cDNA was synthesized using 2 μg of total RNA with iScript Select cDNA synthesis kit (Bio-rad Co.) in a 20-μl reaction using oligo-dT's according to manufacturer's instructions.

Construction of RNAi cassette

A 470-bp fragment of the *BnSHP1* cDNA (Accession, AY036062) without MADS-box region was amplified by PCR using specific primers; F: 5'-<u>ATACTAGTGGCGCGC CCCGTTAACCC</u>TCCACTG-3' and R: 5'-<u>GCCTTAATT AAATTTAAAT</u>TTGAAGAGGAGGTTGGTC-3' containing restriction enzyme digestion sites for *Asc*1, *Aws*1, *Spe*1 and *Pac*1 (underlined), for cloning the sense and antisense fragments in the above sites in pGSA1252 behind the CaMV35S promoter. The RNAi cassette was removed with *Pst1* digestion and sub-cloned in the *Pst1* site in pCAMBIA3301 to make pCAMRNAi.

Production of transgenic plants

Agrobacterium tumefaciens strain AGL0 containing the plasmid pCAMBIA3301 was used for transformation. Cotyledon explants of rapeseed cv. SLM046 were inoculated and co-cultivated with Agrobacterium inoculum on MS medium containing 1 mg/l 2,4-D and 4.5 mg/l BAP. After co-cultivation, cotyledonary explants were transferred to MS selection medium, containing 4.5 mg/l BAP and 4 mg/l phosphinothricine 400 mg/l cefotaxime and 300 mg/l carbeniciline. The regenerated plants were analyzed by histochemical GUS assay according to the method reported by Jefferson et al. (1987).

The rooted transgenic plants were transferred into a mixture of peat and perlite (1:1, v/v), and they were grown in the greenhouse conditions. At five-leaf stage, the plants were incubated at 4 °C for 8 weeks to vernalize, and then they were moved to 25 °C for 16 h in light and 8 h in dark till maturity. The presence of transgene in T1 transgenic plants was confirmed by amplification of *BnSHP* sense, antisense cassette (F: 5'-<u>AATACTAGTGGCGCGCCCCG TTAACCC TCCTACTG</u>-3', R: 5'-<u>GCCTTAATTAAATT TAAATTTGAAGAGGAGGTTGGTC</u>-3', underlined part is a tail segment) and *bar* (F: 5'-ATCTCGGTGACGGG CAGGAC-3', R: 5'-CGCAGGACCCGCAGGAGTG-3') by PCR. A T$_2$ transgenic line (cultured seeds from T1 line) was used for gene expression evaluation.

To study the expression of *BnSHP* alleles (*SHP1, SHP2-a* and *SHP2-b*), pod samples were taken from three transgenic plants of a T$_2$ line and one non-transgenic plants (two replications from each plant) for RNA extraction. RT- and Real-time PCR was conducted with two specific primer pairs:

P14 F: 5'-TGAACTAGTCCATGGAGATCTTCTTCTC ATGATCAGTCGCAGCATT-3', P14 R: 5'-AGCTTAA TTAAATTTAAATTTAAACAAGTTGAAGAGGAGGT TGG-3' (producing a 151-bp fragment from three alleles, underlined part is a tail segment); P19 F: 5'-GAACAA GGCGCGAGATTGAATCC-3' and P19 R: 5'-GATCATG AGAAGAAGACAGACCGG-3' (producing a 94-bp fragment from *SHP1*). The *GAPDH* gene-specific primers: F: 5'-AGAGCCGCTTCCTTCAACATCATT-3' and R: 5'-TG GGCACACGGAAGGACATACC-3' (producing a 112-bp fragment) were used as a reference gene. The relative gene expression data were analyzed using the $2^{-\Delta\Delta CT}$ method as described previously (Schmittgen and Livak 2008). The amount of *BnSHP* gene expression in transgenic and non-transgenic plants was calculated using the threshold data by $2^{-\Delta\Delta CT}$ method (Pfaffl et al. 2002), and the data were statistically analyzed, and the graphs were drawn by Bio-Rad software package.

Results and discussion

Construction of RNAi cassette and transgenic plant production

To construct the RNAi cassette, a 506-bp fragment (containing a 470-bp sequence downstream the MADS-box region of the *BnSHP1* cDNA) was amplified and cloned in the sense and antisense direction in either side of a *GUS* intron in plasmid vector pGAS1252 (Fig. 1). The cassette was then sub-cloned in the *Pst1* site in pCAMBIA3301, and the recombinant plasmid was designated as pCAMR-NAi. The RNAi cassette was transferred into rapeseed, and

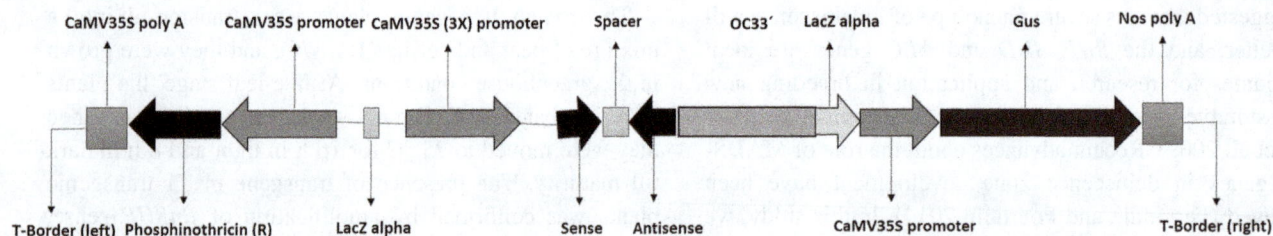

Fig. 1 Schematic representation of the silencing cassette for the *BnSHP* gene. A 470-bp fragment of the *BnSHP1* cDNA was cloned into pGSA1252 vector as sense and antisense segments, and the cassette was sub-cloned in pCAMBIA3301 transformation vector. CaMV35S: 35S promoter of a cauliflower mosaic virus, OCS3′: octopine synthase gene, GUS: β-glucuronidase gene, NOS A: terminator of the nopaline synthase gene

Fig. 2 Development of transgenic *B. napus*. **a** Explants on shoot induction medium supplemented with 400 mg/l cefotaxime and 300 mg/l carbeniciline. **b** Non-transformed shoot on selection medium **c, d** Putative transgenic shoot on MS medium supplemented with 4.5 mg/l BAP, 4 mg/l phosphinothricine, 400 mg/l cefotaxime and 300 mg/l carbeniciline). **e** Flowering and fruiting (pod development) of transgenic plants in the greenhouse (*Bar*: 10 cm). **f** Ripening pods. *Bars* **a**, **b**, **c**, **d** and **f**: 1 cm

putative transgenic plants were regenerated on medium containing phosphinotricin (Fig 2). Transformation efficiency of 3.7 % was obtained in SLM046 cultivar. Histochemical GUS assay in transgenic plants was done using Jefferson's method (Jefferson et al. 1987), and blue leaf samples were observed in transgenic plants (Fig. 3). PCR analysis on T0 putative transgenic and wild-type plants showed the insertion of 501-bp antisense segment only in putative transgenic plants (Fig. 4). Seeds of putative T1 transgenic plants were sown in soil in pots, and growing plants at 5–6 leaf stage were subjected to 1.5 % basta herbicide. Three herbicide resistant lines were selected, and the insertion of 1,000-bp fragment of the *GUS* gene was confirmed by PCR (Fig. 5).

Fig. 3 Histochemical GUS assay in transgenic plant. Leaf of non-transformed plant (**a**) and stable GUS expression (**b**) of *B. napus* transgenic plant. *Bars 5 mm*

were grown in greenhouse conditions and showed normal vegetative and reproductive characteristics compared with control wild-type plants. The presence of the transgene was confirmed by amplification of a 400 bp of the *bar* gene in transgenic plants (Fig. 6).

Expression of the *BnSHP* gene was analyzed using RT-PCR. A sharp 100-bp band for *BnSHP1* and *BnSHP2* gene expression was observed in wild-type plants, while transgenic plants showed a weak band as was expected (Fig. 5). The primer P14 amplifies all three *BnSHP* alleles (*SHP1*, *SHP2-a* and *SHP2-b*), and P19 amplifies *BnSHP1*. A very slight difference in gene expression was observed in transgenic plants. Expression level for housekeeping *GAPDH* gene as control was the same in both transgenic and control plant, indicating that expression of *BnSHP* alleles was decreased in transgenic plants as judged with the *GAPDH* gene expression (Fig. 7).

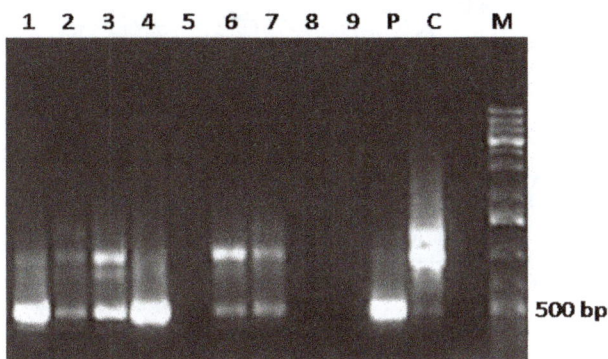

Fig. 4 Detection of the antisense segment of *BnSHP*RNAi cassette in T0 putative transgenic *B. napus* plants. A 501-bp antisense fragment of the *BnSHP* gene was amplified using PCR. *Lanes 1–4 and 6–7* produced clear bands of 501 bp for the transgene (higher band is related to the internal *SHP* gene); *5–8–9* blank; *P* positive control (plasmid vector of transformation); *C* control (non-transgenic plant); *M* 1 kb Plus DNA Ladder

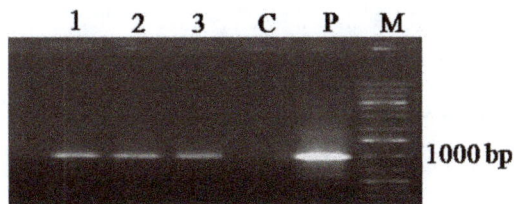

Fig. 6 Detection of the *bar* gene in T2 transgenic *B. napus* plants by PCR. Amplification of a 400-bp fragment of the *bar* gene: *1* negative control, *c* wild type, *2–5* transgenic plants; *M* DNA marker

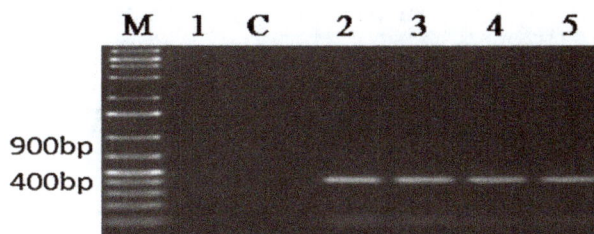

Fig. 5 Detection of the *GUS* gene in T1 transgenic *B. napus* plants. A 1000-bp fragment of the *GUS* gene was amplified using PCR. *Lanes 1–3* transgenic plants; *C* control (non-transgenic plant); *P* positive control (plasmid vector of transformation); *M* 1 kb Plus DNA Ladder

Analysis of *BnSHP1* gene in wild-type and transgenic plants

The effect of silencing cassette on gene expression was evaluated using T2 transgenic lines. The transgenic plants

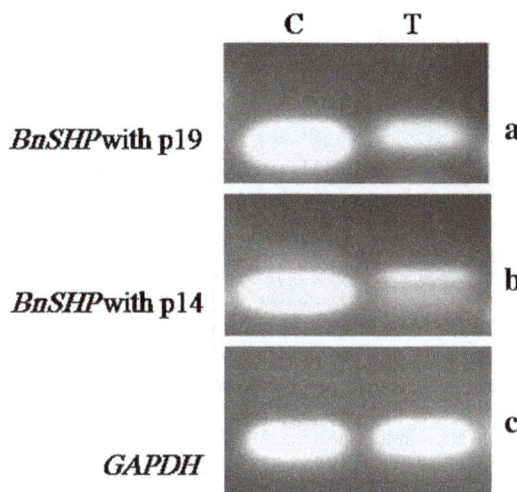

Fig. 7 Expression analysis of *BnSHP* gene in transgenic *B. napus* plants and wild type by RT-PCR: **a** amplification of a 105-bp band using the p14 specific primers, **b** amplification of a 94-bp band using the p19 specific primers and **c** amplification of a 100-bp band using the *GAPDH* specific primers. *C* non-transgenic control plant, *T* transgenic plant

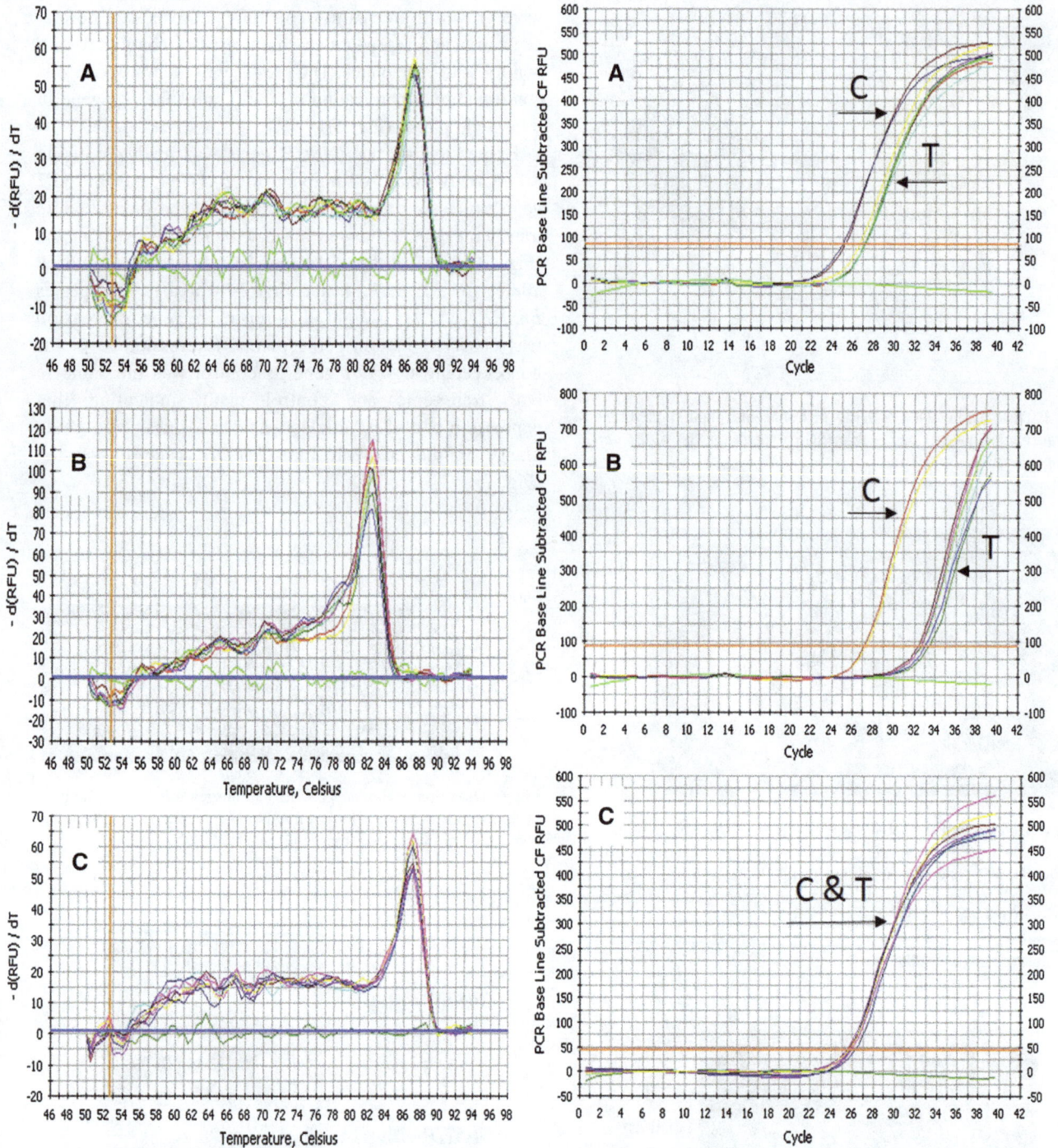

Fig. 8 Real-time analysis of the *BnSHP* gene and *GAPDH* expression in transgenic and non-transgenic *B. napus* plant using the P19 (**a**), P14 (**b**) *BnSHP* specific primers and *GAPDH* (**c**) primers. For *BnSHP*: delay in amplification in transgenic plant indicating the reduced level of the *BnSHP* RNAs and for *GAPDH*: no difference in amplification curves, indicating the same amount of RNA in the samples. High PCR specificity is shown by melting curve analysis

Real-time PCR

To analyze the relative changes in gene expression, a quantitative Real-time PCR using two specific primers for *BnSHP* alleles and *GAPDH* gene was applied (Fig. 8). The efficiency of amplification using dilution series was 95 and 94 % for the *GAPDH* and the *BnSHP* genes. A higher C_T value was observed for transgenic plants than non-transgenic plants, implying that the *BnSHP* gene expression has been decreased in transgenic plants. The data were statistically analyzed, and the related vertical graph showed the 97 % reduction in gene expression (Fig. 9).

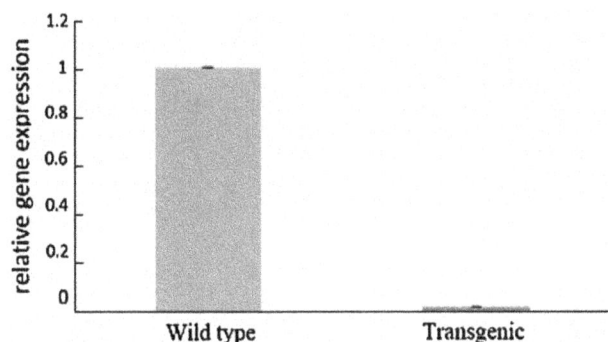

Fig. 9 The high reduction in the *BnSHP* gene expression in *B. napus* transgenic plant compared with wild-type plant using Real-time PCR analysis

The genetic variation in shatter resistance has been reported in Brassica species, including *B. rapa* L., *B. juncea* L., *B. hirta* L. and within wild relatives of Brassica (Kadkol et al. 1985; Wang et al. 2007, Bagheri et al. 2012). However, to avoid windrowing of crops on a routine basis, the level of resistance is insufficient (Raman et al. 2011). Random impact test (RIT) on 229 accessions of *B. napus* for shatter resistance detected only two shatter resistant lines (Wen et al. 2008). Using introgression method, shatter resistance could be improved in *B. juncea* (Kadkol 2009; Raman et al. 2011). However, unwanted traits could be present too as reported by Summers et al. (2003). The DK142 line (resynthesized *B. napus* using *B. oleracea* alboglabra and *B. rapa* chinensis) showed higher shatter resistance in all locations, but significantly lower seed yield than commercial variety (Summers et al. 2003).

Indehiscent and harder siliques were obtained by constitutive MADSB expression in winter rape lines compared with wild-type winter rape plants, and precocious seed dispersal was prevented (Chandler et al. 2005). However, for unclear reason, the transgenic summer rape lines did not show a non-opening silique phenotype. This indicated that the tender rape cultivars differences could affect the coordination of signaling events involved in fruit dehiscence. A number of genes have been shown to involve in fruit dehiscence and seed shattering. Among the reported genes, *SHP*, *IND* and *ALC* genes have been suggested as ideal candidates for manipulation in breeding shatter resistant lines (Liljegren et al. 2000; Tan et al. 2009). In *B. napus*, three *SHP* alleles have been reported, and due to more probable redundancy of these alleles, resistance to shattering may need control of all these alleles simultaneously (Tan et al. 2009).

In Arabidopsis, double mutant of both *SHATTER-PROOF1* (*SHP1*) and *SHATTERPROOF2* (*SHP2*) produced indehiscent pods (Liljegren et al. 2000). Pod shatter resistance was also observed in mutants of the *INDEHIS-CENT* gene (Liljegren et al. 2004; Wu et al. 2006), and the

ALCATRAZ gene in Arabidopsis (Rajani and Sundaresan 2001). Sorefan et al. (2009) showed that the *INDEHIS-CENT (IND)* mutant produced indehiscent fruits by preventing differentiation of tissue in dehiscence zone into an abscission layer.

Over-expression of *FRUITFUL* (*FUL*), a repressor of *SHP* and *IND*, produced indehiscent siliques in Arabidopsis (Ferrandiz et al. 2000). Ectopic expression of *FUL* gene in *B. juncea* produced indehiscent fruits, but they could not be threshed in a combine harvester without seed damage (Ostergaard et al. 2006). In the present study, introduction of the *BnSHP* gene-silencing cassette into *B. napus* showed a drastic reduction in *BnSHP* expression. Morphologically, transgenic plants were normal and set seeds at greenhouse conditions. At ripening stage, pods failed to shatter. Further phenotypic evaluation of shatter resistance such as the cantilever test, pendulum test (Kadkol et al. 1984) and the RIT are needed to be done in produced transgenic plants.

Acknowledgments This research was supported by Agricultural Research, Education and Extension Organization (AREEO) of Iran.

Conflict of interest The authors certify that there is no conflict of interest with any financial organization regarding the material discussed in our manuscript.

References

Bagheri H, El-Soda M, Oorschot I, Hanhart C, Bonnema G, Jansen-van den Bosch T, Mank R, Keurentjes J, Meng L, Wu J, Koornneef M, Aarts M (2012) Genetic analysis of morphological traits in a new, versatile, rapid-cycling *Brassica rapa* recombinant inbred line population. Front Plant Sci 3:183.

Basalma D (2008) The correlation and Path analysis of yield and yield components of different winter rapeseed (*Brassica napus* ssp. oleifera L.) cultivars. Res J Agric Biol Sci 4:120–125

Chandler J, Corbesier L, Spielmann P, Dettendorfer J, Stahl D, Apel K, Melzer S (2005) Modulating flowering time and prevention of pod shatter in oilseed rape. Mol Breed 15:87–94

Child RD, Evans DE (1989) Improvement of recoverable yields in oilseed rape (*Brassica napus*) with growth retardants. Asp Appl Biol 23:135–143

Daminato M, Guzzo F, Casadoro G (2013) A SHATTERPROOF-like gene controls ripening in non-climacteric strawberries, and auxin and abscissic acid antagonistically affect its expression. J Exp Bot.

Dellaporta SL, Wood J, Hickes JB (1983) A plant DNA miniprep-aration: version II. Plant Mol Biol Rep 1:19–21

Ferrándiz C, Fourquin C (2014) Role of the FUL–SHP network in the evolution of fruit morphology and function. J Exp Bio.

Ferrandiz C, Liljegran SJ, Yanofsky MF (2000) Negative regulation of the *SHATTERPROOF* genes by *FRUITFULL* during Arabidopsis development. Science 280:436–438

Flanagan CA, Hu Y, Ma H (1996) Specific expression of the AGL1 MADS-box gene suggests regulatory functions in *Arabidopsis gynoecium* and ovule development. J Plant 10:343–353

Jefferson RA, Kavanagh TA, Bevan MW (1987) GUS fusion. B-glucuronidase as a sensitive and versatile gene fusion marker in higher plants. J EMBO 6:3901–3907

Kadkol GP (2009) Brassica shatter-resistance research update. In: Proceedings of the 16th Australian research assembly on brassicas conference, Ballarat Victoria, pp 104–109

Kadkol GP, MacMillan RH, Burrow RP, Halloran GM (1984) Evaluation of *Brassica* genotypes for resistance to shatter. I. Development of a laboratory test. Euphytica 33:63–73

Kadkol GP, Halloran GM, MacMillan RH (1985) Evaluation of *Brassica* genotypes for resistance to shatter. II. Variation in siliqua strength within and between accessions. Euphytica 34:915–924

Liljegren SJ, Ditta GS, Eshed Y, Savidge B, Bowman JL, Yanofsky MF (2000) *SHATTERPROOF* MADS-box genes control seed dispersal in *Arabidopsis*. Nature 404:766–770

Liljegren SJ, Roeder AH, Kempin S, Gremski K, Ostergaard L, Guimil S, Yanofsky MF (2004) Control of Fruit Patterning in *Arabidopsis* by INDEHISCENT. Cell 116:843–853

Macleod J (1981) Harvesting in oilseed rape. Cambridge Agricultural Publishing, Cambridge, pp 107–120

Meakin PJ, Roberts JA (1990) Dehiscence of fruit in oilseed rape (*Brassica napus* L.), II. The role of cell wall degrading enzymes and ethylene. J Exp Bot 41:1003–1011

Ostergaard L, Kempin SA, Bies D, Klee HJ, Yanofsky MF (2006) Pod shatter-resistant *Brassica* fruit produced by ectopic expression of the *FRUITFULL* gene. J Plant Biotechnol 3:1–7

Petersen M, Sander L, Child R, Onckelen H, Ulvskov P, Boekhardt B (1996) Isolation and characterization of a pod dehiscence zone-specific polyglactronase from *Brassica napus*. Plant Mol Biol 31:517–527

Pfaffl WM, Horgan GW, Dempfle L (2002) Relative expression software tool for group-wise comparison and statistical analysis of relative expression results in Real-time PCR. Nucleic Acids Res 9:1–10

Prakash S, Chopra VL (1990) Reconstruction of allopolyploid Brassicas through non-homologous recombination introgression of resistance to pod shatter in *Brassica napus*. Genet Res 26:1–2

Price JS, Neale MA, Hobson RN, Bruce DM (1996) Seed losses in commercial harvesting of Oilseed Rape. Eng J Agric Res 80:343–350

Rajani S, Sundaresan V (2001) The *Arabidopsis* myc/bHLH gene *ALCATRAZ* enables cell separation in fruit dehiscence. Curr Biol 11:1914–1922

Raman R, Raman H, Kadkol GP, Coombes N, Taylor B, Luckett D (2011) Genome-wide association analyses of loci for shatter resistance in Brassicas. In: Proceedings of the 11th Australian research assembly on brassicas (ARAB) conference, WaggaWagga, NSW, pp 36–41

Savidge B, Rounsley SD, Yanofsky MF (1995) Temporal relationship between the transcription of two Arabidopsis MADS box genes and the floral organ identity genes. Plant Cell 7:721–733

Schmittgen TD, Livak KJ (2008) Analyzing Real-time PCR data by the comparative CT method. Nat Protoc 3:1101–1108

Sorefan K, Girin T, Liljegren SJ, Ljung K, Robles P, Galvan-Ampudia CS, Offringa R, Friml J, Yanofsky MF, Ostergaard L (2009) A regulated auxinminimumis required for seed dispersal in *Arabidopsis*. Nature 459:583–587

Summers JE, Bruce DM, Vancanneyt G, Redig P, Werner CP, Morgan C, Child RD (2003) Pod shatter resistance in the resynthesised *Brassica napus* line DK142. J Agric Sci 140:43–52

Tan X, Xia Z, Zhang L, Zhang Z, Guo Z, Qi C (2009) Cloning and sequence analysis of oilseed rape (*Brassica napus*) *SHP2* gene. Bot Stud 50:403–412

Wang R, Ripley VL, Rakow G (2007) Pod shatter resistance evaluation in cultivars and breeding lines of *B. napus, B. juncea* and *Sinapis alba*. Plant Breed 126:588–595

Wen YC, Fu TD, Tu JX, Ma CZ, Shen JX, Zhang SF (2008) Screening and analysis of resistance to siliquae shattering in rape (*Brassica napus* L.). Acta Agronomic Sinica 34:163–166

Wu H, Mori A, Jiang X, Wang Y, Yang M (2006) The INDEHIS-CENT protein regulates unequal cell divisions in *Arabidopsis* fruit. Planta 224:971–979

16

Insights into controlling role of substitution mutation, E315G on thermostability of a lipase cloned from metagenome of hot spring soil

Pushpender Kumar Sharma · Rajender Kumar ·
Prabha Garg · Jagdeep Kaur

Abstract Rational mutagenesis was performed (at the vicinity of the active site residues D317 and H358 of a mature polypeptide) to investigate the role of amino acids in the thermostability/activity of a lipase enzyme. The single variant enzyme created with E315G (lip M2) mutation near one of the active site residue (D317) found to be an important residue in controlling the thermal stability, the variant with E315G mutation demonstrated biochemical properties similar to that of native lipase. However, we found that this mutation strongly affected the activity and stability of the lip M1 mutant, reported in our previous study (Sharma et al. in Gene 491:264–271, 2012b). The dual mutant with E315G/N355K mutation in the Wt showed small increase in the protein thermostability compared to the native lipase, however, the thermostability of the mutant lip M1 was reduced several fold. Presumably, E315G (lip M2) mutation reverted the thermostability evolved by N355K (lip M1). The native and variant enzymes also displayed large variation in enzyme kinetics and their preference for pNP-esters (substrates). We further generated 3D models and studied the loop modelling of the WT and variants. Interestingly, loop region Leu314-Asn321 showed structural flexibility on introducing E315G mutation in the native lipase. On the other hand, lysine in mutant N355K exhibited side chain conformational changes in the loop Thr353-His358 which resulted in its H-bonding with Glu284. In addition, replacing glutamic acid by glycine at 315 position in lip M3 distorted the electrostatic interactions between Glu315 and Lys355 in the flexible loop region Leu314-Asn321.

Mutants reported in the study are, single mutant N355K, pronounced as lip M1, E315G, pronounced as lip M2 and E315G/N355K, pronounced as lip M3 respectively, the native lipase is pronounced as WT.

P. K. Sharma and R. Kumar contributed equally to this study.

P. K. Sharma · J. Kaur (✉)
Department of Biotechnology, Panjab University,
Sector 14, Chandigarh 160014, India
e-mail: jagsekhon@yahoo.com

P. K. Sharma
Department of Biotechnology, Sri Guru Granth Sahib World
University, Fatehgarah Sahib, India
e-mail: pushpg_78@rediffmail.com

R. Kumar
Department of Pharmacoinformatics, National Institute
of Pharmaceutical Education and Research (NIPER),
S.A.S. Nagar, Mohali 160062, Punjab, India

R. Kumar · P. Garg
Computer Centre, National Institute of Pharmaceutical
Education and Research (NIPER), S.A.S. Nagar,
Mohali 160062, Punjab, India

Keywords Lipase · Mutagenesis · Thermostability ·
Conformation · α/β hydrolase fold · Homology modeling

Introduction

Recent advances in genetic engineering methods offers valuable tools to modify enzyme activity with desired functions (Kazlauskas and Bornscheuer 2009; Lutz and Bornscheuer 2009). Despite these advances, the intrinsic properties of the enzymes are difficult to explain or predict, as even slight change in the altered structure can affect both enzyme activity and stability (Bloom et al. 2006;

Spiller et al. 1999; Shimotohno et al. 2001). Therefore, understanding structure function relation at molecular level is multifaceted and raised many question i.e. why certain protein catalyzes a biochemical reaction more efficiently, and why certain proteins are more thermostable than the other (Bouzas et al. 2006). In the past several years, many researchers have employed a variety of molecular tools to improve the catalytic function of the enzymes (Magnusson et al. 2005; Reetz 2000; Koga et al. 2003). Typically, the factors which are considered important in enhancing protein thermostability are increased hydrophobicity, rigidity, compactness in the structure, decrease percentage age of thermolabile residues, increased hydrogen bonding and presence of salt bridge etc. (Haney et al. 1997; Sadeghi et al. 2006; Russel et al. 1997; Bogin et al. 1998; Gromiha 2001; Kumar et al. 2000), whilst, reverse may be true for less thermostable proteins. It is also suggested that the energy difference between a stable versus unstable proteins is very small (Tokuriki and Tawfik 2009). In addition, it was also stated that thermostability of enzyme is related to the rigidity of a protein structure and can affect protein function to much more extent. In this particular case, we are studying the effect of substitution mutation on the activity and stability of a lipase that had shared more than 90 % homology with thermostable enzyme of *Bacillus* species. Lipases belong to a group of enzyme that catalyzes the synthesis and hydrolysis of long chain of fatty acids. These are important biocatalysts and find numerous applications in different industries that mainly include food, leather, pharmaceutical, dairy and detergent making industry (Jaeger et al. 1999). They belong to serine proteases and include a conserved Gly-X-Ser-X-Gly motif, near the catalytic residue serine (Kim et al. 1998). In addition to serine, their catalytic triad contains aspartate, and histidine (Choi et al. 2005). Structurally, lipase belongs to α/β hydrolase fold that comprises parallel β-strands surrounded by α-helices (Schrag and Cygler 1997; Cherukuvada et al. 2005). These enzymes had been modified previously for enhancing protein thermostability, catalytic function, and for the development of enantiomeric pure compounds etc. Recent studies have shown that the stabilizing mutation which conferred thermostability had not resulted in loss of the structural rigidity in lipase mutants derived from the native lipases of a mesophilic *Bacillus* sp. (Acharya et al. 2004; Ahmad et al. 2008). Previously, we reported cloning and characterization of a gene encoding extracellular lipase in detail (Sharma et al. 2012a, b). Recently, we described a highly thermostable mutant lip M1, carrying mutation N355K close to the active site of the WT enzyme (Sharma et al. 2012a). Interestingly, during multiple sequence alignment, we notice that lysine at 355 position was critical in determining protein thermostability and was conserved in homologous proteins. Furthermore, we observed another

alteration in amino acid sequence of this polypeptide mutation close to aspartate (a catalytic residue at 317 position), where the conserved glycine at 315 position was replaced to glutamic acid. Therefore, we set out our objective to mutate glutamic acid to the conserved glycine, in both native and N355K. All purified enzymes (WT and variants) characterized biochemically for various properties demonstrated great variations (Table 1). We, next performed molecular dynamics in the loop region to shed light into structural plasticity at three dimensional levels.

Materials and methods

Reagents/kits/plasmids

pGEM-T easy vector (Promega, USA) was used for cloning and pQE30-UA plasmid (Quiagen, Germany) was used for expression purpose. Gel extraction kit was purchased from MOBIO (USA). Taq DNA polymerase (5 U/µl), dNTPs mix, each was purchased from Fermentas (Germany). Substrates (pNP-esters and tributyrin), used for biochemical assays and screening, were purchased from Sigma Aldrich (USA). All other chemicals were procured from Merck (Germany). The WT and lip M1 plasmid DNA with mutation (N355K) used in this study were cloned earlier in the lab.

Site specific mutagenesis and molecular manipulations

Site directed mutagenesis was carried out by overlap extension PCR. The full length primer used were as follows: 5'-TGATGAARGGNTGYAGRGTNCC-3' (forward) and 5'-TTANGGNCGNA (A/G) N(C/G) (T/A) NGCNA (G/A) (T/C) TGNCC-3' (reverse). Primer sequences used for the amplification of single mutant encoding E315G were as follows: 5'-ATTGGCTTGG**G**GAACGACGG-3' (forward) and 5'-CCGTCGTTCC**C**CAAGCCAAT 3' (reverse). Double mutant encoding E315G/N355K was generated using the plasmid DNA extracted from clone E315G. Oligonucleotide sequences used for amplification mutant encoding N355K were as follows: 5'-TGG AAT GAC ATG GGA ACG TAC AA**G**GTC GAC CAT TTG G 3' (forward) and 5'-CCA AAT GGT CGA C**C**T TGT ACG TTC CCA TGT CAT TCC A-3' (reverse). PCR (gradient) reaction was performed in a Bio-Rad thermal cycler as follows: 94 °C for 4 min, followed by 30 cycles at 94 °C for 1 min, 55/59.5 °C for 50 s and 72 °C for 2 min, with a final extension of 10 min at 72 °C. Full length amplified gene product was cloned in pGEM-T easy vector, as per manufacturer's instructions. Mutations were confirmed by sequencing, and all the genes (WT and variants) were submitted to gene bank. Following confirmation of the sequence, genes encoding WT and variants were sub-cloned in pQE-30 UA expression vector and transformed in *Escherichia coli* M15 cells. Transformed cells were selected on LB agar medium plates containing ampicillin (100 µg) and kanamycin (35 µg).

Expression and purification of recombinant protein

E. coli M15 cells harboring recombinant pQE-30 UA plasmids were cultivated overnight at 37 °C in 5 ml LB media having antibiotics as mentioned above. Next day, 1 % overnight grown culture was inoculated into 500 ml media and the expression was induced by addition of 1 mM IPTG when OD was reached ∼0.4–0.6. Cells were harvested after 6 h growth and the extracellularily secreted protein (WT and variants) was purified from the supernatant as reported previously (Sharma et al. 2012a). Recombinant protein samples from WT and variants were purified separately at 4 °C, unless and otherwise stated.

Enzyme assay

Enzyme assays were carried out according to the Sigurgisladottir et al. (1993). Phosphate buffer 0.8 ml (0.05 M, pH 8.0) was premixed with 0.1 ml enzyme (appropriately diluted) and 0.1 ml of 0.002 M *p*-nitrophenyl laurate. The reaction mixture was incubated at 50 °C for 10 min. Reaction was stopped by adding 0.1 M Na_2CO_3 (0.25 ml). Reaction mixture was centrifuged and supernatant was used to determine the enzyme activity. Enzyme activity was measured at 420 nm in UV/Vis spectrophotometer (JENWAY 6505, UK). One unit of enzyme activity is defined as the amount of enzyme, which liberates 1 µmole of *p*-nitrophenol from pNP-laurate as substrate/min under standard assay conditions. The total enzyme activity was expressed in U/ml, whereas the specific activity was expressed as U/mg of protein.

Effect of temperature on enzyme activity and stability

To calculate temperature optima, the purified lipase proteins (WT and variants) were assayed at different temperatures (20–80 °C). To study the thermal inactivation, the enzymes were incubated at different temperatures (20–80 °C) for 30 min. After heat treatment the reaction tubes were kept in ice for 15 min and assayed for enzyme activity. Enzyme without incubation was taken as control (100 %). Similarly, the thermal inactivation was also studied at 55 and 60 °C for different time intervals. The enzyme activity at the start of the experiment was taken as 100 %, and the residual lipase activity after incubation was determined. Reaction mix without enzyme served as blank.

Effect of pH on enzyme activity and stability

Optimum pH for the purified lipase (WT and variants) was determined by assaying these enzymes at various pH i.e. sodium acetate—pH 5.0, sodium phosphate—pH 6.0–8.0, Tris.HCl—pH 9.0, Glycine NaOH—pH 10.0–11.0, at 50 °C and 40 °C respectively. The pH stability assays of the lipases were performed by pre-incubating these enzymes in presence of 0.05 M buffer of different pH (5.0–11.0) for 1 h at room temperature, followed by enzyme assay.

Specificity

Substrate specificity for WT and its variants were determined using pNP ester (final concentration 0.2 mM) of following chain lengths: pNP-acetate (C_3), pNP-butyrate (C_4), pNP-caprylate (C_8), pNP-deconate (C_{10}), pNP-laurate (C_{12}), pNP-myristate (C_{14}), pNP-palmitate (C_{16}), pNP stearate (C_{18}) from Sigma (USA) were dissolved in absolute alcohol, and used in enzyme assay reaction according to standard assay method.

Kinetic study

Enzyme activity of WT and its variants were determined as a function of range of substrate concentration (0.01–2.5 mM of pNP laurate). The Michaelis–Menten constant (K_m) and maximum velocity for the reaction (V_{max}) with pNP-laurate as substrate were calculated by Lineweaver–Burk plot. The k_{cat} and k_{cat}/K_m were also calculated and the results were compared.

Gene submission

All genes (WT and variants) were submitted in gene bank (NCBI) with accession numbers WT-FJ392756.1, lip M1-GU292533, lip M2-GU292534, and lip M3-GU292535.

Molecular modeling

We employed MODELLER 9v7 program to build the homology models, using highly identical templates, for which crystal structure was reported in Protein Data Bank (PDB). The crystal structure of *Bacillus stearothermophilus* L1 lipase (PDB ID: 1KU0, sequence identity 96 %) (Jeong et al. 2002) and *Geobacillus thermocatenulatus* (PDB ID: 2W22, sequence identity 94 %) (Carrasco-López et al. 2009) showed high structural identity, when BLASTP (Basic Local Alignment Search Tool for Protein at NCBI) was performed. Furthermore, *Bacillus stearothermophilus* L1 lipase (PDB ID: 1KU0) was used as template for construction of a homology models, and all 3-D models of the lipase (WT and variants) were built, starting from the template. Energy minimizations of the modeled structures were carried out using forces field GROMOS96 43a1 and steepest descent method for 1,000 steps in the GROMACS 3.3.1 (Sali et al. 1995; Lindahl et al. 2001). Additionally, the loop regions (Leu314-Asn321 and Thr353-His358) were also modeled using loop script in MODELLER 9v7 and eventually verified and validated on Structure Analysis and Verification Server (http://nihserver.mbi.ucla.edu/SAVES). The final loop refinement structures were used for structural comparison analysis. Additionally, *Eris* server was used to calculate the changes in protein stability induced by mutations ($\Delta\Delta G$) utilizing the recently developed Medusa modeling suite. The server is freely accessible online (http://eris.dokhlab.org) (if $\Delta\Delta G <0$: stabilizing mutations $\Delta\Delta G >0$: destabilizing mutations) (Ding and Dokholyan 2007).

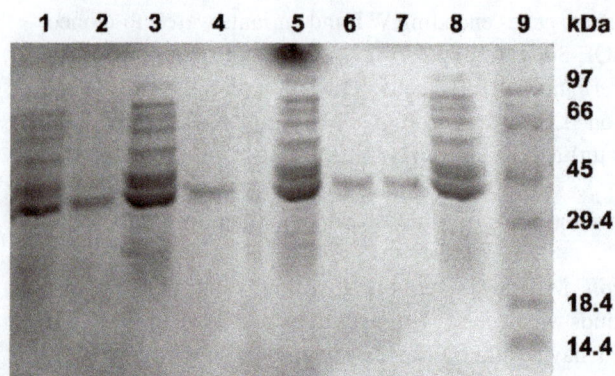

Fig. 2 SDS-PAGE analysis for wild type and mutant lipase. *Lane 1, 3, 5, 8*: induced culture from WT and variants. *Lane 2, 4, 6, 7*: purified protein sample from WT and variants after Phenyl-Sepharose chromatography (*Lane 1, 2* (WT), *3, 4* (lip M1) *5, 6* (lip M2) *7, 8* (lip M3). *Lane 9*: protein molecular weight marker

Results and discussion

Protein engineering method offers valuable tools to modify various properties of the biocatalysts that include e.g. the operational stability under denaturing conditions, enantioselectivity and chain length specificity etc. (Fujii et al. 2005; Magnusson et al. 2005; Reetz 2000; Koga et al. 2003). Here, we created mutant lip M2 and lip M3 from a native (WT) lipase, cloned from a metagenomic DNA extracted and purified from hot spring soil (Sharma et al. 2007), as depicted in Fig. 1. The results were compared with lip M1 and WT.

Protein expression, purification and Biochemical properties

Protein purified from WT and its variants showed single band of expected molecular weight on 12 % SDS-PAGE, as deduced from number of amino acids present in the mature polypeptide (Fig. 2). We further compared purification profile of WT and enzyme variants (data not shown) and observed that variant lip M1 had highest specific enzyme

activity i.e. 3,090 ± 14 U/mg compared to the WT, lip M2 and lip M3 whose specific enzyme activity was calculated to be 2,022 ± 31, 1,816.2 ± 18 and 1,972 ± 50 U/mg protein respectively. All enzymes demonstrated virtually a comparable activity and stability over wide range of pH (data not shown). Furthermore, all of them had displayed enzyme activity over broad range of temperature, with optimum enzyme activity observed at 50 °C, except to the lip M1 that displayed optimum enzyme activity at 40 °C (data not shown). Additionally, the thermostability of WT and variants tested over a range of temperature i.e. 20–80 °C for 30 min did not show loss in the enzyme activity until 50 °C. However at 60 °C, all of them showed a decrease in enzyme activity, apart from lip M1 (Fig. 3a). Thermal stability assays were also performed at 55 °C for varying time points (Fig. 3b). Various thermostability assays revealed following order for the thermal denaturation i.e. lip M1 > lip M3 > WT > lip M2. In addition, half life of all the proteins was also calculated and compared at 60 °C (Table 1).

Biochemical kinetics study

Kinetic parameters determined for the WT and variants using *p*-nitrophenyl laurate as substrate, demonstrated great variation (supplementary figures, S1–S4). The mutant lip M1 displayed $k_{cat} \sim 9$ times higher than WT enzyme, whereas lip M2 and lip M3 showed ~ 2 fold less k_{cat}. The overall catalytic efficiency i.e. k_{cat}/K_m of lip M1 showed ~ 20 fold increase, whereas lip M2 and lip M3 illustrated ~ 4 fold decrease in the catalytic efficiency than the WT, which is ~ 85 and ~ 99 folds lower than lip M1 respectively (Table 1). Information gathered from the biochemical kinetics data suggested that mutation near the active site might affect the binding of the substrate to the active site, and may be attributed to more flexibility and

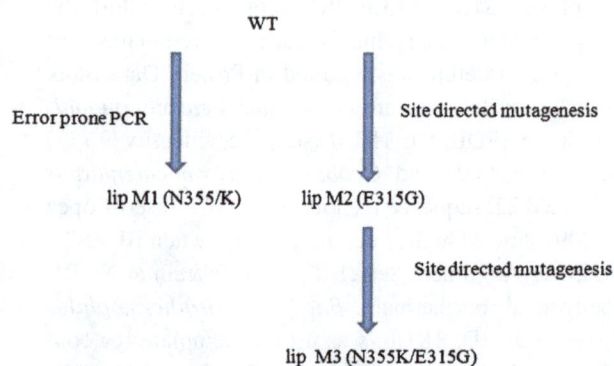

Fig. 1 Schematic presentation for the generation of variants

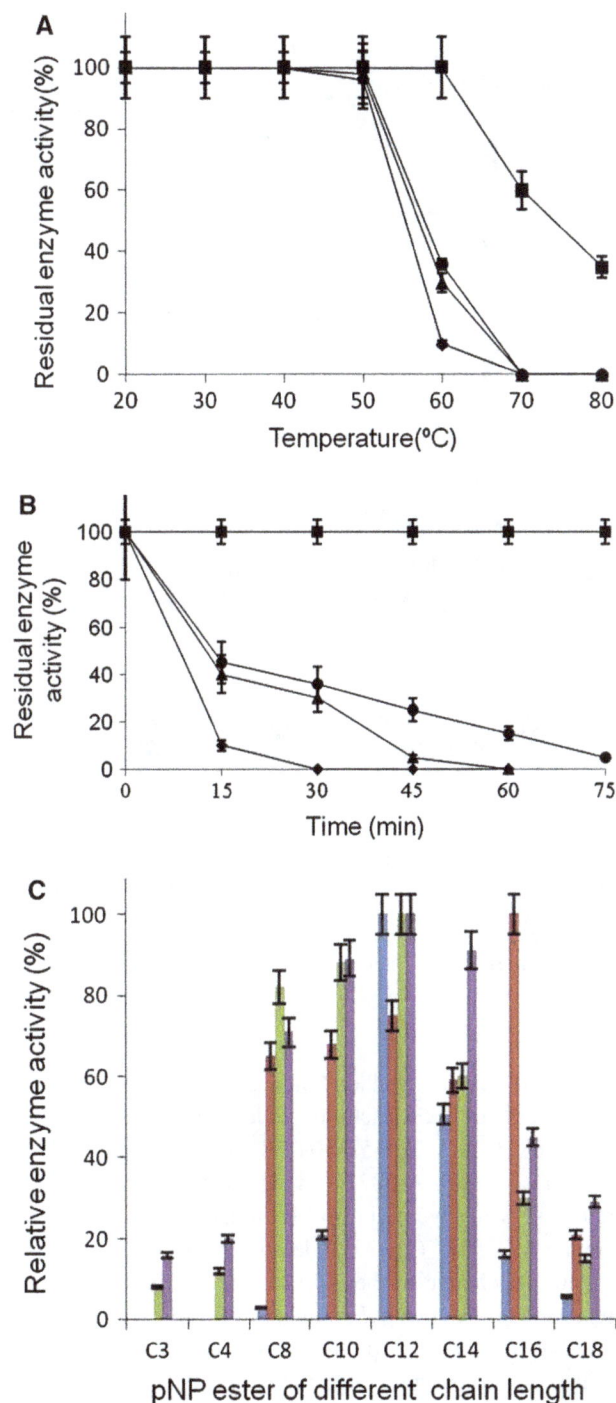

Table 1 Comparative biochemical studies for the WT and variant enzymes

Biochemical properties	WT	Lip M1	Lip M2	Lip M3
Specific activity	$2,022 \pm 31$	$3,090 \pm 14$	$1,816 \pm 18$	$1,972 \pm 50$
Temperature optimum (°C)	50	40	50	50
pH optimum	9	9	7–8	9
Half life at 60 °C	5 min	14 h	<5 min	>5 min
pH stability	8–9	7–9	7–8	8–9
k_m (µM)	0.73	0.33	1.33	1.18
V_{max} (µmol/ml/min)	239 ± 16	312	28	23
k_{cat} (s^{-1})	569	5,199	249	189
k_{cat}/K_m (µM^{-1} s^{-1})	779	15,754	187	160
Preferred substrate	pNP-laurate	pNP-palmitate	pNP-laurate	pNP-laurate

distortion in the structure on bringing in E315G mutation. All over it appears that the mutation has either affected the binding of the substrate molecule in the catalytic site or resulted in dispersion of the enzymatic product away from the micro-catalytic environment.

Effect of p-NP ester chain on enzyme activity

Substrate specificity of the WT and variants was evaluated and compared by testing enzyme activity in presence of pNP-esters of varying carbon chain length C_3–C_{18} (Fig. 3c). It is evident from the figure that WT, lip M2 and lip M3 displayed maximum enzyme activity with pNP-laurate while lip M1 showed maximum activity towards C_{16} (pNP–palmitate). In addition, lip M2 and lip M3 also demonstrated lipase activity towards short chain pNP-esters. Alteration of the specificity may be the result of changed conformation that has affected the catalytic pocket of the enzyme.

Structural implications and molecular modeling

Mutation near the active site residues can influence both hydrogen bonding and conformational stability (Offman et al. 2011; Morley and Kazlauskas 2005). Previous studies also established that mutation altering enzyme structure and functions exists at the vicinity of the active site (Takase 1993; Ollis et al. 1992). In the present investigation, we report that less thermo stability and catalytic efficiency of lip M2 and lip M3 by E315G mutation may be attributed to its presence at vicinity of the active site. Substituting a negatively charged polar amino acid (glutamic acid) to a non polar and flexible amino acid (glycine)

Fig. 3 a Thermostability assay for WT and variant proteins at different temperature performed by pre-incubating all enzymes at different temperature (20–80 °C) for 30 min followed by cooling of enzymes for 10 min on ice prior to enzyme assay. (*Filled square*) lip M1, (*filled circle*) lip M3, (*filled triangle*) WT, (*filled diamond*) lip M2. **b** Thermal denaturation assays for the WT and variant proteins at 55 °C for varying time periods, (*filled square*) lip M1, (*filled circle*) lip M3, (*filled triangle*) WT, (*filled diamond*) lip M2. **c** Substrate specificity of the WT and variant proteins (*blue*) WT, (*brown*) lip M1, (*purple*) lip M2, (*green*) lip M3

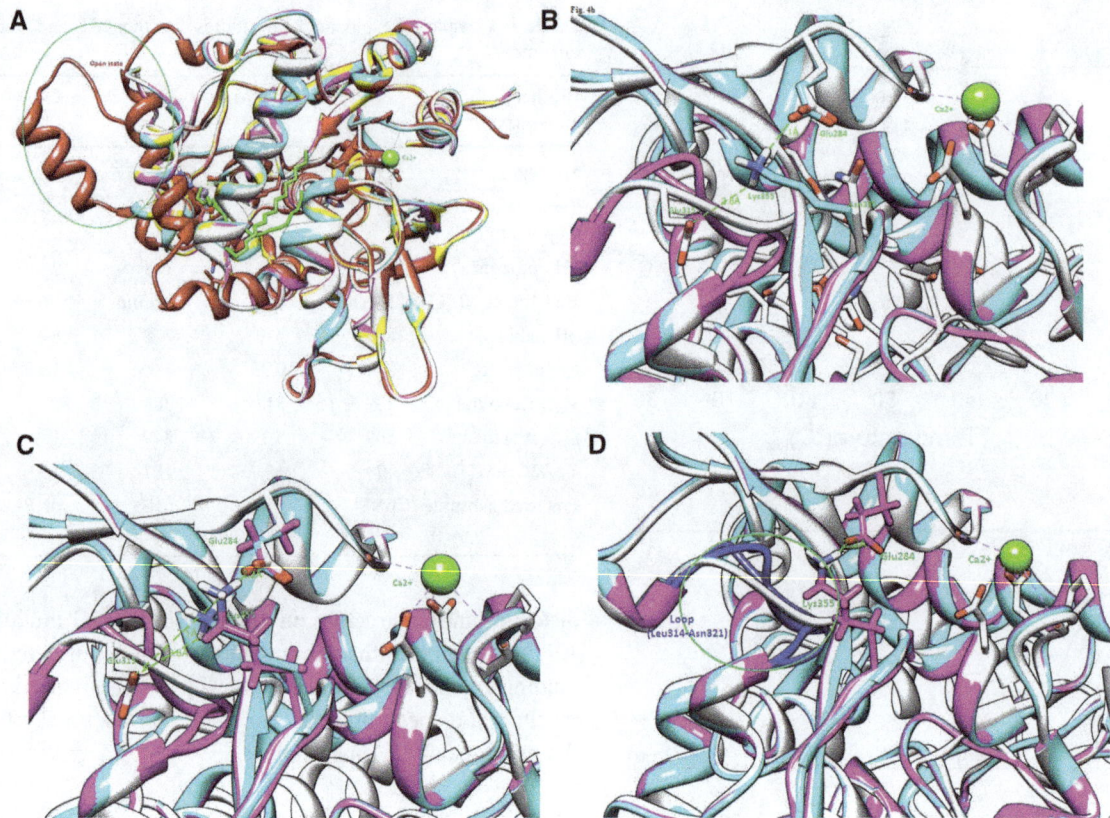

Fig. 4 Modeled structures of WT (*white color*) and mutants lipases (*cyan* and *magenta color*), superimposed on crystal structure 1KU0 (*maroon color*) in various color representation depicted in **a**; modeled structure of variant N355K (*cyan color*) showing H-bonding between Lys355 and Glu284 with a distance, 2.1 Å (Lys355 NH–O=C Glu284) and between Lys355 and Glu3153.8 Å (Lys355 NH–O=C Glu315) depicted in **b**; double mutated model (*magenta color*) showing H-bonding interaction between Lys355 and Glu284 with distance 2.1 Å (Lys355 NH–O=C Glu284), superimposed on mutant N355K showing distance 7.1 Å (Lys355 NH–O=C Glu315) with WT amino acid Glu315, as depicted in **c**; the overall structural changes in loop region (*blue color*) of double mutated and WT form, after loop refinements shown in **d**

in mutant lip M2 and lip M3 may affect the conformational plasticity of the enzyme. In general, glycine is reported to be present in those part of the protein structures which are forbidden to other amino acids, while glutamic acid commonly exist on the protein surface, and can interact with other amino acids. Furthermore, in order to demonstrate the effect of these mutations at three dimensional structural levels, we performed three dimensional modellings. To verify models, they were energy minimized using GROMACS 3.3.1 program and further validated by Ramachandran plot. The plot revealed presence of ∼91.1 % amino acids in the allowed core region, while ∼8.3 % were located in the additionally allowed region, and no residues were observed in the disallowed region, hence illustrated their best fit. Further, all these models exhibited superimposed view of the template IKU0, and hence high structural similarity (rmsd values for the model struture was found to be, C-α 0.181 with reference IKU0 and C-α 4.76 with reference 2W22), except in the loop region Leu314-Asn321 (Fig. 4a). In addition, all of them exhibited typical α/β hydrolase fold, a characteristics feature of lipases with conserved metal ion binding sites (Zn^{2+}, Ca^{2+}). Further refining of the structural models in loop regions Leu314–Asn321 and Thr353–His358 showed structural flexibility in Leu314–Asn321, owing to replacement of glutamic acid with glycine in lip M2 mutant. On the other side, mutant N355K (lip M1) demonstrated side chain conformational change in the loop region Thr353–His358 and resulted in extensive H-bonding of Lys 355 with Glu284 that renders the protein more thermostable as mentioned in our previous study (Sharma et al. 2012b). Computational molecular modeling experiment further demonstrated compact packing of positively charged amino acid lysine in between two negatively charged amino acids Glu284 and Glu315 and the distance measured was predicted to be 2.1 and 3.8 Å respectively (Fig. 4b), which may be another rationale for the improved protein thermostability in lip M1. Consequently, when both these mutations were brought together (lip M3), mutation E315G resulted in disruption of the electrostatic interactions between Glu315 and Lys355. To know whether this altered structural plasticity (conformation) has affected the distance between

Lys355 and Glu284, and Lys355 and Glu315, we superimposed lip M3 modeled structure on WT model. In fact, we observed an increase in the distance between Lys355 and Glu284 (2.1–2.2 Å) and Lys355 and Glu315 (3.8–7.1 Å) in lip M3 (Fig. 4c) which may be accredited to augmented flexibility in the loop region Leu314–Asn321 (Fig. 4d) as predicted. Our observations were further strengthened by calculating change in protein stability induced by mutations ($\Delta\Delta G$). The values obtained for different proteins were as follow e.g. E315G showed +4.39, for N355K +1.02, whereas dual mutant (G315E/N355K) it was −4.87, respectively (Ding and Dokholyan 2007). Therefore, our present investigation provide strong evidence that several loop conformational changes accompanied by distortion in electrostatic interactions resulted in loss of protein thermo stability and enzyme activity in the lip M3.

Conclusion

Altogether, our biochemical, circular dichroism and computational studies provide strong evidence that an amino acid substitution at positions 355 is critical for lipase stability and activity, whereas, residue at position 315 had only a marginal effect on its own. The mutation at 315 was able to nullify the enhanced thermostability and catalytic efficiency acquired by mutant N355K, might be due to altered loop conformations and disruption of the electrostatic interactions. We strongly feel that data in the manuscript can contribute in understanding the structural and functional of proteins.

Acknowledgments Independent senior research fellowship to PKS by CSIR and financial assistance to JK, by CSIR and DST, New Delhi, INDIA is duly acknowledged.

Conflict of interest None.

References

Acharya P, Rajakumara E, Sankaranarayanan R, Rao NM (2004) Structural basis of selection and thermostability of laboratory evolved *Bacillus subtilis* lipase. J Mol Bio 341:1271–1281

Ahmad S, Zahid Kamal M, Sankaranarayanan R, Rao NM (2008) Thermostable *Bacillus subtilis* lipases: *in vitro* evolution and structural insight. J Mol Bio 381:324–340

Bloom JD, Labthavikul ST, Otey CR, Arnold FH (2006) Protein stability promotes evolvability. Proc Natl Acad Sci USA 103:5869–5874

Bogin O, Peretz M, Hacham Y, Korkhin Y, Frolow F, Kalb AJ, Burstein Y (1998) Enhanced thermal stability of *Clostridium beijerinckii* alcohol dehydrogenase after strategic substitution of amino acid residues with prolines from the homologous thermophilic *Thermoanaerobacter brockii* alcohol dehydrogenase. Protein Sci 7:1156–1163

Bouzas TD, Barros-Velazquez JT, Villa G (2006) Industrial applications of hyperthermophilic enzymes: a review. Prot Pept Lett 13:445–451

Carrasco-López C, Godoy C, Rivas B, Fernández-Lorente G, Palomo JM, Guisán JM, Fernández-Lafuente R, Martínez-Ripoll M, Hermoso JA (2009) Activation of bacterial thermoalkalophilic lipases is spurred by dramatic structural rearrangements. J Biol Chem 284:4365–4372

Cherukuvada SL, Seshasayee ASN, Raghunathan K, Anishetty S, Pennathur G (2005) Evidence of a Double-Lid Movement in *Pseudomonas aeruginosa* lipase: insights from molecular dynamics simulations. PLoS Comput Biol 1:e28

Choi WC, Kim MH, Ro HS, Sang RR, Oh TK, Lee JK (2005) Zinc in lipase L1 from *Geobacillus stearothermophilus* L1 and structural implications on thermal stability. FEBS Lett 579:3461–3466

Ding YF, Dokholyan NV (2007) Eris: an automated estimator of protein stability. Nat Methods 4:466–467

Fujii R, Nakagawa Y, Hiratake J, Sogabe A, Sakata K (2005) Directed evolution of *Pseudomonas aeruginosa* lipase for improved amide-hydrolyzing activity. Protein Eng Des Sel 18:93–101

Gromiha MM (2001) Important inter-residue contacts for enhancing the thermal stability of thermophilic proteins. Biophy Chem 91:71–77

Haney P, Konisky J, Koretke KK, Luthey-Schulten Z, Wolynes PG (1997) Structural basis for thermostability and identification of potential active site residues for adenylate kinases from the archaeal genus *Methanococcus*. Proteins 28:117–130

Jaeger KE, Dijkstra BW, Reetz MT (1999) Bacterial biocatalysts; molecular biology, three dimensional structures, and biotechno-biogical applications of lipases. Ann Rev Microbiol 53:315–351

Jeong ST, Kim HK, Kim JS, Chi SW, Pan JG, Oh TK, Ryu SE (2002) Novel zinc-binding center and a temperature switch in the *Bacillus stearothermophilus* L1 lipase. J Biol Chem 277:17041–17047

Kazlauskas RJ, Bornscheuer UT (2009) Improving enzyme properties: when are closer mutations better? Nat Chem Biol 5:526–529

Kim HK, Park SY, Oh TK, Lee JK (1998) Gene cloning and characterization of thermostable *lipase* from *Bacillus stearothermophilus* L1. Biosci Biotechnol Biochem 62:66–71

Koga Y, Kato K, Nakano H, Yamane T (2003) Inverting enantioselectivity of *Burkholderia cepacia* KWI-56 lipase by combinatorial mutation and high-throughput screening using single-molecule PCR and in vitro expression. J Mol Bio 331:585–592

Kumar S, Ma B, Tsai CJ, Nussinov R (2000) Electrostatic strengths of salt bridges in thermophilic and mesophilic glutamate dehydrogenase monomers. Prot Str Funct Bioinfo 38:368–383

Lindahl E, Hess B, Van Der DS (2001) GROMACS: a package for molecular simulation and rajectory analysis. J Mol Model 7:306–317

Lutz S, Bornscheuer UT (2009) Protein engineering handbook, 2nd edn. Wiley VCH, Weinheim

Magnusson O, Takwa M, Harnberg A, Hult K (2005) An S-selective lipase was created by rational redesign and the enantioselectivity increased with temperature. Angew Chem Int Ed 44:4582–4585

Morley KL, Kazlauskas RJ (2005) Improving enzyme properties: when are closer mutations better? Trends Biotechnol 23:231–237

Offman MN, Krol M, Patel N, Krishnan S, Liu J, Saha V, Bates PA (2011) Rational engineering of L-asparaginase reveals importance of dual activity for cancer cell toxicity. Blood 117:1614–1621

Ollis DL, Cheah E, Cygler M, Dijkstra B, Frolow F, Franken SM, Harel M, Remington SJ, Silman I, Schrag J, Sussman JL, Verschueren KHG, Goldman A (1992) The alpha/beta hydrolase fold. Protein Eng 5:197–211

Reetz MT (2000) Evolution in the test tube as a means to create enantioselective enzymes for use in organic synthesis. Sci Progr 83:157–172

Sadeghi M, Naderi-Manesh H, Zarrabi M, Ranjbar B (2006) Effective factors in thermostability of thermophilic proteins. Biophy Chem 119:256–270

Sali A, Potterton L, Feng Y, Herman V, Martin K (1995) Evaluation of comparative protein modeling by Modeller. Protein: Str Funct Genet 23:318–326

Schrag JD, Cygler M (1997) Lipases and alpha/beta hydrolase fold. Methods Enzymol 284:85–107

Sharma PK, Capalash N, Kaur J (2007) An improved method for single step purification of metagenomic DNA. Mol Biotechnol 36:61–63

Sharma PK, Singh K, Singh R, Capalash N, Ali A, Mohammad O, Kaur J (2012a) Characterization of a thermostable lipase showing loss of secondary structure at ambient temperature. Mol Biol Rep 39:2795–2804

Sharma PK, Kumar R, Kumar R, Mohammad O, Singh R, Kaur J (2012b) Engineering of a metagenome derived lipase towards thermal tolerance: effect of asparagines to lysine mutation on the protein surface. Gene 491:264–271

Shimotohno A, Oue S, Yano T, Kuramitsu S, Kagamiyama R (2001) Demonstration of the importance and usefulness of manipulating non-active-site residues in protein design. J Biochem 129: 943–948

Sigurgisladottir S, Konraosdottir M, Jonsson A, Kristjansson JK, Matthiasson E (1993) Lipase activity of thermophilic bacteria from icelandic hot springs. Biotechnol Lett 15:361–366

Spiller B, Gershenson A, Arnold F, Stevens RA (1999) Structural view of evolutionary divergence. Proc Natl Acad Sci USA 96:12305–12310

Takase K (1993) Effect of mutation of an amino acid residue near the catalytic site on the activity of Bacillus stearothermophilus amylase. Eur J Biochem 211:899–902

Tokuriki N, Tawfik D (2009) Protein dynamism and evolvability. Science 324:203–207

17

Optimization of amorphadiene production in engineered yeast by response surface methodology

Rama Raju Baadhe · Naveen Kumar Mekala ·
Sreenivasa Rao Parcha · Y. Prameela Devi

Abstract Isoprenoids are among the most diverse bio-active compounds synthesized by biological systems. The superiority of these compounds has expanded their utility from pharmaceutical to fragrances, including biofuel industries. In the present study, an engineered yeast strain *Saccharomyces cerevisiae* (YCF-AD1) was optimized for production of Amorpha-4, 11-diene, a precursor of anti-malarial drug using response surface methodology. The effect of four critical parameters such as KH_2PO_4, methionine, pH and temperature were evaluated both qualitatively and quantitatively and further optimized for enhanced amorphadiene production by using a central composite design and model validation. The "goodness of fit" of the regression equation and model fit (R^2) of 0.9896 demonstrate this study to be an effective model. Further, this model will be used to validate theoretically and experimentally at the higher level of amorphadiene production with the combination of the optimized values of KH_2PO_4 (4.0), methionine (1.49), pH (5.4) and temperature (33 °C).

Keywords Response surface methodology · *S. cerevisiae* · Amorphadiene · Isoprenoids

R. R. Baadhe (✉) · N. K. Mekala · S. Rao Parcha
Department of Biotechnology, National Institute of Technology, Warangal 506004, India
e-mail: ramarajub@nitw.ac.in; baadheramaraju@gmail.com

Y. Prameela Devi
Department of Zoology, Kakatiya University, Warangal 506009, India

Introduction

Isoprenoids (terpenoids) are the most structurally diverse class of natural compounds commonly produced in plants (Croteau et al. 2000). Terpenoids are classified according to their carbon number (basic isoprene (C_5) unit) as mono (C_{10}), sesqui (C_{15}), di (C_{20}), sester (C_{25}), tri (C_{30}), tetra (C_{40}) and polyterpenoids (C_n) (Ruzicka 1959). More than 55,000 terpenes have been isolated and characterized, consistently doubling in their numbers each decade (Breitmaier 2006; McGarvey and Croteau 1995). Isoprenoids have diverse functional roles in plants such as growth, defense and development (McGarvey and Croteau 1995). Based on these characteristic features, terpenoids have prominence in pharmaceutical, fragrances and biofuel industries (for e.g. bisabolene is an alternative source for jet fuel (Breitmaier 2006; Peralta-Yahya et al. 2012).

Artemisinin is a well-known sesquiterpene lactone peroxide, extracted from the shrub *Artemisia annua*. 'Artemisininins' (artemisinin and its derivatives) are recommended by the World Health Organization (WHO) in combination with other effective anti-malarial drugs, known as artemisinin-based combination therapy (ACT) for malarial treatment (Bloland 2001). Since then, the incompetence in large-scale chemical synthesis of artemisinin and enormous demand and price directed the scientific world towards the semi-synthesis of artemisinin followed by microbial production of the precursor amorpha-4,11-diene. Heterologous production of amorpha-4, 11-diene was first established in *Escherichia coli* by the expression of the mevalonate pathway from yeast and amorpha-4, 11-diene synthase (ADS) from *A. annua* (Martin et al. 2003). The production of amorpha-4, 11-diene from *Saccharomyces cerevisiae* revealed that cytochrome P450 enzyme was responsible for the production of artemisinic acid (Mercke et al. 2000; Martin et al. 2003; Ro et al. 2006). Artemisinic acid was produced from yeast by a series of alterations and adjustments to the endogenous mevalonate pathway, such as high-level expression of ADS, overexpression of farnesyl diphosphate synthase (FDPS), expression of the catalytic domain of HMG-CoA reductase(HMGCR), reduced expression of squalene synthase (SQS) and increased expression of *UPC2* allele transcription factor (Ro et al. 2006). Artemisinic acid was produced by a three-step oxidation of amorphadiene, by cytochrome P450 reductase (*A. annua*) (Ro et al. 2006). However, cytochrome P450 reductase instability and lower yields of artemisinic acid compared to amorphadiene drew attention towards improving the production of amorphadiene, the precursor of artemisinic acid in *S. cerevisiae*. (Westfall et al. 2012). In combination with traditional metabolic engineering, we also applied enzyme fusion technology for improved production of amorphadiene in *S. cerevisiae* (YCF-AD-1) (unpublished data). Our previous observations show that in engineered yeast, the mevalonate pathway is tightly regulated by methionine and phosphate levels along with other physical parameters such as pH and temperature. Optimization of these parameters by classical experimental optimization is difficult because it involves changing one variable at a time while keeping the others constant. In addition, it is not practical to carry out experiments with every possible factorial combination of the test variables, because of the large number of experiments required to be done and/or evaluated (Akhnazarova and Kafarov 1982; Myers and Montgomery 1995) which does not emphasize the effect of interactions among various parameter. Besides this, it will be a tedious and time-consuming process, especially when there are a large number of parameters to take into consideration. An alternative and more efficient approach is the use of the statistical method to resolve this kind of practical hurdles. Response surface methodology (RSM) has been widely used to evaluate and understand the interactions between different process parameters (Khuri et al. 1987). RSM was applied successfully for optimizing process parameters for various processes in biotechnology, from biological treatment of toxic wastes (Ravichandra et al. 2008a, b) to enzyme production (Doddapaneni et al. 2007; Tatineni et al. 2007; Ravichandra et al. 2008a, b; Chennupati et al. 2009) including recombinant products (Vellanki et al. 2009; Farhat-Khemakhem et al. 2012). Till date, studies with statistical optimization of parameters for production of amorphadiene have not been reported elsewhere. Our present work emphasizes the key parameters (KH_2PO_4, methionine, pH and temperature) affecting amorpha-4,11-diene production in engineered *S. cerevisiae* strain (YCF-AD-1), optimized using RSM.

Materials and methods

Microbial strain and inoculum preparation

The yeast strain *S. cerevisiae* (YCF-AD-1) used in this study was developed in our previous studies (unpublished data) and originated from *S. cerevisiae* MTCC 3157. The strain was cultured in 250 mL Erlenmeyer flasks containing 100 mL medium with the following composition (g/L): galactose, 20; $(NH_4)_2.SO_4$, 7.5; $MgSO_4.7H_2O$, 0.5; trace metals solution, 2 mL; vitamins solution, 1 mL and 50 µl/L silicone anti-foam. The pH of the media was adjusted to 5.0 using 1 M NaOH and further autoclaved. Filter-sterilized vitamin solution and galactose solution were aseptically added to the sterile medium. The flasks were incubated for 24 h at 28 ± 2 °C at 150 rpm.

Amorphadiene production

The media components KH_2PO_4 and methionine were added according to experimental designs (Table 2) to the minimal medium (Verduyn et al. 1992) which consisted of (g/L): galactose, 20; $(NH_4)_2SO_4$, 5; $MgSO_4.7H_2O$, 0.5; EDTA, 0.015; $ZnSO_4.7H_2O$, 0.0045; $CoCl_2.6H_2O$, 0.0003; $MnCl_2. 4H_2O$, 0.001; $CuSO_4.5H_2O$, 0.0003; $CaCl_2.2H_2O$, 0.0000045; $FeSO_4.7H_2O$, 0.0003; $NaMoO_4.2H_2O$, 0.0004; H_3BO_3, 0.001; KI, 0.0001; 25 µl/L silicone anti-foam (Merck). It was autoclaved and cooled to room temperature. The filter solution was added to this sterile medium (Dynesen et al. 1998). The pH was adjusted according to the experimental design (Table 2). Aseptically, 1 % of inoculum was added to the flask, mixed thoroughly and incubated at the temperature specified in the experimental designs (Table 1) for 80 h at 150 rpm. After cells reached OD600 value of 1.0, 20 % (v/v) of isopropyl myristate (Merck Millipore, Germany) was added aseptically to the media. This isopropyl myristate layer was sampled and diluted with ethyl acetate for determination of amorphadiene by gas chromatography coupled with mass spectrometry GC–MS (Agilent Technologies, USA).

Analytical methods

Amorpha-4, 11-diene analysis

Amorpha-4, 11-diene was analysed by gas chromatography with flame-ionization detection (GC–FID). Samples from flasks were centrifuged at 5,000 rpm for 5 min and diluted directly into ethyl acetate and mixed for 30 min on a vortex mixer. After phase separation, 0.6 mL of the ethyl acetate layer was transferred to a capped vial for analysis. The ethyl acetate-extracted samples were analysed using the GC–FID

Table 1 Range and levels of the variables in coded units for response surface methodology studies

Variables	−2	−1	0	+1	+2	ΔX
KH_2PO_4 (x_1)	0	4	8	12	14	4
Methionine (x_2)	0	1	2	3	4	1
pH, 5.5 (x_3)	4.0	4.5	5.0	5.5	6.5	0.5
Temperature, °C (x_4)	25	27	32	37	39	2

with a split ratio of 1:20 and separated using a DB-WAX column (50 m × 200 μm × 0.2 μm) with hydrogen as carrier gas with a flow rate of 1.57 mL/min. The temperature program for the analysis was as follows: the column was initially held at 150 °C for 3 min, followed by a temperature gradient of 5 °C per min to a temperature of 250 °C. Amorpha- 4, 11-diene peak areas were converted to concentration values from external standard calibrations using *trans*-caryophyllene standard (Westfall et al. 2012).

Experimental design and response optimization

Response optimization method was used to increase the yield of amorphadiene by using RSM. On the basis of previous experience (unpublished data), four critical parameters for amorphadiene production were selected and further evaluated for their interactive behaviour by using statistical approach. The levels of the four medium variables, KH_2PO_4, 6.5(x_1); methionine, 1.5(x_2); pH, 5.5(x_3); and temperature, 32 °C (x_4), were selected as central points, and each variable was coded at five levels, −2, −1, 0, +1 and +2, using Eq. (1). For statistical calculations, the centre variable X_i was coded as x_i according to the following transformation. The range and levels of the variables in coded units for RSM studies are given in Table 1.

$$x_i = X_i - X_0/\Delta X \tag{1}$$

where x_i is the dimensionless coded value of the variable X_i, X_0 represents the value of X_i at the centre point and ΔX the step change. The behaviour of the system is explained by the following quadratic model [Eq. (2)].

$$Y = \beta_0 + \sum \beta_i X_i + \sum \beta_{ii} X_i^2 + \sum \beta_{ij} X_i X_j \tag{2}$$

where Y is the predicted response, β_0 is the intercept term, β_i the linear effect, β_{ii} the squared effect and β_{ij} the interaction effect. The full quadratic equation for four factors is given by the following model [Eq. (3)].

$$Y = \beta_0 + \beta_1 X_1 + \beta_2 X_2 + \beta_3 X_3 + \beta_4 X_4 + \beta_{11} X_1^2 + \beta_{22} X_2^2$$
$$+ \beta_{33} X_3^2 + \beta_{44} X_4^2 + \beta_{12} X_1 X_2 + \beta_{13} X_1 X_3 + \beta_{14} X_1 X_4$$
$$+ \beta_{23} X_2 X_3 + \beta_{24} X_2 X_4 + \beta_{34} X_3 X_4 \tag{3}$$

Previous experimental studies have considered such models using central composite design (CCD) (Cochran

and CoxIn 1957; Montgomery 2001). In this study, a 2^4 full-factorial design with eight star points and six replicates at the central points were employed to fit the second-order polynomial model, where we carried out a set of 30 experiments. Data obtained in the above experiments were analysed for regression, and graphical analysis using Design Expert® software (Stat-Ease Inc, USA) was used for regression and graphical analysis of the data obtained. The optimal combination of variables for the amorphadiene production were analysed using CCD experiments and were tabulated in Table 2. Table 2 shows the results of CCD experiments used for studying the effect of four independent variables along with the mean predicted and experimental responses. Each response was analysed, and a second-order regression model was developed. The model was validated in each case, and a set of optimal values were calculated.

Results and discussion

Multiple responses optimization and building model

RSM is a sequential and effective procedure where the primary objective of the methodology is to run rapidly and efficiently along the path of enhancement towards the general vicinity of the optimum, identifying the optimal region for running the process (Mekala et al. 2008; Chennupati et al. 2009; Potumarthi et al. 2012). The four independent variables such as KH_2PO_4, methionine, pH and temperature were chosen for optimized production of amorphadiene and experiments were performed according to the given CCD experimental design (Table 2), to obtain optimal combination of variables for the process. Thirty experimental runs with different combinations of four factors were carried out. For each run, the experimental responses along with the predicted response were calculated from the regression Eq. (4).

$$Y = 190.777 - 2.867X_1 - 1.756X_2 - 0.123X_3 + 6.121X_4$$
$$- 0.0719X_1X_2 + 1.4744X_1X_3 - 1.1194X_1X_4$$
$$- 0.3944X_2X_3 - 2.243X_2X_4 + 0.0956X_3X_4 - 3.481X_1^2$$
$$- 111.521X_2^2 - 13.075X_3^2 - 14.7455X_4^2 \tag{4}$$

where, Y is the predicted response, and x_1, x_2, x_3 and x_4 are coded values of KH_2PO_4, methionine, pH and temperature, respectively. The regression equation was used to calculate the predicted responses given in Table 2, and assessment of the predicted values with the experimental values indicated that these data were in reasonable agreement. The maximum response (205.34 mg/L) was obtained in run number 7, and in general all the runs with middle levels of parameters gave higher yields compared to other

Table 2 Design of experiments by central composite design for response surface methodology studies

Std. order	Run order	x_1	x_2	x_3	x_4	Coefficients assessed by	Amorphadiene (mg/L) Experimental	Amorphadiene (mg/L) Predicted
1	14	−1	−1	−1	−1	Full-factorial 2^4 design (16 expts)	41.98	44.31
2	10	1	−1	−1	−1		40.12	38.01
3	22	−1	1	−1	−1		46.24	46.22
4	8	1	1	−1	−1		42.37	39.63
5	30	−1	−1	1	−1		48.24	41.71
6	2	1	−1	1	−1		39.21	41.31
7	29	−1	1	1	−1		46.21	42.04
8	9	1	1	1	−1		40.35	41.35
9	26	−1	−1	−1	1		68.24	63.09
10	1	1	−1	−1	1		48.25	52.31
11	18	−1	1	−1	1		58.23	56.02
12	3	1	1	−1	1		42.58	44.96
13	21	−1	−1	1	1		58.24	60.87
14	11	1	−1	1	1		60.12	55.99
15	15	−1	1	1	1		54.27	52.23
16	25	1	1	1	1		49.5	47.06
17	4	−2	0	0	0	Star points (8 expts)	175	190.15
18	17	2	0	0	0		182.54	184.42
19	27	0	−2	0	0		74.21	81.00
20	20	0	2	0	0		67.25	77.49
21	19	0	0	−2	0		174.35	177.81
22	16	0	0	2	0		164	177.57
23	24	0	0	0	−2		159.77	169.90
24	28	0	0	0	2		175.24	182.14
25	7	0	0	0	0	Central points (6 expts)	205.34	190.77
26	23	0	0	0	0		201.27	190.77
27	12	0	0	0	0		198.24	190.77
28	6	0	0	0	0		195.28	190.77
29	13	0	0	0	0		197.32	190.77
30	5	0	0	0	0		198.25	190.77

combinations. The data were analysed by regression analysis, and the optimized values to maximize the responses were observed at 4, 1.49, 5.47 and 33.13 for KH_2PO_4, methionine, pH and temperature, respectively.

Suitability of the model was confirmed by the analysis of variance (ANOVA) using Design Expert software and the results are shown in Table 3. ANOVA of the quadratic regression model suggests that the model is significant with a computed F value of 101.6917 and a $P > F$ value less than 0.05. A lower value for the coefficient of variation suggests higher consistency of the experiment, and in this case the obtained CV value of 9.19 % demonstrates a greater reliability of the trials. R^2 is the coefficient of variance of response under test and whose values are always between 0 and 1; closer the value of R^2 to 1, stronger is the statistical model and better is the prediction of response (Myers and Montgomery 1995). The

Table 3 Model summary and analysis of variance for the quadratic model

Source of variations	Sum of squares	Degree of freedom	Mean square	F value	Probability (P)
Regression	132,761.320	14	9,482.95	101.69	<0.0001
Residual	1,398.780	15	93.25		
Total	134,160.099	29			

$R = 0.9947$, $R^2 = 0.9896$, adjusted $R^2 = 0.9798$, CV = 9.19 %

coefficient of determination (R^2) for response of amorphadiene is 0.9896 (Table 3), indicating that the statistical model can explain 98.96 % of variability in the response and only 1.04 % of the variations for amorphadiene not explained by the model. The adjusted R^2 value corrects the R^2 value for the sample size and for the number of terms in the model. The value of the adjusted determination

Table 4 Model coefficients estimated by multiple linear regressions (significance of regression coefficients)

Model term	Coefficient estimates	Standard error	F value	P value Prob > F
Intercept	190.767	2.99967	101.692	<0.0001
x_1	−2.8672	2.27611	1.58686	0.227
x_2	−1.7561	2.27611	0.59528	0.4524
x_3	−0.1233	2.27611	0.00294	0.9575
x_4	6.12111	2.27611	7.23228	0.0168[a]
x_1x_2	−0.0719	2.41418	0.00089	0.9766
x_1x_3	1.47438	2.41418	0.37297	0.5505
x_1x_4	−1.1194	2.41418	0.21499	0.6495
x_2x_3	−0.3944	2.41418	0.02669	0.8724
x_2x_4	−2.2431	2.41418	0.86332	0.3675
x_3x_4	0.09563	2.41418	0.00157	0.9689
x_1^2	−3.4805	5.99933	0.33658	0.5704
x_2^2	−111.52	5.99933	345.545	<0.0001[a]
x_3^2	−13.076	5.99933	4.75021	0.0456[a]
x_4^2	−14.746	5.99933	6.04108	0.0266[a]

[a] Significant at $P < 0.05$

coefficient (Adj R^2) for amorphadiene (0.9798) is also good, supporting the significance of this developed model (Cochran and CoxIn 1957). The significance of individual variables can be evaluated from their P values, with the more significant terms having a lower P value (Table 4). The values of $P > F$ less than 0.05 indicate that the model terms are significant and in this case X_4, X_2^2, X_3^2 and X_4^2 were found to be significant model terms and there were no significant interactions between the parameters.

Surface plots are generally the graphical representation of the regression equation for identifying the optimal levels of each parameter for attaining the maximum response (amorphadiene) production. Figure 1a–f shows the response surfaces obtained for the interaction effects of tested variables. In each response graph, the effect of the two variables on amorphadiene production was shown when the other two variables were kept constant. Figure 1a shows the interaction relationship between the two independent variables, namely, KH_2PO_4/methionine and their effects on amorphadiene production .

It was observed from Fig. 1a that amorphadiene synthesis was significantly affected by methionine concentration. Amorphadiene synthesis was increased with increase in methionine concentration up to 1.5 mM and further increase in methionine concentration did not show any influence on amorphadiene production, whereas the addition further resulted in decreased production. The same pattern was observed in other graphs (Fig. 1d, e). This indicates that the increase in the methionine concentration tightly regulates the engineered repressible methionine promoter in *S. cerevisiae* by limiting the conversion of farnesyl pyrophosphate into squalene (Asadollahi et al. 2008).

Studies on the effect of varied methionine concentration (0–2 mM) with engineered yeast reported approximately 125 mg/L of amorphadiene with 0.2 mM methionine concentration. In previous studies, 1.5 and 2 mM concentrations of methionine were considered for the production of plant sesquiterpenes in yeast during batch and fed-batch operations, respectively (Asadollahi et al. 2008; Paradise et al. 2008). But these reported studies were not statistically optimized for methionine concentration; in the present work, it was observed that 1.49 mM of methionine was the optimum concentration with combinations of other optimum variables leading to synthesis of 191.5 mg/L of amorphadiene. The effect of KH_2PO_4 did not have significant effect in combination with methionine concentration, but there was significant effect observed in combination with the other two variables, temperature and pH (Fig. 1a, b and c). There was a significant increase in amorphadiene production with increase in KH_2PO_4 concentration up to 6.5 g/L and further increase in its concentration did not show any significant improvement in amorphadiene production. Previous studies reported that low phosphate concentration improved amorphadiene production, which may be by limiting the growth and channelling the carbon flux towards amorphadiene production (Westfall et al. 2012). In this study, 4.01 g/L of KH_2PO_4 was the recommended concentration for the optimized production of amorphadiene in combination with other optimized parameters.

Figure 1b, d, f shows the effect of pH on amorphadiene production in combination with KH_2PO_4 and temperature. There is increase in amorphadiene production with increase in pH and the maximum production was at pH 5.5. In previous studies, the production of plant sesquiterpenes in yeast was carried out at pH 6.50, 5 ± 0.5, 5.0 for shake flasks, batch and fed-batch cultivation, respectively (Asadollahi et al. 2008), whereas the enzyme responsible for amorphadiene production (amorphadiene synthase) showed

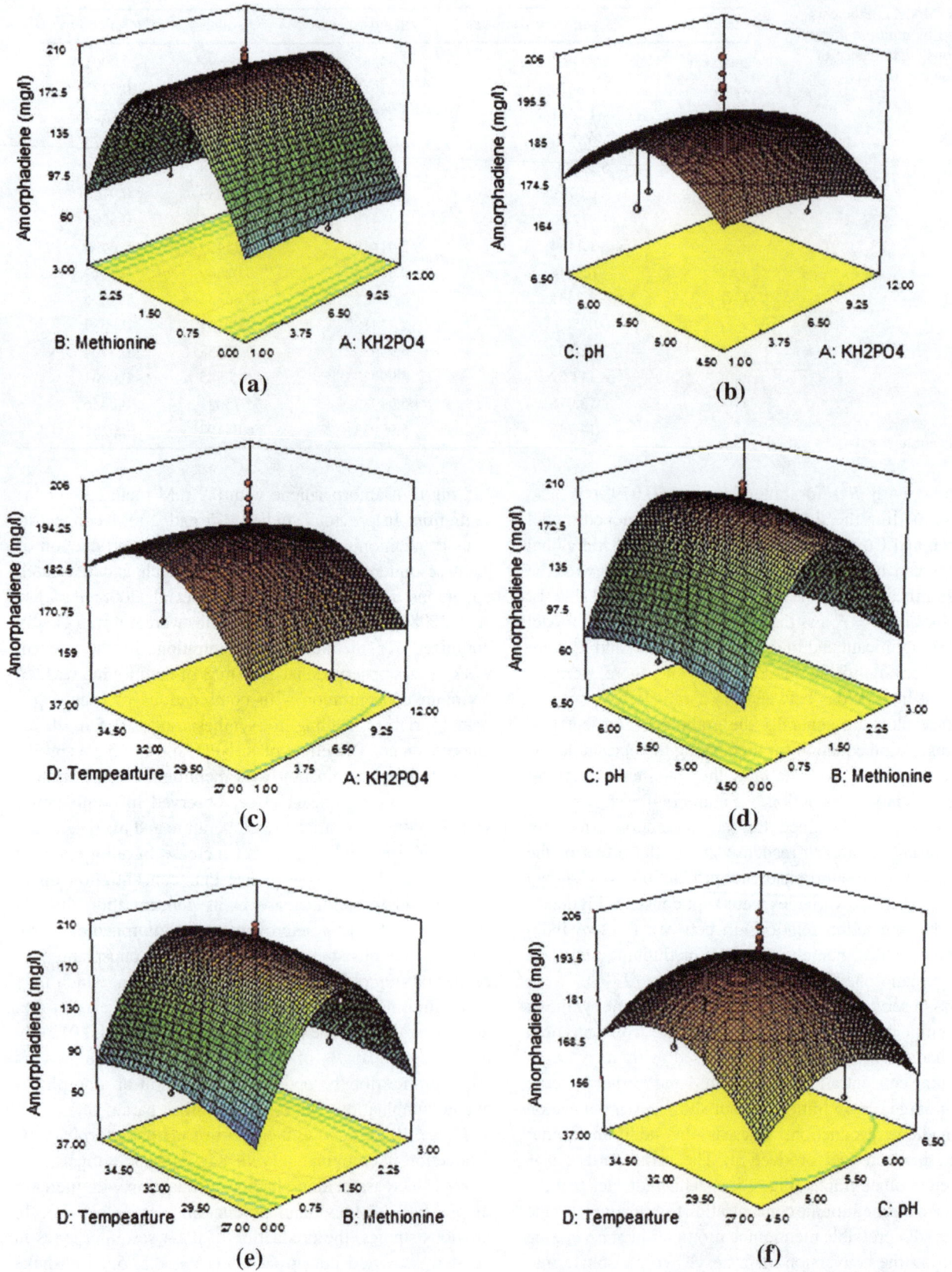

Fig. 1 a–f 3-D surface and contour plot of amorphadiene production by *S. cerevisiae* (mg/L): the effect of two variables while the other two were held at 0 level

optimum activity at varied pH 6.5–7.5 levels in *artemisia annua* (Bouwmeester et al. 1999; Mercke et al. 2000; Picaud et al. 2005; Picaud et al. 2007). In this study, *S. cerevisiae* showed optimum pH as 5.5 and the present model gave 5.47 as an optimum value along with other optimal parameters.

The effects of temperature in response to combination with other variables, KH_2PO_4, methionine and pH, are shown in Fig. 1c, e, f. At low temperature (27 °C), amorphadiene synthesis was very less and increased with increment in temperature up to 33 °C. There was a rapid increase in amorphadiene production in combination with KH_2PO_4 and pH, whereas in combination with methionine the effect of temperature was not significant. Based on this model, the optimal combination of all parameters is KH_2PO_4, 4.01; methionine, 1.49; pH, 5.47; temperature 33.13 °C with a predicted response value of 192.119 mg/L. Experiments conducted with the same optimal conditions, such as KH_2PO_4, 4.0; methionine, 1.49; pH, 5.4; temperature 33 °C, yielded 191.5 mg/L of amorphadiene, which resembles closely the predicted response. Finally, these results suggest that methionine has a high significant effect on amorphadiene production compared to other variables. Hence, the maximum amorphadiene production can be achieved with a relatively limited number of experimental runs using the appropriate statistical design and optimization technique.

Conclusion

The use of RSM with a full-factorial rotatable CCD for determination of optimal medium and physical parameters for amorphadiene production was demonstrated using the essential parameters. The use of this methodology will be successful for any combinational analysis, in which an analysis of the effects and interactions of many experimental factors are required. Rotatable central composite experimental design maximizes the amount of information that can be obtained while limiting the number of individual experiments. Thus, smaller and less time-consuming experimental designs could generally be sufficient for optimization of many such fermentation processes (Tatineni et al. 2007). The superiority of terpenoids has expanded their utility from pharmaceutical to fragrances, including biofuel industries. Significant efforts have been made for establishing microbial cell factories for the production of a wide variety of high value-added chemicals. However, there are some difficulties for the large-scale production of these chemicals. In addition to the synthetic biology and metabolic engineering approaches, statistical optimization methods will provide insights into the production of high value-added chemicals. In the present study, the overall

view on the optimization of the process using essential parameters for amorphadiene production provides insights into the process development and further scaling-up process. The results of ANOVA and regression of the second-order model showed that the linear effects of temperature and the interactive effects of the three variables, methionine, pH and temperature, were significant for amorphadiene production. Among these three variables, methionine has a more significant interactive effect. Finally, we conclude our study by stating that the optimization of amorphadiene production was by the second-order model, and ANOVA requires optimal conditions of: KH_2PO_4, 4.0; methionine, 1.49; pH, 5.4; temperature 33 °C.

Acknowledgments The authors express their deep sense of gratitude to the Head, Department of Biotechnology, and Director, NIT, Warangal for all the support and constant encouragement in carrying out this work. One of the authors, RR Baadhe, acknowledges M.H.R.D, India, for the Ph.D. fellowship.

Conflict of interest The authors confirm that this article content has no conflict of interest.

References

Akhnazarova S, Kafarov V (1982) Experiment optimization in chemistry and chemical engineering. Mir Publishers, Moscow

Asadollahi MA, Maury J, Møller K, Nielsen KF, Schalk M, Clark A, Nielsen J (2008) Production of plant sesquiterpenes in *Saccharomyces cerevisiae*: effect of ERG9 repression on sesquiterpene biosynthesis. Biotechnol Bioeng 99:666–677

Bloland PB (2001) Drug Resistance in Malaria. World Health Organization, publications http://www.who.int/csr/resources/publications/drugresist/malaria.pdf. Accessed 26 June 2011

Bouwmeester HJ, Wallaart TE, Janssen MH, van Loo B, Jansen BJ, Posthumus MA, Schmidt CO, De Kraker JW, König WA, Franssen MC (1999) Amorpha-4,11-diene synthase catalyses the first probable step in artemisinin biosynthesis. Phytochemistry 52:843–854

Breitmaier, E (2006) Terpenes: flavors, fragrances, pharmaca, pheromones. Wiley-VCH, Germany

Chennupati S, Potumarthi R, Gopal Rao M, Manga PL, Sridevi M, Jetty A (2009) Multiple responses optimization and modeling of lipase production by *Rhodotorula mucilaginosa* MTCC-8737 using response surface methodology. Appl Biochem Biotechnol 159:317–29

Cochran WG, CoxIn GM (1957) Experimental design. Wiley, New York

Croteau R, Kutchan TM, Lewis NG (2000) Natural products (secondary metabolites). In: Buchanan B, Gruissem W, Jones R (eds) Biochemistry and molecular biology of plants. ASPB publications, Mary land, USA, pp 1250–1318

Doddapaneni KK, Tatineni R, Potumarthi R, Mangamoori LN (2007) Optimization of media constituents through response surface methodology for improved production of alkaline proteases by *Serratia rubidaea*. J Chem Technol Biot 82:721–729

Dynesen J, Smits HP, Olsson L, Nielsen J (1998) Carbon catabolite repression of invertase during batch cultivations of *Saccharomyces cerevisiae*: the role of glucose, fructose, and mannose. Appl Microbiol Biotechnol 50:579–582

Farhat-Khemakhem A, Farhat MB, Boukhris I, Bejar W, Bouchaala K, Kammoun R, Maguin E, Bejar S, Chouayekh H (2012) Heterologous expression and optimization using experimental designs allowed highly efficient production of the PHY US417 phytase in *Bacillus subtilis* 168. AMB Express 2:1–11

Khuri AI, Cornell JA, Dekker M (1987) Response surfaces: design and analysis. Dekker, New York

Martin VJ, Pitera DJ, Withers ST, Newman JD, Keasling JD (2003) Engineering a mevalonate pathway in *Escherichia coli* for production of terpenoids. Nat Biotechnol 21:796–802

McGarvey DJ, Croteau R (1995) Terpenoid metabolism. Plant Cell 7:1015–1026

Mekala NK, Singhania RR, Sukumaran RK, Pandey A (2008) Cellulase production under solid-state fermentation by *Trichoderma reesei* RUT C30: statistical optimization of process parameters. Appl Biochem Biotechnol 151:122–131

Mercke P, Bengtsson M, Bouwmeester HJ, Posthumus MA, Brodelius PE (2000) Molecular cloning, expression, and characterization of amorpha-4,11-diene synthase, a key enzyme of artemisinin biosynthesis in *Artemisia annua* L. Arch Biochem Biophys 381:173–180

Montgomery D (2001) Design and analysis of experiments. Wiley, New York

Myers RH, Montgomery DC (1995) Response surface methodology: process and product optimization using designed experiments. Wiley-Interscience, New York

Paradise EM, Kirby J, Chan R, Keasling JD (2008) Redirection of flux through the FPP branch-point in *Saccharomyces cerevisiae* by down-regulating squalene synthase. Biotechnol Bioeng 100:371–378

Peralta-Yahya PP, Zhang F, Del Cardayre SB, Keasling JD (2012) Microbial engineering for the production of advanced biofuels. Nature 488:320–328

Picaud S, Olofsson L, Brodelius M, Brodelius PE (2005) Expression, purification, and characterization of recombinant amorpha-4,11-diene synthase from *Artemisia annua* L. Arch Biochem Biophys 436:215–226

Picaud S, Olsson ME, Brodelius PE (2007) Improved conditions for production of recombinant plant sesquiterpene synthases in *Escherichia coli*. Protein Expr Purif 51:71–79

Potumarthi R, Jacques L, Harry W, Michael D (2012) Surface immobilization of *Rhizopus oryzae* (ATCC 96382) for enhanced production of lipase enzyme by multiple responses optimization. Asia-Pac J Chem Eng 7:S285–S295

Ravichandra P, Gopal M, Annapurna J (2008a) Biological treatment of toxic petroleum spent caustic in fluidized bed bioreactor using immobilized cells of *Thiobacillus* RAI01. Appl Biochem Biotech 151:532–546

Ravichandra P, Subhakar C, Pavani A, Jetty A (2008b) Evaluation of various parameters of calcium-alginate immobilization method for enhanced alkaline protease production by *Bacillus licheniformis* NCIM-2042 using statistical methods. Bioresour Technol 99:1776–1786

Ro DK, Paradise EM, Ouellet M, Fisher KJ, Newman KL, Ndungu JM, Ho KA, Eachus RA, Ham TS, Kirby J, Chang MC, Withers ST, Shiba Y, Sarpong R, Keasling JD (2006) Production of the antimalarial drug precursor artemisinic acid in engineered yeast. Nature 440:940–943

Ruzicka L (1959) The isoprene rule and the biogenesis of terpenic compounds. Experientia 9:357–367

Tatineni R, Doddapaneni KK, Potumarthi RC, Mangamoori LN (2007) Optimization of keratinase production and enzyme activity using response surface methodology with *Streptomyces sp7*. Appl Biochem Biotech 141:187–201

Vellanki RN, Potumarthi R, Mangamoori LN (2009) Constitutive expression and optimization of nutrients for streptokinase production by *Pichia pastoris* using statistical methods. Appl Biochem Biotech 158:25–40

Verduyn C, Postma E, Scheffers WA, Van Dijken JP (1992) Effect of benzoic acid on metabolic fluxes in yeasts: a continuous-culture study on the regulation of respiration and alcoholic fermentation. Yeast 8:501–517

Westfall PJ, Pitera DJ, Lenihan JR, Eng D, Woolard FX, Regentin R, Horning T, Tsuruta H, Melis DJ, Owens A, Fickes S, Diola D, Benjamin KR, Keasling JD, Leavell MD, McPhee DJ, Renninger NS, Newman JD, Paddon CJ (2012) Production of amorphadiene in yeast, and its conversion to dihydroartemisinic acid, precursor to the antimalarial agent artemisinin. Proc Natl Acad Sci USA 109:E111–E118

18

Molecular modeling and expression analysis of a *MADS-box* cDNA from mango (*Mangifera indica* L.)

Magda A. Pacheco-Sánchez · Carmen A. Contreras-Vergara ·
Eduardo Hernandez-Navarro · Gloria Yepiz-Plascencia · Miguel A. Martínez-Téllez ·
Sergio Casas-Flores · Aldo A. Arvizu-Flores · Maria A. Islas-Osuna

Abstract *MADS-box* genes are a large family of transcription factors initially discovered for their role during development of flowers and fruits. The MADS-box transcription factors from animals have been studied by X-ray protein crystallography but those from plants remain to be studied. In this work, a *MADS-box* cDNA from mango encoding a protein of 254 residues was obtained and compared. Based on phylogenetic analysis, it is proposed that the MADS-box transcription factor expressed in mango fruit (MiMADS1) belongs to the SEP clade of MADS-box proteins. *MiMADS1* mRNA steady-state levels did not changed during mango fruit development and were up-regulated, when mango fruits reached physiological maturity as assessed by qRT-PCR. Thus, MiMADS1 could have a role during development and ripening of this fruit. The theoretical structural model of MiMADS1 showed the DNA-binding domain folding bound to a double-stranded DNA. Therefore, MiMADS1 is an interesting model for understanding DNA-binding for transcriptional regulation.

Keywords *Mangifera indica* L. · Gene expression · MADS-box · Molecular modeling · Transcription factor

Introduction

Gene expression associated to specific inductive events on fruit development is poorly characterized in fruits of commercial importance. In a plant cell, the temporal and spatial organization of these events is mediated by transcription factors required for gene expression. The MADS-box proteins comprise one of the largest groups of plant transcription factors with diverse functions during important events of plant development. *MADS-box* genes represent a highly conserved gene family encoding transcription factors in plants, which contain a conserved sequence called MADS-box as an acronym of the first four identified members of the family (MADS: MCM-AGAMOUS-DEFICIENS-SRF) (Liu et al. 2009), and have regulatory functions in flower and fruit development (Theißen and Saedler 2001). More recently, studies of ripening-inhibited spontaneous tomato mutants have shown that *MADS-box* genes play a major role in the molecular regulation of development in tomato fruit (Elitzur et al. 2010), aside of their role in organ and flower development initially described in *Arabidopsis* (Weigel and Meyerowitz 1994).

The function of MADS-box protein family on regulation of gene expression have been extensively studied, mainly the floral organ identity in *Arabidopsis* (Ito et al. 2008). Using information of mutants in *Arabidopsis thaliana* and *Antirrhinum majus* in which some identity of floral organs is changed, the ABC model is the most simple to explain how identity arises during development (Theißen and Saedler 2001). Tomato MADS-RIN protein regulates tissue differentiation acting as a transcription factor.

M. A. Pacheco-Sánchez · C. A. Contreras-Vergara ·
E. Hernandez-Navarro · G. Yepiz-Plascencia ·
M. A. Martínez-Téllez · M. A. Islas-Osuna (✉)
Plant Molecular Biology Lab, Centro de Investigación en Alimentación y Desarrollo, A.C., Carretera a la Victoria Km 0.6, Apartado Postal 1735, 83304 Hermosillo, Sonora, Mexico
e-mail: islasosu@ciad.mx

A. A. Arvizu-Flores
Departamento de Ciencias Químico Biológicas, Universidad de Sonora, Blvd. Luis Encinas y Blvd. Rosales S/N, 83000 Hermosillo, Sonora, Mexico

S. Casas-Flores
División de Biología Molecular, IPICYT, Camino a la Presa San José No. 2055, Lomas 4a sección, 78216 San Luis Potosí, Mexico

Development of fleshy fruits such as tomato involves cell division and expansion of the ovary tissues. Recent studies in *A. thaliana* and tomato have been important to elucidate the function of MADS-box transcription factors on ethylene biosynthesis, perception and signaling pathways. Mango (*Mangifera indica* L.) is one of the most important tropical fruits and it could serve as a model to study development and ripening of this fruit.

The *MADS-box* gene family encodes transcription factors present in a variety of organisms in diverse kingdoms. An exclusive family for plant MADS proteins belong to the MICK-type that includes a MADS (M), intervening (I), keratin like (K) and C-terminal (C) domains (Kaufmann et al. 2005). MADS-box proteins have in common a highly conserved DNA-binding domain located at the N-terminal, this sequence of approximately 60 amino acids is common to all MADS type proteins. The MADS-box domain binds to a conserved DNA motif know as CArG box [CC(A/T)$_6$GG]. In terms of structure, the DNA-binding domain of MEF2A has been determined by X-ray diffraction for the myocyte enhancer factor 2 (MEF2), which is involved in muscle development (Han et al. 2005; Wu et al. 2010), and also by NMR (Huang et al. 2000). In this work, a *MADS-box* cDNA was cloned, its deduced amino acid sequence was compared and modeled bound to DNA, and *MiMADS1* mRNA levels were assayed at different mango fruit developmental stages.

Materials and methods

cDNA isolation and analysis of MiMADS1 amino acid sequence

Total RNA from mango pulp was isolated as described elsewhere (Lopez-Gomez and Gomez-Lim 1992), and purified by Oligotex Direct mRNA Mini Kit (QIAGEN). The cDNA was synthesized with the SMART cDNA library construction kit (Clontech) following manufacturer's recommendations. Three independent *MADS-box* cDNA clones were sequenced thoroughly at the Genomic Analysis and Technology Core at the University of Arizona (Tucson, AZ, USA). These three clones were piled up to form a contiguous unambiguous clone and its identity was obtained after a BLASTX search. The cDNA contig was named MiMADS1 and deposited in GenBank. The deduced amino acid sequence was obtained and compared to other MADS-box transcription factors with CLUSTAL W using the T-Coffee server (Notredame et al. 2000).

MiMADS1 phylogenetic analysis

A phylogenetic analysis was done using the Maximum Likelihood algorithm. The bootstrap consensus tree inferred

from 1,000 replicates was taken to represent the evolutionary history of the analyzed taxa (Felsenstein 1985). The analyses were conducted with MEGA5 software (Tamura et al. 2011).

Molecular modeling

A molecular model of MiMADS1 was obtained by homology modeling using the MOE v2012.10 software. The crystallographic structure used, as template was the MEF2A bound to DNA deposited with the PDB code 1EGW (Santelli and Richmond 2000). A multiple sequence alignment was done previously to match the conserved residues from MiMADS1 and MEF2A N-terminal domain. Then, 25 intermediate models were constructed under the CHARMM27 force-field and DNA bounded from the template coordinates was included for induced fit simulation on MiMADS1 model. The final model was further refined under the default parameters of MOE. Figures from the molecular model were done with MOE software.

MiMADS1 gene expression

Mango (*Mangifera indica* L.) fruits cv. Keitt were hand harvested during four different developmental stages in a commercial packinghouse at El Porvenir, Sinaloa, México (25°55′56.83″N, 109°5′40.11″W) and transported to the laboratory. Fruits were selected and washed with chlorinated water (200 ppm sodium hypochlorite), numbered and weighed. Fruits were taken and the peel was removed and cut to obtain small pieces of the pulp and were immediately frozen with liquid nitrogen.

Total RNA extraction and cDNA synthesis was done using frozen mango mesocarp tissue as described (1992). The extracted RNA was cleaned with RNAse free DNase I (Roche). RNA quantity was estimated at 260 nm using a NanoDrop ND-1000 ultraviolet–visible spectrophotometer (NanoDrop Technologies Inc., Wilmington, DE, USA). The integrity of total RNA was evaluated by formaldehyde-agarose gel electrophoresis (Sambrook and David 2001). The cDNAs were synthesized from 5 μg of total RNA using a cDNA synthesis kit (Invitrogen) that was prepared from the pool of five experimental units.

Relative expression of *MiMADS1* was evaluated by qRT-PCR and the *18S* rRNA gene was used for data normalization. Real-time PCR was realized in a StepOne Real-Time PCR (Life Technology, CA) using iQSYBR Green Supermix (BIO-RAD). For each gene, a standard curve was generated using cDNA serial dilutions and PCR efficiencies were calculated from slope according to the manufacturer's instructions. The specific primers for *MiMADS1* were MiMADS1-Fw: 5′-ATGGGATCAAGCTGGAGTAC-3′ and MiMADS1-Rv: 5′TCAAAGCATCCATCCTGG-3′, primers for *18S* rRNA were Mi18S-Fw: 5′-GGTGACGGAGAATTAGGGTTC-3′ and Mi18S-Rv: 5′-CCGTGTCAGGATTGGGTAAT-3′. qRT-PCR reactions were carried out using 4 ng of cDNA under the following conditions: 94 °C for 30 s as denaturation step, followed for 35 cycles of: 94 °C for 30 s, 60 °C for 30 s, 72 °C for 1 min, and 72 °C for 5 min as final extension. The data were analyzed using the $2^{-\Delta\Delta Ct}$ method for calculation of relative changes in gene expression (Livak and Schmittgen 2001) where the fold change was calculated as follows:

$$\text{Fold change} = 2^{-\Delta\Delta Ct} = [(\text{Ct}_{\text{Target}} - \text{Ct}_{18S})\text{Time}_x] \\ - [(\text{Ct}_{\text{Target}} - \text{Ct}_{18S})\text{Time}_0)].$$

Thus, the relative expression levels reported here were calibrated and normalized to the constitutive gene *18S* rRNA.

Statistical analysis

The data were analyzed as a completely randomized design with three replicates using the GLM procedure of NCSS 2007. Comparison of mean was performed using the multiple range test of Tukey–Kramer with a significance level of 0.05.

Results and discussion

cDNA and deduced amino acid sequence of MiMADS1

The complete sequence of the *MiMADS1* cDNA (GenBank KF214778) was 1,062 bp with a 62 bp 5′-untranslated region, a 765 bp Open Reading Frame including the initial methionine and the stop codon and a 3′-untranslated region of 195 nt and a 24 nt poly-A$^+$ tail (Fig. 1). The encoded protein of 254 residues has a theoretical pI of 9.1 and a molecular weight of 29.7 kDa. Multiple amino acid sequence alignment showed that MiMADS1 is similar to MADS-box proteins from other fruits (Fig. 2) for example it is 70 % identical to PpMADS5 from peach (*Prunus persica*, AAZ16241) (Xu et al. 2008) and 68 % identical to AcSEP4 from kiwifruit (*Actinidia chinensis*, ADU15479) (Varkonyi-Gasic et al. 2011).

MiMADS1 has a MADS-box domain of 57 residues based on sequence comparison to other MADS-box transcription factors. The I domain (linker region) of MiMADS1 has a length of 33 residues, which is similar in length to the MICK protein QUAMOSA (SQUA) that has an I domain of 35 residues (Henschel et al. 2002). Also, the conserved K domain of MiMADS1 consists of three subdomains that form alpha helixes with hydrophobic amino acids (Kaufmann et al. 2005). The structure of the K domain is important for protein–protein interactions since point mutations made in this domain have shown its importance. Furthermore, the coiled-coil structure of K domain is important for protein–protein interaction. The linker region is variable and located between the MADS- and K-domains, which shows sequence and structural similarity to the coiled-coil domain of keratin, likely forms amphipathic α-helices. Both the I and K-domains, are implicated in determining protein–protein dimerization specificity (Cseke and Podila 2004). The C region is not required for dimerization but it is needed for the formation of ternary complexes (Aswath and Kim 2005). Moreover, the C domain, located at the C terminus is the domain that has the capacity of transcriptional activation (Immink et al. 2009). In fruits like apple, grapes and peach, expression of *MADS-box* genes has been observed during early stages of fruit development (Yao et al. 1999; Choudhury et al. 2012). MADS-box proteins can form homo or heterodimers as it has been demonstrated by a large number of in vitro and yeast two-hybrid experiments. Structure of these proteins will also help to elucidate functional properties related to their structure (Smaczniak et al. 2012).

MiMADS1 phylogenetic analysis

Sequence analysis showed that MiMADS1 protein primary structure is similar to members of the SEP clade. SEP are transcription factors that facilitate the formation of complexes along with other MADS-box proteins and activate the potential of these complexes due to the fact that members of the SEP subfamily have a role as mediators of higher order complex formation (Ito et al. 2008; Smaczniak et al. 2012). The amino acid sequence of MiMADS1 was aligned with other MADS-box proteins and it is more similar to proteins from the SEP and AGL families.

Therefore, phylogenetic analysis was performed using proteins from the SEP and AGL subfamilies from several fruits; the resultant dendrogram showed that MiMADS1 is more similar to AcSEP4 expressed in ripe kiwifruit (Fig. 3). Thus, MiMADS1 transcription factor could be involved in gene expression regulation during development and ripening of mango fruit. Several *MADS*-box genes were discovered in other fruits, and most of them are expressed in early stages of fruit development. In peach, an increase in expression was detected for two *MADS-box* genes during fruit development (Tadiello et al. 2009). Also, expression analysis of *MADS-box* transcripts in grapes was detected during fruit development (Elitzur et al. 2010). *MADS-box* genes in kiwifruit were required for floral meristem and floral organ specification (Varkonyi-Gasic et al. 2011). Due to the similarity between MADS1 form mango and other MICK-type proteins belonging to the SEP clade, we did a molecular modeling of MiMADS and

```
              aaccaagaccagaaacagcttaaagaaagggaaatcagagaaacaaaaataaaaaaaaat

     1    M  G  R  G  R  V  E  L  K  R  I  E  N  K  I  N  R  Q  V  T
    61    ATGGGAAGAGGAAGAGTTGAACTGAAGAGGATAGAGAACAAAATTAATAGGCAGGTTACA

    21    F  A  K  R  R  N  G  L  L  K  K  A  Y  E  L  S  V  L  C  D
   121    TTTGCTAAGAGAAGAAATGGGTTGCTTAAAAAAGCTTATGAACTCTCTGTTCTTTGTGAT

    41    A  E  V  A  L  I  I  F  S  N  R  G  K  L  Y  E  F  C  S  S
   181    GCTGAAGTTGCTCTCATCATTTTCTCCAACCGGGGGAAACTATATGAATTTTGTAGCAGT

    61    S  S  M  M  K  T  L  E  R  Y  Q  R  C  S  Y  G  A  L  D  T
   241    TCTAGCATGATGAAAACTCTTGAGCGATACCAAAGGTGTAGCTATGGTGCACTGGATACT

    81    N  R  T  G  N  E  S  Q  G  T  Y  Q  E  Y  L  K  L  K  T  T
   301    AACCGTACAGGGAATGAGTCACAGGGAACTTACCAGGAGTATTTGAAGCTGAAAACAACA

   101    V  E  V  L  Q  R  S  Q  R  N  L  L  G  E  D  L  G  P  L  S
   361    GTTGAGGTCCTGCAGCGATCTCAAAGAAACCTTTTGGGAGAAGATCTGGGTCCATTGAGC

   121    T  K  E  L  D  Q  L  E  N  Q  L  E  T  S  L  K  H  I  R  T
   421    ACAAAAGAGCTTGACCAACTTGAGAATCAGCTAGAGACATCCTTGAAGCATATCAGAACT

   141    T  K  T  Q  F  M  V  D  Q  L  S  E  L  Q  K  R  E  Q  M  L
   481    ACAAAGACCCAATTTATGGTTGATCAGCTTTCCGAACTTCAAAAGAGGGAACAAATGCTT

   161    V  E  T  N  K  A  L  R  K  K  F  E  E  T  G  A  Q  V  P  F
   541    GTTGAAACTAACAAGGCTCTGAGAAAGAAGTTTGAGGAAACCGGTGCTCAAGTTCCTTTT

   181    R  L  A  W  D  Q  A  G  V  Q  N  M  A  Y  N  T  R  L  P  G
   601    CGACTGGCATGGGATCAAGCTGGAGTACAAAACATGGCATACAACACTCGTCTCCCCGGT

   201    H  S  E  G  F  F  Q  P  L  G  A  N  S  T  I  N  M  G  Y  N
   661    CATTCAGAAGGGTTTTTCCAGCCCCTGGGAGCCAACTCCACCATTAACATGGGATACAAT

   221    P  M  A  V  V  A  S  S  G  A  E  E  V  V  S  F  S  V  A  G
   721    CCAATGGCGGTGGTGGCCTCTTCTGGTGCGGAGGAGGTGGTGAGTTTCAGCGTTGCAGGG

   241    Q  T  Q  T  V  N  G  Y  I  P  G  W  M  L  *
   781    CAAACTCAGACTGTCAATGGATACATTCCAGGATGGATGCTTTGAtaatggagtcttttt

   841    gcttgcttggcctctatggtgattttggtcatcttcttcatgcagcatgtaattaattat

   901    ttatatatatatgtatgtgcaaactttgagtattatttgtaatttgtgtagtggaacaaa

   961    tgatgatgcgtttattttattttattttatatatatgtatgctctgtacttcatgtaaag

  1021    cctctaaccctttcaacaaaaaaaaaaaaaaaaaaaaaaaaaaa
```

Fig. 1 The cDNA and deduced amino acid sequence of MiMADS1. The lower case letters indicate the 5′-untranslated region and the 3′-untranslated region

docking with DNA. It is also known that SEP-like proteins interact with MADS by interaction with bridging proteins like TM5 (Leseberg et al. 2008). These studies in tomato suggest that two-hybrid approaches may help in identifying interactions during development and ripening in mango.

MiMADS structural model and DNA interaction

The N-terminus of the amino acid sequence of MiMADS1 was readily modeled using MOE 2012.10, due to its identity to the crystal structure of MEF2A (Wu et al. 2010) (PDB 1EGW, 3KOV, 1C7U, 1TQP, 1N6J). The model obtained was a homodimer that also was docked to double strand DNA fragment (5′-TAAGCTAATAATAGCTT-3′) as reported for MEF2A (Fig. 4). The molecular model includes the first 72 residues of the MiMADS1 amino acids sequence, since the remainder of the molecule, probably involved in transcriptional activation, is novel and does not have significant identity with known structures. The amino acid identity with the template was 55 %, ensuring that the

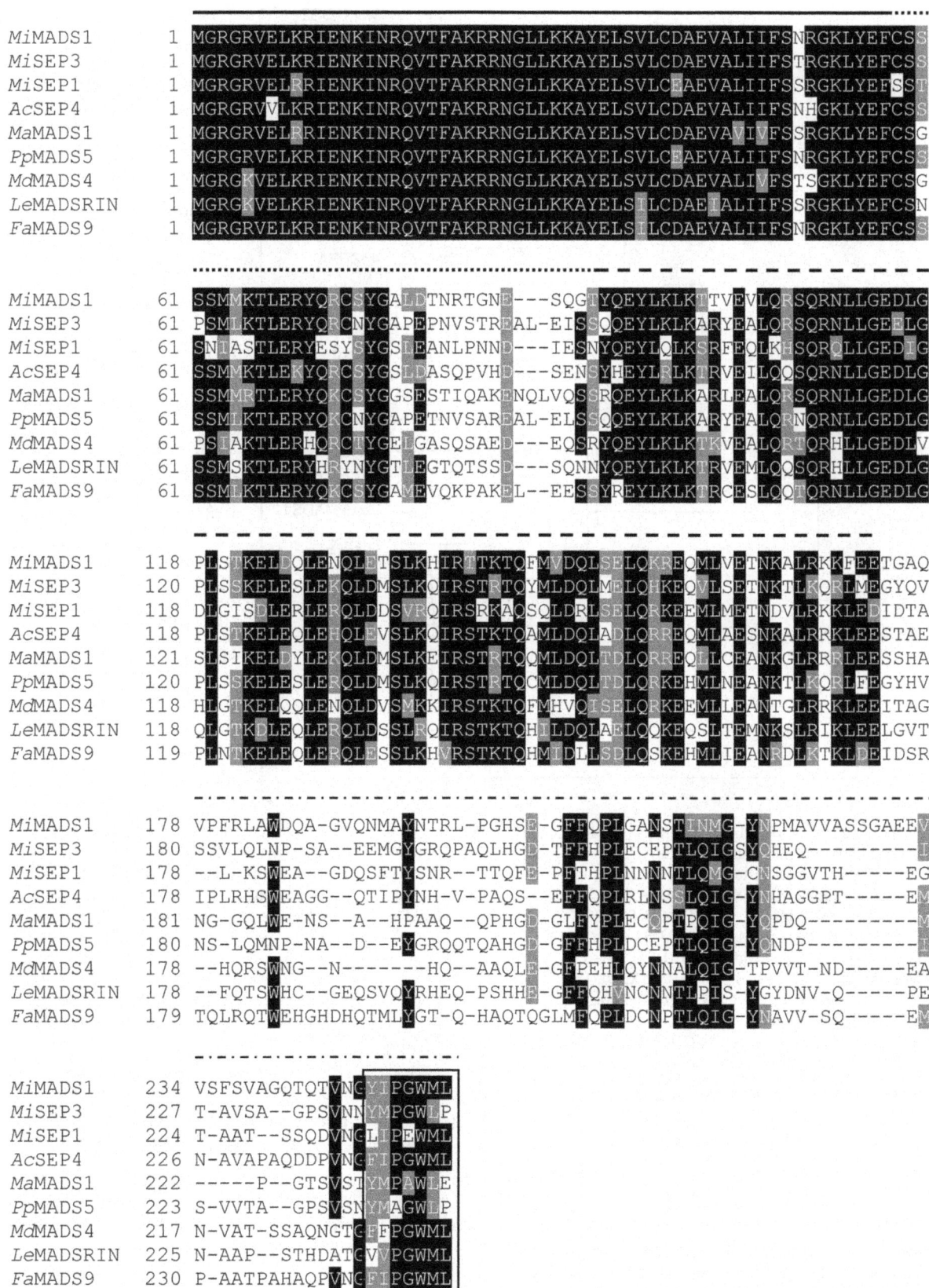

Fig. 2 Sequence alignment of selected MADS-box proteins: MiMADS (KF214778) this work, MiSEP3 (AEO45959) (mango), MiSEP1 (ADX97328) (mango), AcSEP4 (ADU15479) (kiwifruit), MaSEP3MADS1 (ACJ64679) (banana), PpMADS5 (AAZ16241) (peach), MdSEP4MADS4 (AAD51423) (apple), LeMADSRINSEP4 (NP_001234670) (tomato), FaMADS9 (XP_004304204) (woodland strawberry)

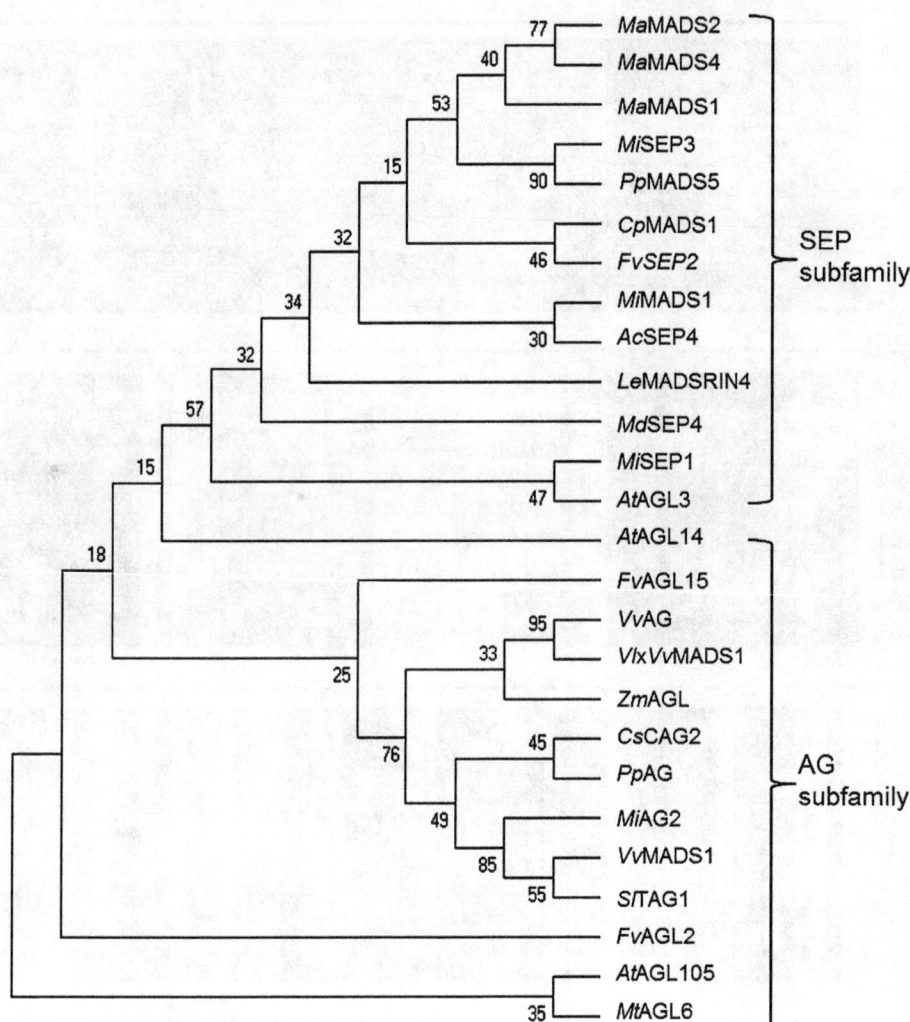

Fig. 3 Phylogenetic analysis of MADS proteins by Maximum Likelihood method. The protein sequences of MADS-box used for the construction of the tree are listed in the GenBank under the following accession numbers: *Mi*SEP3 (AEO45959), *Mi*SEP1 (ADX97328), AcSEP4 (ADU15479), MaSEP3MADS1 (ACJ64679), MaSEP3MADS2 (ACJ64678), MaSEP3MADS4 (ACJ64681), PpSEP3MADS5 (ABO27621), MdSEP4MADS4 (AAD51423), LeMADSRINSEP4 (NP_001234670), FvSEP2 (XP_004304204), CpMADS1 (ACD39982), VvMADS1 (XP_002283924), AtAGL3 (NP_849930), VlxVvAG (ABN46892), SlAGL1 (NP_001234187), AtAGL105 (NP_001119325), AtAGL14, (NP_192925), CsCAG2 (XP_004147393), VlxVvMADS1 (ABN46892), ZmAGAMOUS-like (NP_001105946), FvAGL62-like (XP_004305781), FvAGL15-like (XP_004305660), MtAGL6 (XP_003636892), MiAG2 (AER34989), PpAG (ABU41518). *Mi*, mango, *Mangifera indica; Ac*, kiwi fruit, *Actinidia chinensis; Ma*, banana, *Musa acuminata; Pp*, peach, *Prunus persica; Md*, apple, *Malus domestica; Le*, tomato, *Lycopersum esculentum; Cp*, papaya, *Carica papaya; At*, *Arabidopsis thaliana; Sl*, tomato, *Solanum lycopersicum; Cs*, cucumber, *Cucumis sativus; Vl × Vv*, grape, *Vitis labrusca × Vitis vinifera; Zm*, maize, *Zea mays; Fv*, woodland strawberry *Fragaria vesca; Mt*, *Medicago truncatula; Vv*, grape, *Vitis vinifera*

fold is conserved. Also, the conserved residues R3, V6, N16, K23, R24, K30, K31 and E34 are present and can establish specific contacts to DNA. The cognate fold serum-response factor (SRF) is comprised of a $\alpha - \beta\beta - \alpha$ dimer with an antiparallel sheet dimerization domain (Fig. 4).

The N-terminus was predicted as a loop, which has several residues that interacts with the DNA minor groove and the phosphate backbone. In a lateral view (Fig. 4a), the loops appear bound to the DNA, and in Fig. 4b the loops appear in the shape of a hook contacting the bottom of the minor groove with its N-term amine group. SRF-fold is different to other DNA-binding motifs such as the helix-loop-helix since in a small stretch of amino acids DNA is bound and dimerization is also achieved. A top view (Fig. 4c) shows the interaction between the antiparallel beta sheets, which are locked by one alpha helix from each monomer at 90° from the sheets orientation on top of them. The presence of the dimer has been reported by multi-angle light scattering (Wu et al. 2010) and by gel-filtration chromatography. A rotated lateral view (Fig. 4d) shows very clear the internal symmetry of the MiMADS dimer and the hooks interacting with the minor groove.

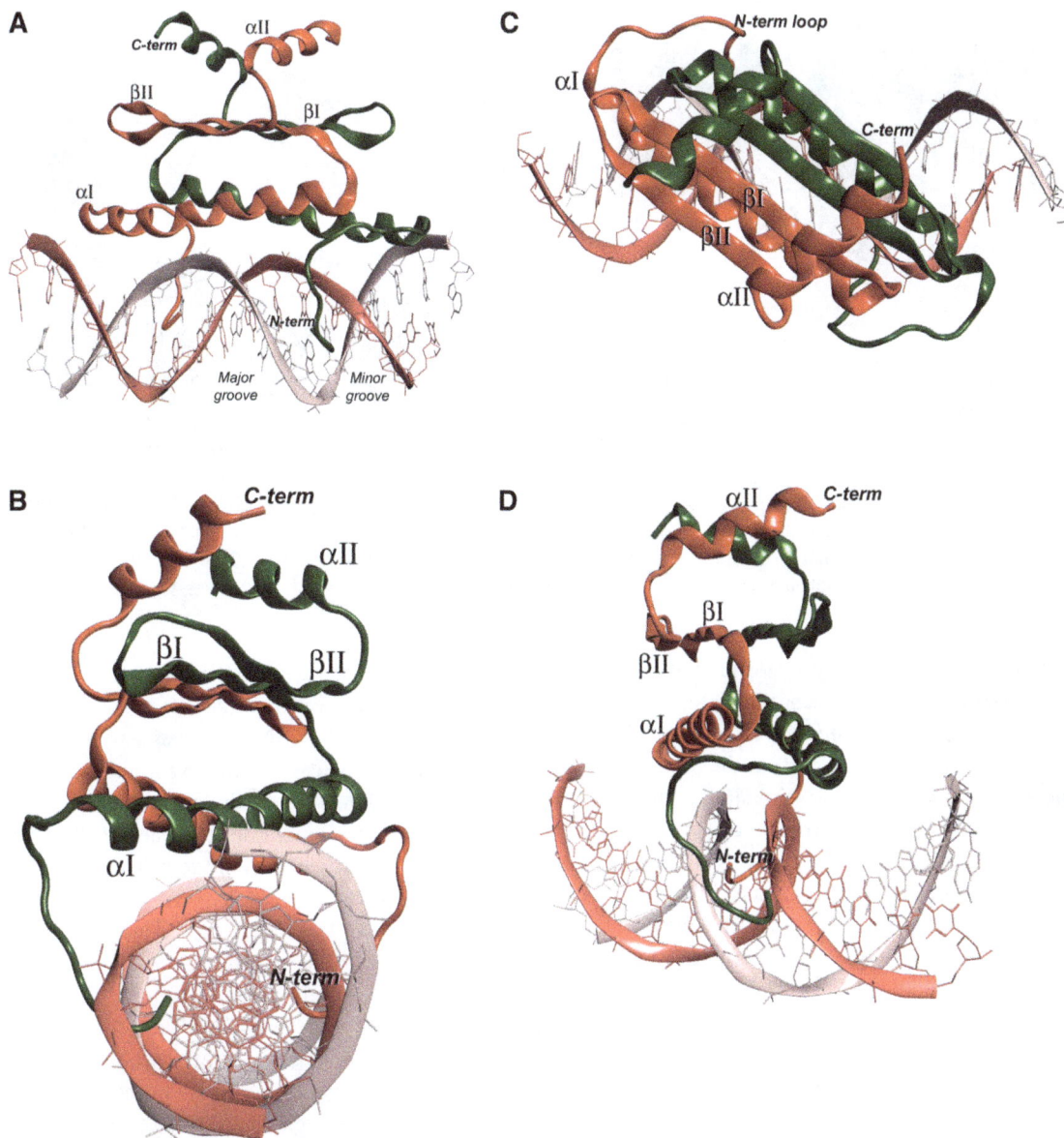

Fig. 4 Ribbon representation of the DNA-binding domain of Mi-MADS1 bound to a double-stranded fragment. Each protein monomer is represented in *red* and *green* and helices and strands are numbered, and DNA on *pink*. **a** Side view of the complex. **b** Lateral view, where the N-terminus of MiMADS1 contacts the DNA **c** Top view, depicting the β-strand interaction between monomers. **d** Rotated lateral view showing the interaction of N-terminus with minor groove

It is known that MEF2A interacts with DNA mainly to the minor groove by contacts mediated from the N-terminal loop and helix αI (Fig. 5). The interactions between Mi-MADS1 and the two strands of DNA are mainly to the phosphate backbone, as found for the basic side chain of R5, R24, K30 and K31. The K31 amine side chain group established two H-bond contacts to phosphate groups at the two DNA strands, narrowing the minor groove cavity. The two K31 residues from the two monomers of MiMADS1 are adjacent from each other, possibly making a strong contribution to the DNA backbone stretching. E34 is other conserved residue found in MADS-box proteins. Its

function is to coordinate an H-bond network through its carboxyl side chain that anchors K30, K31 and R24′ (from the other monomer) side chains in place, so it may be important for MiMADS1 dimerization and DNA backbone stretching. At the center of the helix αI is located the invariant G27, that it is important for preventing the steric hindrance at the close proximity to the DNA backbone (Fig. 5).

The conserved R3 side chain extends at the minor groove bottom and interacts to the nitrogen bases of DNA. The main interactions are through H-bonding to two adjacent pyrimidine rings at one strand of DNA and a ribose ring. At the same time, K23 contacts the purine rings

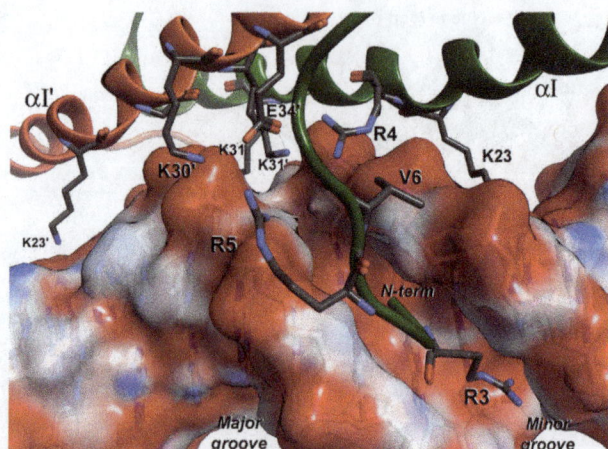

Fig. 5 Details of the protein–DNA interactions predicted by molecular modeling. Arginines (R) and lysines (K) are critical for interaction with the phosphate DNA backbone. Prime numbering corresponds to the second monomer. *Red* color on the surface of DNA corresponds to negative charges of the phosphodiester bonds and *blue* corresponds to positive charges

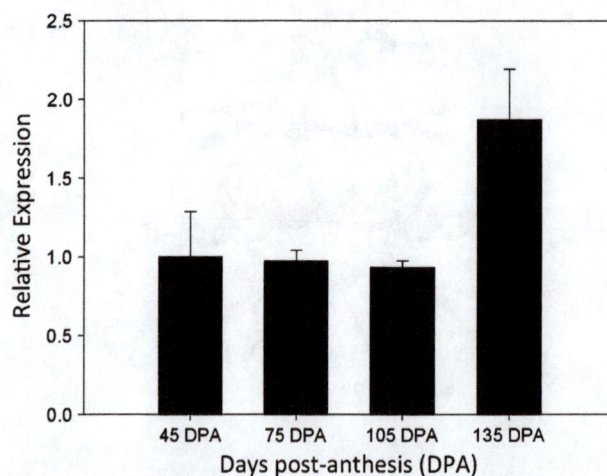

Fig. 6 Expression of *MiMADS1* at different developmental stages of mango fruit (45, 75, 105 and 135 days post-anthesis). Fruits reached their physiological maturity at 135 DPA. *Bars* show the mean ± ES of three measurements ($n = 3$)

from the complementary strand at the major groove. Both R3 and K23 are the main residues that contribute to binding specificity at the consensus sequence. The N-term G2 residue is another invariant residue that lie at the bottom of the minor groove and contacts to pyrimidine rings in the consensus sequence. V6 makes hydrophobic contacts to methylene carbons from the DNA backbone (Fig. 5). We should emphasize that this is a theoretical model and not an experimental crystal structure, but the model is consistent with the ionic charge interactions and this will wait to be demonstrated by the crystallization of MiMADS with a cognate DNA box.

Expression of *MiMADS1* during mango development

The expression pattern of *MiMADS*1 remained constant during development of mango and increased at mango physiological maturity (135 days post harvest, DPA) (Fig. 6). These results suggest a role of MiMADS1 for development of mango fruits and also a function during ripening. MiMADS1 could be up-regulating expression of genes that encode enzymes involved in ethylene biosynthesis based on the mango fruit from the developmental stage that showed higher expression levels (135 DPA). The rise in mRNA levels of *MiMADS1* at onset of ripening could be explained for auto regulation of its own expression. Fujisawa et al. (2011) showed that tomato MADS-RIN controlled its own expression and also they observed an increase of its expression at onset of ripening. *MADS-box* gene expression studies concerning development of fruits are scarce; nevertheless, there are reports of expression of *MADS-box* from banana mesocarp where it was found that *MADS-box* genes regulate ethylene synthesis

(Liu et al. 2009). These results suggest that *MADS-box* genes have a function on fruit climatery and therefore in fruit ripening. However, it is necessary to test this hypothesis using tools like chromatin immunoprecipitation to confirm the regulation of ethylene pathway genes for MiMADS1. Also, it is important to evaluate the levels of the MiMADS protein since they are not necessarily correlated with the mRNA levels (Vogel and Marcotte 2012) and that specific post transcriptional or translational regulation exists during mango ripening.

Until now, we found this cDNA encoding a MADS-box protein from mango fruit cv. "Keitt" and in the future it will be possible to pursue some of the above mentioned experiments. It is also possible to study the mRNA levels of *MiMADS1* during organ and flower development in mango in future studies, as initially described in *Arabidopsis*.

Conclusion

We have identified a MADS-box transcription factor that could be important for mango fruit development, MiMADS1. Molecular modeling predicts that all-important residues are present for interactions in the minor groove of cognate double-stranded DNA. Future research on DNA-protein recognition and interaction will help to determine the MiMADS1 role within the transcriptional machinery acting during mango fruit development and ripening.

Acknowledgments Dr. Islas-Osuna is very grateful for financial support for this study by the CONACYT (Mexico's Consejo Nacional de Ciencia y Tecnología) grants CB2012-01-178296. Dr. Arvizu-Flores thanks grant PROMEP/103.5/12/3590-PTC-155 for supporting computational and molecular modeling software. M.A. Pacheco-Sánchez thanks for a PhD scholarship to CONACYT. Authors thank Emmanuel Aispuro-Hernández, M.Sc. for technical support and Drs. Rogerio R. Sotelo-Mundo and Adriana Muhlia-Almazan for helpful revisions and suggestions.

Conflict of interest The authors declare that they have no competing interests.

References

Aswath CR, Kim SH (2005) Another story of MADS-box genes: their potential in plant biotechnology. Plant Growth Regul 46:177–188

Choudhury SR, Roy S, Nag A, Singh SK, Sengupta DN (2012) Characterization of an AGAMOUS-like MADS Box protein, a probable constituent of flowering and fruit ripening regulatory system in banana. PLoS ONE 7:e44361

Cseke L, Podila G (2004) MADS-box genes in dioecious aspen II: a review of MADS-box genes from trees and their potential in forest biotechnology. Physiol Mol Biol Plants 10:7–28

Elitzur T, Vrebalov J, Giovannoni JJ, Goldschmidt EE, Friedman H (2010) The regulation of MADS-box gene expression during ripening of banana and their regulatory interaction with ethylene. J Exp Bot 61:1523–1535

Felsenstein J (1985) Confidence limits on phylogenies: an approach using the bootstrap. Evolution 38:783–791

Fujisawa M, Nakano T, Ito Y (2011) Identification of potential target genes for the tomato fruit-ripening regulator RIN by chromatin immunoprecipitation. BMC Plant Biol 11:26

Han A, He J, Wu Y, Liu JO, Chen L (2005) Mechanism of recruitment of class II histone deacetylases by myocyte enhancer factor-2. J Mol Biol 345:91–102

Henschel K, Kofuji R, Hasebe M, Saedler H, Münster T, Theißen G (2002) Two ancient classes of MIKC-type MADS-box genes are present in the moss *Physcomitrella* patens. Mol Biol Evol 19:801–814

Huang K, Louis JM, Donaldson L, Lim FL, Sharrocks AD, Clore GM (2000) Solution structure of the MEF2A-DNA complex: structural basis for the modulation of DNA bending and specificity by MADS-box transcription factors. EMBO J 19:2615–2628

Immink RG, Tonaco IA, de Folter S, Shchennikova A, van Dijk AD, Busscher-Lange J et al (2009) SEPALLATA3: the 'glue' for MADS box transcription factor complex formation. Genome Biol 10:R24

Ito Y, Kitagawa M, Ihashi N, Yabe K, Kimbara J, Yasuda J et al (2008) DNA-binding specificity, transcriptional activation potential, and the rin mutation effect for the tomato fruit-ripening regulator RIN. Plant J 55:212–223

Kaufmann K, Melzer R, Theißen G (2005) MIKC-type MADS-domain proteins: structural modularity, protein interactions and network evolution in land plants. Gene 347:183

Leseberg CH, Eissler CL, Wang X, Johns MA, Duvall MR, Mao L (2008) Interaction study of MADS-domain proteins in tomato. J Exp Bot 59:2253–2265

Liu J, Xu B, Hu L, Li M, Su W, Wu J et al (2009) Involvement of a banana MADS-box transcription factor gene in ethylene-induced fruit ripening. Plant Cell Rep 28:103–111

Livak KJ, Schmittgen TD (2001) Analysis of relative gene expression data using real-time quantitative PCR and the $2^{-\Delta\Delta CT}$ method. Methods 25:402–408

Lopez-Gomez R, Gomez-Lim M (1992) A method for extracting intact RNA from fruits rich in polysaccharides using ripe mango mesocarp. HortScience 27:440–442

Notredame C, Higgins DG, Heringa J (2000) T-Coffee: a novel method for fast and accurate multiple sequence alignment. J Mol Biol 302:205–217

Sambrook JR, David W (2001) Molecular cloning. A laboratory manual, 3rd edn. Cold Spring Harbor Laboratory, NY

Santelli E, Richmond TJ (2000) Crystal structure of MEF2A core bound to DNA at 1.5 A resolution. J Mol Biol 297:437–449

Smaczniak C, Immink RG, Angenent GC, Kaufmann K (2012) Developmental and evolutionary diversity of plant MADS-domain factors: insights from recent studies. Development 139:3081–3098

Tadiello A, Pavanello A, Zanin D, Caporali E, Colombo L, Rotino GL et al (2009) A PLENA-like gene of peach is involved in carpel formation and subsequent transformation into a fleshy fruit. J Exp Bot 60:651–661

Tamura K, Peterson D, Peterson N, Stecher G, Nei M, Kumar S (2011) MEGA5: molecular evolutionary genetics analysis using maximum likelihood, evolutionary distance, and maximum parsimony methods. Mol Biol Evol 28:2731–2739

Theißen G, Saedler H (2001) Plant biology: floral quartets. Nature 409:469–471

Varkonyi-Gasic E, Moss SM, Voogd C, Wu R, Lough RH, Wang YY et al (2011) Identification and characterization of flowering genes in kiwifruit: sequence conservation and role in kiwifruit flower development. BMC Plant Biol 11:72

Vogel C, Marcotte EM (2012) Insights into the regulation of protein abundance from proteomic and transcriptomic analyses. Nat Rev Genet 13:227–232

Weigel D, Meyerowitz EM (1994) The ABCs of floral homeotic genes. Cell 78:203–209

Wu Y, Dey R, Han A, Jayathilaka N, Philips M, Ye J et al (2010) Structure of the MADS-box/MEF2 domain of MEF2A bound to DNA and its implication for myocardin recruitment. J Mol Biol 397:520–533

Xu Y, Zhang L, Xie H, Zhang Y-Q, Oliveira MM, Ma R-C (2008) Expression analysis and genetic mapping of three SEPALLATA-like genes from peach (*Prunus persica* (L.) Batsch). Tree Genet Genomes 4:693–703

Yao J-L, Dong Y-H, Kvarnheden A, Morris B (1999) Seven MADS-box genes in apple are expressed in different parts of the fruit. J Am Soc Hortic Sci 124:8–13

19

Molecular authentication of *Cissampelos pareira* L. var. *hirsuta* (Buch.-Ham. ex DC.) Forman, the genuine source plant of ayurvedic raw drug *'Patha'*, and its other source plants by ISSR markers

Deepu Vijayan · Archana Cheethaparambil · Geetha Sivadasan Pillai · Indira Balachandran

Abstract *Cissampelos pareira* L. var. *hirsuta* (Buch.-Ham. ex DC.) Forman belongs to family Menispermaceae. The roots of this taxon are used in the treatment of various diseases like stomach pain, fever, skin disease, etc., in Ayurveda and is commonly known as *Patha*. Two other species, viz., *Cyclea peltata* (Lam.) Hook.f. & Thomson and *Stephania japonica* (Thunb.) Miers of the same family are being used as the source of this drug in various parts of India. This type of substitution or adulteration will ultimately affect the therapeutic efficacy of the medicines adversely. ISSR profiles of all the three taxa are generated and analyzed to assess the genetic relationships among these three species. The profiles of all the three species displayed a high level of polymorphism among them. ISSR markers developed can be used in authenticating and validating the exact species discrimination of the genuine raw drug of *'Patha'* from its substitutes/adulterants to guarantee the quality and legitimacy of this drug in the market.

Keywords Adulteration · Substitution · Molecular markers · *Cyclea peltata* · *Stephania japonica*

Abbreviations
API   Ayurvedic Pharmacopoeia of India
CTAB   Cetyl trimethyl ammonium bromide
DNA   Deoxyribonucleic acid
ISSR   Inter simple sequence repeats
PCR   Polymerase chain reaction

A. Cheethaparambil · G. S. Pillai · I. Balachandran
Crop Improvement and Biotechnology Division, Centre for Medicinal Plants Research, Arya Vaidya Sala, Kottakkal, Malappuram 676503, Kerala, India

Present Address:
D. Vijayan (✉)
Botanical Survey of India, Eastern Regional Centre, Shillong 793003, India
e-mail: deepundd@gmail.com

Introduction

The roots of *Cissampelos pareira* L. var. *hirsuta* (Buch.-Ham. ex DC.) Forman, *Cyclea peltata* (Lam.) Hook.f. & Thomson and *Stephania japonica* (Thunb.) Miers of family Menispermaceae are known as *Patha* in Ayurveda. They are used in the treatment of various diseases like stomach pain, fever, skin conditions, cardiac pain, etc., among which *C. pareira* var. *hirsuta* is the accepted source in Ayurveda (API 2001; Yoganarsimhan 1996). However, the plants of *C. peltata* are used as *Patha* in Kerala (Warrier et al. 1994). Authentication of raw medicinal plants is a fundamental requirement for quality assurance in herbal drug markets. Certain rare and expensive medicinal plant species are often adulterated or substituted by morphologically similar, easily available or less expensive species. Pharmaceutical companies procure plant materials from traders, who gather them from untrained collectors in the rural and forest areas. This has given rise to widespread adulteration or substitution, leading to poor quality of herbal formulations (Mehrotra and Rawat 2000). Herbal medicinal products may vary in composition and properties, unlike conventional pharmaceutical products, which are usually prepared from synthetic, chemically pure materials by means of reproducible manufacturing techniques and procedures. Correct identification and quality assurance of the starting material is, therefore, an essential prerequisite to ensure reproducible quality of herbal medicine, which contributes to its safety and efficacy (De Smet 2002; Straus 2002).

The morphological, biochemical or histological characteristics employed in the identification are prone to different environmental conditions (Kiran et al. 2010). Limitations of these markers for authentication of herbal drugs have generated a need to develop more reproducible molecular markers for quality control of these medicinal herbs. In view of these limitations, there is need for a new approach that can complement or, in certain situations, serve as an alternative. DNA markers are reliable for informative polymorphisms as the genetic composition is unique for each species and is not affected by age, physiological conditions as well as environmental factors (Chan 2003). DNA-based molecular markers have proved their utility in various fields of science and recently researchers have tried to explore the application of these markers in pharmacognostic characterization of herbal medicine. DNA-based techniques have been widely used for authentication of plant species of medicinal importance. This is especially useful in case of those that are frequently substituted or adulterated with other species or varieties that are morphologically and/or phytochemically indistinguishable. Attempts have been made to compare the source plants of *Patha* using anatomical and phytochemical markers (Hullatti and Sharada 2007, 2010). But, so far nobody has attempted to characterize these three species at the molecular level. The main objective of the present study is to evaluate the source plants of *Patha* using inter

simple sequence repeat (ISSR) markers for the accurate identification of *C. pareira* var. *hirsuta, C. peltata* and *S. japonica*.

Materials and methods

Plant materials and DNA extraction

The total genomic DNA was extracted from the fresh leaves of *C. pareira* var. *hirsuta, C. peltata* and *S. japonica* grown at the Centre for Medicinal Plants Research, Arya Vaidya Sala, Kottakkal, Kerala, India, using a modified CTAB method (Doyle and Doyle 1987). The leaf samples (0.2 g) were ground in liquid nitrogen using mortar and pestle and re-suspended in 2 mL of DNA extraction buffer [2 % CTAB, 1 % PVP, 1.4 M NaCl, 20 mM EDTA (pH 8.0), 100 mM Tris–HCl (pH 8.0) and 0.2 % β-mercaptoethanol]. The mixture was incubated at 65 °C for 30 min and centrifuged with equal volume of chloroform:isoamylalcohol (24:1) at 8,000 rpm for 10 min. The supernatant was transferred to fresh tubes and DNA was precipitated by adding equal volume of ice cold isopropanol. The DNA precipitate was washed with 70 % ethanol, air dried and stored in 500 μL TE buffer. RNA was eliminated by treating the samples with RNase A (10 mg/mL) at 37 °C for 30 min. Again, it was extracted with equal volume of

Table 1 Total number of bands and percentage of polymorphism amplified by 19 ISSR primers

Sl. no.	Primer	Annealing temperature	Total number of bands	Number of polymorphic bands	Number of monomorphic bands	Polymorphism (%)
1	(ACTG)$_4$	49	11	10	1	90.91
2	UBC 807	50	9	8	1	88.89
3	UBC 808	52	17	17	0	100
4	UBC 810	50	16	16	0	100
5	UBC 825	50	14	13	1	92.86
6	UBC 840	54	17	17	0	100
7	UBC 841	54	12	12	0	100
8	UBC 842	54	10	8	2	80
9	UBC 847	52	16	13	3	81.25
10	UBC 851	54	16	14	2	87.5
11	UBC 854	56	13	12	1	92.31
12	UBC 855	52	19	19	0	100
13	UBC 857	54	15	14	1	93.33
14	UBC 859	54	8	7	1	87.50
15	UBC 862	55	13	13	0	100
16	UBC 873	52	12	9	3	75
17	UBC 881	59	19	19	0	100
18	UBC 890	59	20	18	2	90
19	UBC 900	59	13	13	0	100
		Total	270	252	18	93.33

chloroform:isoamylalcohol (24:1). Two volumes of cold ethanol were added in aqueous layer and centrifuged at 12,000 rpm for 10 min. Pellets were washed with 70 % alcohol, air dried and dissolved in TE buffer. DNA quantification as well as quality assessment was carried out spectrophotometrically using Biophotometer (Eppendorf—AG 22331, Germany). The DNA sample was also quantified on 0.8 % agarose gel electrophoresis.

ISSR-PCR

Nineteen ISSR primers (Eurofins, Bangalore, India) were selected for the present investigation. PCR for amplifying the DNA preparations was carried out in a 25-μl volume of reaction mixture. A reaction tube contained 25 ng of DNA, 1 U of *Taq* DNA polymerase enzyme, 2.5 mM of each dNTPs, 1× *Taq* buffer with 25 mM $MgCl_2$ and 25 pmol primers. Amplifications were carried out in a DNA thermal cycler (Eppendorf, mastercycler gradient) using following parameters: 94 °C for 5 min; 35 cycles at 94 °C for 1 min, 49–59 °C for 1 min, and 72 °C for 2 min; and a final extension at 72 °C for 10 min. The annealing temperature was adjusted to a range of 49–59 °C depending on GC content and length of the primers. PCR products were subjected to agarose gel [1.5 % (w/v)] electrophoresis in 1× TBE buffer, along with 1 kb DNA ladder (Fermentas Life Sciences) as size markers. DNA was stained with ethidium bromide and electrophoretic profile was photographed on gel documentation system (Alpha Innotech, USA).

Results and discussion

ISSR fingerprinting of 19 primers generated 270 scorable fragments, out of which 252 were polymorphic (93.33 % polymorphism) (Table 1). The primers viz., UBC 808, 810, 840, 841, 855, 862, 881 and 900 exhibited 100 percent polymorphism. The number of polymorphic fragments for each primer varied from 7 to 19 with an average of 13.26 polymorphic fragments (Fig. 1). ISSR banding pattern exhibited a maximum of 133 fragments in *S. japonica* followed by *C. pareira* var. *hirsuta* (132 fragments) and *C. peltata* (119). A maximum of eight unique bands were observed in *C. pareira* var. *hirsuta* while screening with the primer UBC 851. Interestingly, the number of unique bands in *C. peltata* was less compared to other plants. But in *S. japonica*, six and seven unique bands were observed while screening with the primers UBC 890 and UBC 900, respectively. Here also, the total number of unique bands was highest in *C. pareira* var. *hirsuta* (Table 2).

　　Morpho-anatomical studies of roots of *C. pareira* var. *hirsuta*, *C. peltata* and *S. japonica* clearly indicate the

significant differences among themselves (Hullatti and Sharada 2007). Hullatti and Sharada (2010) reported the HPTLC and HPLC fingerprinting of *Patha* plants and their studies clearly indicate the significant differences among the three plant materials. They concluded that all the studied parameters clearly indicate that roots of *C. pareira* var. *hirsuta* are the genuine source of *Patha* as it fulfills the ayurvedic claims of this drug. The present investigation confirms that there is no genetic relationship among *C. pareira* var. *hirsuta*, *C. peltata* and *S. japonica*. The high level of polymorphism and unique fragments detected in ISSR profiles authenticate that there exists a wide genetic disparity among the selected plants. The findings of the present study substantiate the reports on the

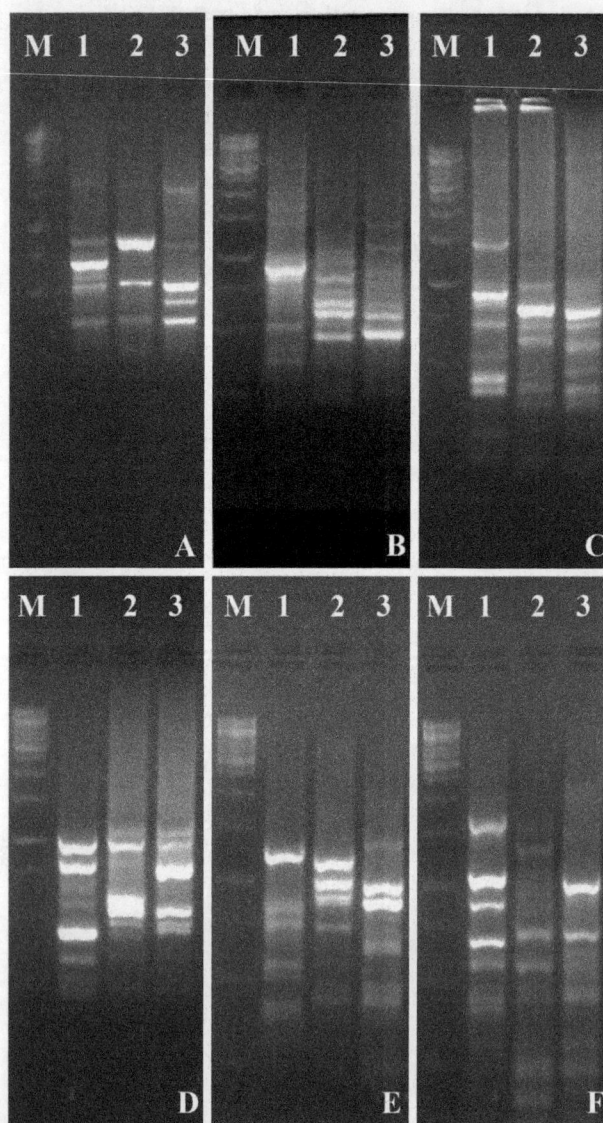

Fig. 1 Representative ISSR profiles of *Patha* plants using primers **a** (ACTG)₄, **b** UBC 840, **c** UBC 847, **d** UBC 857, **e** UBC 881 and **f** UBC 890. *M* represents 1 kb ladder, *lane 1 Cissampelos pareira* var. *hirsuta, lane 2 Cyclea peltata* and *lane 3 Stephania japonica*

Table 2 Primerwise ISSR banding pattern of *Patha* plants

Primer	Cissampelos pareira var. hirsuta				Cyclea peltata				Stephania japonica			
	TB	PB	MB	UB	TB	PB	MB	UB	TB	PB	MB	UB
(ACTG)₄	6	5	1	3	4	3	1	1	6	5	1	3
UBC 807	4	3	1	0	7	6	1	4	3	2	1	1
UBC 808	9	9	0	5	4	4	0	1	9	9	0	5
UBC 810	7	7	0	4	10	10	0	5	6	6	0	0
UBC 825	9	8	1	6	6	5	1	1	6	5	1	1
UBC 840	9	9	0	3	10	10	0	3	7	7	0	2
UBC 841	3	3	0	2	4	4	0	4	6	6	0	5
UBC 842	4	2	2	0	7	5	2	2	7	5	2	2
UBC 847	9	6	3	5	7	4	3	2	9	6	3	3
UBC 851	11	9	2	8	7	5	2	1	6	4	2	1
UBC 854	5	4	1	2	7	6	1	3	6	5	1	4
UBC 855	9	9	0	7	6	6	0	1	10	10	0	5
UBC 857	9	8	1	5	7	6	1	1	8	7	1	1
UBC 859	4	3	1	3	4	3	1	1	4	3	1	1
UBC 862	6	6	0	4	6	6	0	3	5	5	0	2
UBC 873	7	4	3	4	5	2	3	2	6	3	3	3
UBC 881	9	9	0	6	8	8	0	3	9	9	0	3
UBC 890	9	7	2	5	7	5	2	3	12	10	2	6
UBC 900	3	3	0	3	3	3	0	2	8	8	0	7
Total	132	114	18	75	119	101	18	43	133	115	18	55

TB number of total bands, *PB* number of polymorphic bands, *MB* number of monomorphic bands, *UB* number of unique bands

applicability of ISSR markers to authenticate the herbal medicinal materials from its adulterants (Shen et al. 2006; Wang 2011). This investigation is the first report on the study of genetic relationships of *Patha* group of plants using ISSR markers.

Acknowledgments Authors are grateful to the Management, Arya Vaidya Sala, Kottakkal and TATA Trust, Mumbai, for providing the facilities for taking up the programme. Technical help rendered by Mr. P. Vipin is also gratefully acknowledged.

Conflict of interest The authors declare that there is no conflict of interest.

References

API (2001) The ayurvedic pharmacopoeia of India PART-I vol I. Department of ISM & H, Ministry of Health and Family Welfare, Govt. of India, pp 122–123

Chan K (2003) Some aspects of toxic contaminants in herbal medicines. Chemosphere 52:1361–1371

De Smet PAGM (2002) Herbal remedies. N Engl J Med 347:2046–2056

Doyle JJ, Doyle JL (1987) A rapid DNA isolation procedure for small quantities of fresh leaf tissue. Phytochem Bull 19:11–15

Hullatti K, Sharada MS (2007) Morpho-anatomical studies of roots of three species of Menispermaceae. J Trop Med Plants 8(1):71–77

Hullatti K, Sharada MS (2010) Comparative phytochemical investigation of the sources of ayurvedic drug Patha: a chromatographic fingerprinting analysis. Indian J Pharm Sci 72(1):39–45

Kiran U, Khan S, Mirza KJ, Ram M (2010) SCAR markers: a potential tool for authentication of herbal drugs. Fitoterapia 81:969–976

Mehrotra S, Rawat AKS (2000) Standardization and quality evaluation of herbal drugs. In: Chauhan DK (ed) Recent trends in botanical research. Dikshit Press, Allahabad, pp 313–332

Shen J, Ding X, Liu D, Ding G, He J, Li X, Tang F, Chu B (2006) Inter simple sequence repeats (ISSR) molecular fingerprinting markers for authenticating populations of *Dendrobium officinale* Kimura et Migo. Biol Pharm Bull 29:420–422

Straus SE (2002) Herbal medicines—what's in the bottle? N Engl J Med 347:1997–1998

Wang X (2011) Inter-simple sequence repeats (ISSR) molecular fingerprinting markers for authenticating the genuine species of rhubarb. J Med Plants Res 5(5):758–764

Warrier PK, Nambier VPK, Ramankutty C (1994) Indian medicinal plants, a compendium of 500 species, vol 2. Orient Longman Private Limited, Arya Vaidya Sala, Kottakkal, pp 277–280

Yoganarsimhan SN (1996) Medicinal plants of India—Karnataka, vol 1. Interline Publishing Pvt. Ltd, Bangalore, pp 447

20

Hybrid purity assessment in *Eucalyptus* F$_1$ hybrids using microsatellite markers

V. Subashini · A. Shanmugapriya · R. Yasodha

Abstract The worldwide expansion of hybrid breeding and clonal forestry is to meet the demands of paper pulp and bioenergy. Although India was one of the pioneers in hybrid production of eucalypts only recently the hybrid clonal forestry is gaining momentum. Inter-specific hybrids are being produced to exploit the hybrid vigor of F$_1$ individuals. Quality control genotyping for hybrid purity and parentage confirmation at the early stage is one of the essential criteria for clonal propagation and field trails for the assessment of growth performance. Eucalyptus being a obligatory outcrossed species with potential to self pollination, possibilities of pollen contamination are high. Hence, in the present study, *Eucalyptus camaldulensis* × *E. tereticornis* inter-specific hybrids were genotyped using 25 fluorescent labeled microsatellite markers available in public domain. Multiplex loading of PCR products was performed successfully for most of the microsatellite loci. Hybrid purity index was calculated and parentage was confirmed. Hybrid purity values ranged from 85 to 100 % showed the efficiency of controlled pollination techniques. A subset of six fully informative simple sequence repeats was identified for routine quality control genotyping for these hybrids. Detection of non-essential genotypes observed among the hybrid seedlings proved the significance of hybrid purity tests and the false hybrids were removed at the seedling stage. The hybrids with proven hybridity will be used for generation of genetic linkage, discovery of quantitative trait loci and the individuals with high productivity can enter into mass clonal multiplication.

Keywords *Eucalyptus tereticornis* · *Eucalyptus camaldulensis* · Inter-specific hybrids · Simple sequence repeats · Genotyping · Parentage confirmation

Introduction

Hybrid forestry in eucalypts is a success story and currently several eucalypt growing countries are adopting hybrid breeding for industrial plantations. The greatest advance in industrial plantation forestry has undoubtedly been in the clonal deployment of F$_1$ hybrid genotypes. Clonal propagation and hybrid breeding have become a powerful combination of tools for the improvement of production and wood quality in eucalypts (Grattapaglia and Kirst 2008). Reports on controlled inter-specific hybrid eucalypts production and their importance in genetic improvement were available since 1960 in India for the commercially important species like *Eucalyptus camaldulensis*, *Eucalyptus tereticornis* and *Eucalyptus grandis* (Venkatesh and Kedharnath 1965; Venkatesh and Sharma 1977, 1979). However, eucalypts breeding program in India mainly focused on improvement of pure species (Varghese et al. 2009) which is obviously a prerequisite for hybrid development. Many earlier hybrid selections in Congo and Brazil were from sporadic hybrids identified from open pollinated seed orchards and later shifted to planned attempts for production of inter-specific hybrids (Potts and Dungey 2004). The Indian land race of *E. tereticornis* (Mysore gum) was evolved from natural hybrid and suffers from hybrid breakdown or inbreeding depression, necessitating the production of targeted crosses for hybrid vigor. A combination of complementary traits from *E. grandis* and *E. urophylla* has great potential for expanding genetic diversity, for improving pulp properties and for introducing

V. Subashini · A. Shanmugapriya · R. Yasodha (✉)
Division of Plant Biotechnology, Institute of Forest Genetics and Tree Breeding, Coimbatore 641002, India
e-mail: yasodha@icfre.org

adventitious rooting traits in F_1 hybrids. Subsequently, in India attempts are being made to generate inter-specific crosses among *E. camaldulensis*, *E. tereticornis* and *E. grandis* with phenotypically characterized parents (http://ifgtb.icfre.gov.in/news&events/Eucalypts%20Hybrid%20Breeding%20programme.pdf). Such hybrids are essential to map genes controlling various economically important traits for early screening of hybrid progenies with the right combination of desirable characteristics and many other molecular genetic studies. Further, many countries working eucalypt breeding programs also contemplate on marker assisted selection (MAS) in eucalypts, which would aid indirect selection of difficult traits at early stage without waiting for next generation. Thus, MAS speeds up the process of conventional breeding and facilitates improvement of traits which is difficult through conventional breeding. Hybridity validation at the seedling stage is one of the essential criteria for inter-specific hybrid production and establishment of true segregating population. The hybrid authentication procedures identify the pollen contamination, out crossing with foreign pollens and physical admixtures during seed handling at storage and nursery raising stages. Thus use of individuals with erroneous hybridity results in segregation of the traits, lower yields and genetic deterioration of clones. Further, exploitation of heterosis in eucalypts requires sustained production of F_1 hybrids and management of hybrid identity during clonal propagation.

Quality control (QC) genotyping is an important aspect in eucalypt improvement program, however, no well-established methods available for implementation in hybrid breeding program. The use of DNA markers particularly Simple Sequence Repeats (SSRs) are useful for a variety of molecular breeding applications because of their codominance, abundance, high genome coverage and multi allelic nature. SSRs are the most suitable markers for hybrid purity assessment as the heterozygosity of the hybrids can be easily determined by the presence of alleles from both the parents used for controlled pollination. They are already a proven tool for hybrid authentication or hybrid purity assessment and parentage confirmation in many crop species (Bohra et al. 2011). Indeed, molecular markers-based hybrid purity tests have been developed and are in routine use in many crop species such as rice (Yashitola et al. 2002; Sundaram et al. 2008), maize (Asif et al. 2006), cotton (Selvakumar et al. 2010) and safflower (Naresh et al. 2009). Similar to food crops, in forest trees also inter-specific hybrids are generated routinely through hybridization and the heterotic individuals are selected for clonal propagation and mass multiplication (Nikles and Griffin 1992; Stanton et al. 2010). SSRs were recommended for species and hybrid identification, certification of controlled crosses in species

such as *Populus* (Rajora and Rahman 2001), Tsuga (Bentz et al. 2002), Picea (Narendrula and Nkongolo 2012) and *Pinus* (Elliott et al. 2006). Recently, in eucalypts, reasonably good numbers of SSR markers become available (Brondani et al. 2006; Acuna et al. 2012; He et al. 2012; Faria et al. 2010) and these SSRs can be used for developing molecular marker based hybrid purity test. Hence, the objectives of this study were to identify SSR markers for confirmation of parentage and hybrid purity tests in *E. tereticornis* × *E. camaldulensis* and identify a subset of highly informative SSR markers for routine and low-cost quality control genotyping.

Materials and methods

Plant material

Juvenile leaves were collected from the parent trees of *E. camaldulensis* (Clone 7) and *E. tereticornis* (Clone 88) developed through the breeding program of Institute of Forest Genetics and Tree Breeding (IFGTB), Coimbatore. In the case of F_1 inter-specific hybrids (*E. camaldulensis* × *E. tereticornis*), the juvenile leaves were collected from 6-month-old plants. Genomic DNA was extracted from 100 mg of the plant sample using Qiagen DNeasy Plant Mini Kit^{TM}. The extracted DNA quality of each sample was assessed using 0.8 % agarose gel and the quantity of DNA was confirmed using Picodrop spectrophotometer (Picodrop microlitre spectrophotometer version 3.01, UK) and DNA concentrations were normalized at 10 ng/μl.

Polymerase chain reaction (PCR)

SSR locus specific forward primer was designed with a 5'-end M13 tail (5' TGT AAA ACG ACG GCC AGT-3') as reported by Schuelke (2000). Initially, to identify the polymorphic SSRs between the parents, 53 SSR markers located on the 11 linkage groups selected from the SSR set developed and mapped by Brondani et al. (2006) from *E. grandis* and one loci from *E. grandis* mapped by Thamarus et al. (2004) were considered. PCR amplification was carried out in 10 μl volume with 10 × buffer containing 100 mM Tris–HCl (pH 8.3), 500 mM KCl and 15 mM $MgCl_2$, 0.4 pmol of each forward (M13 tailed) and reverse primer, 1 U of *Taq* DNA polymerase, and 100 μM of each dNTPs and 20 ng of template DNA. The PCR amplification was carried out for 5 min at 94 °C, 30 cycles of 1 min at 94 °C, 60 or 30 s at the primer specific annealing temperature, 2 min at 72 °C, and 15 min at 72 °C for final extension. Annealing temperatures varied from 48 to 60 °C, to amplify specific microsatellite marker. PCR products were size-separated using an 5 % denaturing polyacrylamide (PAGE) gels of size 21 × 50 cm (Sequi-Gen GT System, BIO-RAD, USA) containing 7 M urea and 1 × TBE buffer, and visualized by silver staining. SSR loci showing polymorphism between parents were used for genotyping of the F_1 hybrid population. PCR amplification was carried out in 10 µl volume with 10 × buffer containing 100 mM Tris–HCl (pH 8.3), 500 mM KCl and 15 mM $MgCl_2$, 125 µM dNTP mix, 0.1 pmol of M13 tailed forward primer, 0.4 pmol of reverse primer and 0.2 pmol of fluorescently labeled [6-FAM™, NED®, VIC®, or PET® (Applied Biosystems, CA, USA)], 10 ng of template DNA and 1 U of *Taq* DNA polymerase. The touchdown PCR program used was as follows: initial denaturation at 94 °C for 5 min, then 30 cycles of 94 °C for 45 s, annealing temperature varying from 48 to 60 °C for 30 s and 72 °C for 1 min and an elongation step at 72 °C for 15 min. After these cycles, an additional round of 20 cycles with 94 °C for 30 s, annealing of labeled M13 at 50 °C for 45 s and 72 °C for 45 s, and final extension of 72 °C for 30 min.

PCR products were diluted specifically for each dye wherein an aliquot of 1 µl of fluorescently labeled PCR product was mixed with 0.2 µl of GeneScan™ 600-LIZ-Size® Standard V2.0. (Applied Biosystems) and 8.8 µl of Hi-Di formamide (Applied Biosystems). The mixture was electroinjected in ABI 3500 genetic analyzer (Applied Biosystems). Data obtained after 45 min of injection were analyzed with Genemapper™ software version 4.1 (Applied Biosystems) and exported as data table.

Multiplexing in the form of multiloading was carried out by mixing two or three SSRs amplified with different fluorescent dye or two SSRs having different fragment sizes amplified with same dye (Tables 1, 2). While loading more than one PCR product, the concentration of the each product was adjusted (0.5 or 1.0 µl) to avoid off-scale peaks. Only two loci were loaded singly.

Statistical analysis

The allele size data obtained from the Genemapper for all loci were analyzed using Powermarker software to estimate Polymorphic Information Content (PIC). Hybrid purity assessment was carried out wherein the SSR allele data for the hybrid seeds were recorded as "A" [allele of male parent (*E. tereticornis*)], "B" [allele of female parent

Table 1 Details of SSR loci used for genotyping of eucalypt hybrids

S. no.	Microsatellite	Repeat motif	Annealing temperature (in °C)	Size (bp)	Fluorochrome[A]
1	Embra12	$(AG)_{22}$	59	130–150	VIC
2	Embra36	$(AG)_{29}$	55	150–170	FAM[a]
3	Embra98	(AG)8(G)6(AG)3AA… (GA)20	56	240–270	FAM[b]
4	Embra122	$(GA)_{26}$	56	230–280	FAM[a]
5	Embra147	(AG)n	56	184–210	FAM
6	Embra89	(CT)17	54	300–310	FAM[c]
7	Embra33	(AG)19	58	125–160	FAM[b]
8	Embra101	$(AG)_{12}A(AG)_8$	55	130–155	FAM[c]
9	Embra227	$(GA)_{13}$	62	305–340	FAM[d]
10	Embra206	(GA)8AA(GA)11	57	305–335	FAM
11	Embra6	$(AG)_{19}$	58	140–155	FAM[d]
12	Embra186	(GA)27	58	160–190	NED
13	Embra20	$(AG)_{19}$	59	145–170	PET
14	Embra58	$(AG)_{20}$	59	155–180	VIC
15	Embra213	(CT)17	53	220–240	PET
16	Embra28	$(AG)_{25}$	56	190–210	FAM
17	Embra97	(CT)21	58	130–150	NED
18	Eg16	$(GA)_{12}(TGA)_2$	60	245–260	FAM
19	Embra179	$(GA)_9$	56	145–160	VIC
20	Embra303	(CT)20	56	270–300	NED
21	Embra53	(AG)17GT(AG)5	60	130–150	NED
22	Embra304	(GA)35(GGA)4	54	225–245	VIC
23	Embra52	(AG)2T(AG)26	56	120–160	VIC
24	Embra18	$(AG)_3GG(AG_{19})$	52	130–140	NED
25	Embra167	$(TC)_{19}$	48	120–140	FAM

[A] Fluorochrome with same letters was multiplexed together for electrophoresis

Table 2 Combination SSR loci used for multiplex loading of SSR-PCR

S. no.	Fluorochrome			
	FAM	NED	VIC	PET
1	Eg16	Embra97	Embra304	–
2	–	Embra53	Embra179	–
3	–	Embra18	Embra52	–
4	Embra167	Embra303	–	–
5	Embra28	Embra120	–	Embra20
6	–	Embra186	Embra12	Embra213

(*E. camaldulensis*)] and "H" (alleles from both the parents "Hybrid") format. Purity index for each marker was calculated using scored data by applying the following formula (Bohra et al. 2011):

$$\text{Purity index } (\%) = \frac{\text{Number of true hybrids(containing alleles of both the parents)}}{\text{Total number of hybrid seeds tested}} \times 100$$

eucalypts in Brazil wherein for eucalypts clonal protection requests are accompanied by a multilocus DNA profile (DNA fingerprint) of 15–20 microsatellite markers that were recommended based on several aspects such as robustness, polymorphic information content and general availability in the public domain (Grattapaglia 2007) and South Africa (http://www.forestry.co.za/application-of-plant-breeders-rights/).

Preliminary results showed that among the 53 SSR loci analyzed, 25 loci were suitable for further studies and remaining 28 loci showed either large number of non-essential genotypes or no amplification products with more than 20 % of F_1 individuals. These 28 loci required further standardization in fluorescent labeled PCR and hence discarded for further analysis. The details of SSR loci and the information on PCR amplification status, genotype pattern, hybrid purity index and the polymorphic information

Results and discussion

In agricultural crops, hybrid genetic purity is a routine exercise and DNA markers are often used for quality control genotyping (Wu et al. 2006; Selvakumar et al. 2010; Mbanjo et al. 2012; Semagn et al. 2012). In forest tree species, intentional hybrid breeding is restricted to few species like eucalypts (Potts and Dungey 2004; Dickinson et al. 2013), *Acacia* (Griffin et al. 2010), *Populus* (Yu et al. 2001) and *Pinus* (Cappa et al. 2013). Hybrid seed production in eucalypts is generally takes the advantage of the natural protandry and controlled pollination (CP) is usually carried out by hand emasculation and pollination. During this process, the genetic purity of the F_1 seeds is significantly affected by foreign pollen or self pollen. The high cost of performing CP activity and relatively low seed production coupled with large scale clonal propagation of specific hybrids demands the quality control for parentage. Unlike agricultural crops, in tree species, Grow Out Tests (GOTs) for morphological assessment cannot be practiced due to the long duration. The DNA markers such as SSRs could be of excellent choice for hybridity confirmation in eucalypts, which has good resources of SSRs. Hybrid purity assessment would also support the stringent intellectual property requirements for variety registration under plant protection of varieties and farmer's right authority of India (PPV & FRA); and the SSR marker technologies have already been in use for accurate identification of

content for each locus is given in Table 3. Among the selected 25 loci, the SSR primer set Eg16 could not amplify 16 F_1 individuals and EMBRA303 produced maximum number of (27) non-essential genotypes, i.e., non-specific PCR products. The PIC values were ranging from 0.4 to 0.8.

Steenkamp et al. (2003) made an attempt to detect putative hybrids in commercially propagated eucalyptus using 5S rDNA sequence, however, it appears to be unsuitable for constructing species–specific PCR-primers because of the variable sites in the 5S repeat. Similarly, Cupertino et al. (2009) used six SSRs for parentage analysis in multiple hybrid families of eucalypts and found pollen contamination and mislabeling thus warranting parentage confirmation for all controlled crosses. In the present study, hybrid purity assessment was carried out for the F_1 hybrid progenies obtained from *E. camaldulensis* × *E. tereticornis* using the EMBRA SSR markers. The fragment analysis was performed as a multi-loading assay, analyzing two or three loci simultaneously that was labeled by different fluorescence dyes or with loci of different fragment sizes amplified with single fluorescence dye (Fig. 1). This could increase the amount of information generated per assay, and reduced consumable costs. Cost effective genotyping by fluorescence allowed more rapid data collection in many plant breeding experiments compared to earlier methods such as those based on radioactive isotopes and silver staining (Guichoux et al. 2011). Hybrid

Table 3 Hybrid purity index and polymorphic information content of different SSR loci used for genotyping *Eucalyptus* mapping population

Marker name	Genotype of parents	Number of individuals						Purity of designated hybrid (%)	PIC
		Lacking amplified product	Amplified product	Having allele from parent A	Having allele from parent B	Identified as hybrid	Non-essential genotypes		
Embra12	abxac	4	100	0	0	100	0	100.00	0.5
Embra36	abxcd	4	100	0	0	100	0	100.00	0.7
Embra98	abxcd	5	99	0	0	98	1	98.99	0.6
Embra122	abxcc	1	103	0	2	101	0	98.06	0.5
Embra147	abxcc	2	102	2	0	100	0	98.04	0.6
Embra89	abxaa	5	99	0	0	97	2	97.98	0.4
Embra33	abxbc	6	98	0	0	96	2	97.96	0.5
Embra101	abxac	2	102	0	0	101	1	99.02	0.5
Embra227	abxcd	8	96	0	2	94	0	97.92	0.7
Embra206	abxbc	6	98	0	0	98	0	100.00	0.5
Embra6	abxcd	5	99	3	0	95	1	95.96	0.7
Embra186	abxcd	8	96	2	0	94	0	97.92	0.5
Embra20	abxcd	2	102	0	2	100	0	98.04	0.7
Embra58	abxcd	13	91	0	2	89	0	97.80	0.7
Embra213	abxcd	3	101	1	0	100	0	99.01	0.7
Embra28	abxcd	3	101	6	0	90	5	89.11	0.8
Embra97	aaxbb	14	90	2	2	85	1	94.44	0.4
Eg16	abxac	16	88	0	0	88	0	100.00	0.6
Embra179	abxbc	13	91	0	0	91	0	100.00	0.6
Embra303	abxcd	8	96	2	0	90	4	93.75	0.8
Embra53	abxcd	8	96	0	1	93	2	96.88	0.7
Embra304	abxac	8	96	0	0	95	1	98.96	0.5
Embra52	abxcd	14	90	0	2	78	10	86.67	0.8
Embra18	abxbc	12	92	1	0	85	6	92.39	0.6
Embra167	abxcd	8	96	1	0	95	0	98.96	0.7

* Loci marked in bold form a subset for hybrid confirmation and quality control genotyping in clonal propagation

purity values for all 25 loci were >85.0 % and few loci showed up to 10 non-essential genotypes.

Selection of a set of SSRs for hybrid quality control in clonal propagation

To identify a set of informative markers that could be used for routine genotyping of hybrids for QC, minimum number of SSR loci was determined based on the parentage confirmation and amplification pattern. Subset of six fully informative SSRs with four different alleles (Table 3) would be sufficient to verify the hybridity and for quality control during clonal multiplication of these hybrids. Use of fluorescent-based SSR genotyping in hybrid purity assessment generates specific and accurate data for hybrid identification and subsequently when these hybrids enter clonal propagation chain these SSR markers could be useful in confirming the identity.

Conclusion

Generally, F_1 hybrids contain DNA from both the parents and SSR markers identified both male and female parent–specific markers allowing differentiation of true eucalypts hybrids from selfed individuals and outcrossed individuals with foreign pollen. Presence of non-essential genotypes observed among the hybrid seedlings established the importance of hybrid purity tests and those false hybrids could be removed at the early stage of multiplication. The SSR marker information developed through this study will be of immense help for hybrid eucalypts industry to select appropriate marker combinations and assess hybrid purity of the clones. Other possible applications of hybrid purity assessment include development of genetic linkage maps, quantitative analysis of economically important traits, and marker assisted selection where true and pure hybrids are essential.

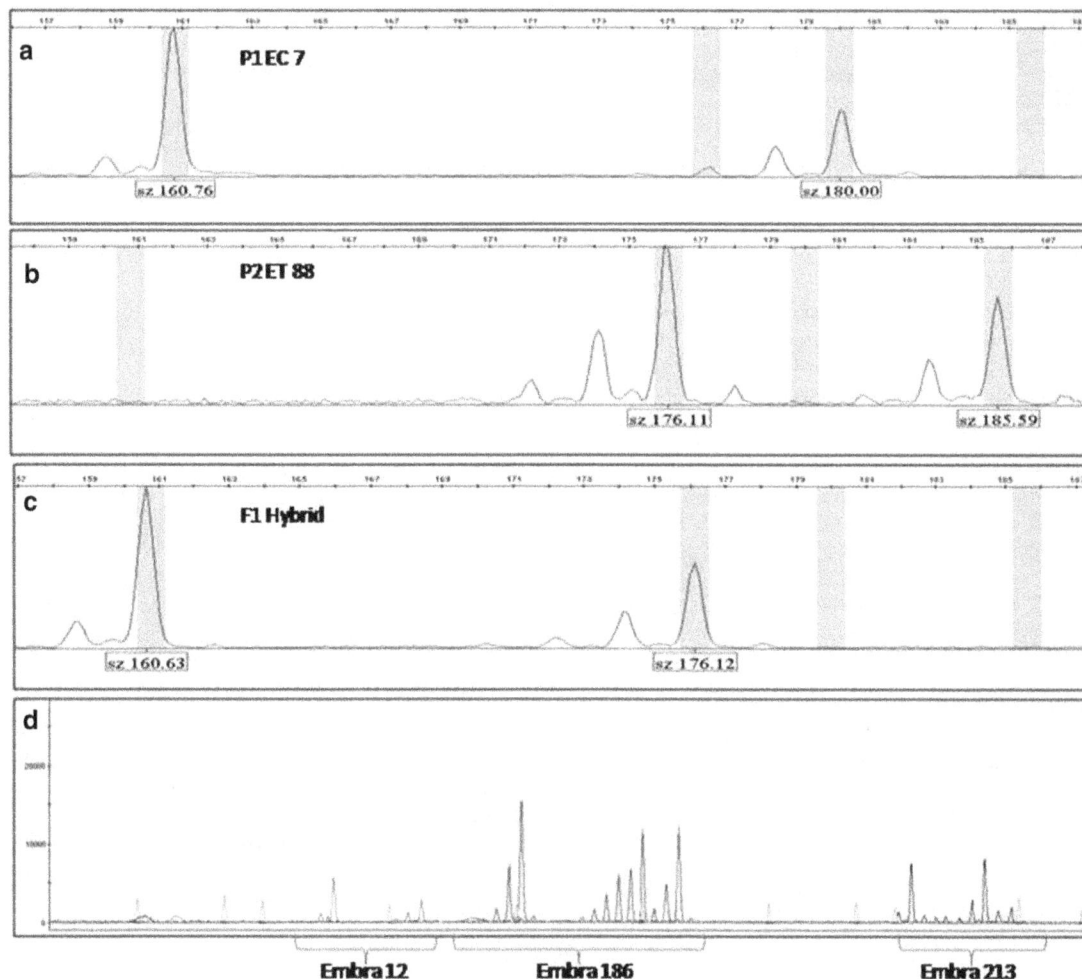

Fig. 1 Electropherogram of SSRs obtained with software Genemapper. Female parent *E. camaldulensis* (**a**), Male parent *E. tereticornis* (**b**) and F1 Hybrid amplified with locus EMBRA186 (**c**). Multi-loading of three loci amplified with different fluorescent dye EMBRA12 (VIC), EMBRA186 (PET) and EMBRA213 (NED) in a single capillary (**d**)

Acknowledgments The authors acknowledge B. Nagarajan, A. Mayavel, S. Chakravarthi, S.P. Subramani, S. Shanmugam, V. Chinnadurai for producing the hybrid seeds. Dr. Sivakumar has raised the hybrid seeds in nursery. V.K.W. Bachpai and M. Ganesan has field planted the F₁ hybrids for further assessment. Authors are grateful to Dr. N. Krishna Kumar, Director, IFGTB for his constant support to carry out this study. The Department of Biotechnology, Government of India is acknowledged for the financial support. The authors, Subashini and Shanmugapriya acknowledge the Indian Council of Forestry Research and Education, Dehradun for proving research fellowship. The authors certify that no actual or potential conflict of interest in relation to this article exists.

References

Acuna CV, Villalba PV, García M, Pathauer P, Hopp HE et al (2012) Microsatellite markers in candidate genes for wood properties and its application in functional diversity assessment in *Eucalyptus globules*. Electronic J Biotech 15:0717–3458

Asif M, Ur-Rahman M, Yusuf Z (2006) Genotyping analysis of six maize (*Zea mays L.*) hybrids using DNA fingerprinting technology. Pak J Bot 38:1425–1430

Bentz SE, Riedel LGH, Pooler MR, Townsend AM (2002) Hybridization and self-compatibility in controlled pollinations of eastern North American and Asian hemlock (*Tsuga*) species. J Arboric 28:200–205

Bohra A, Dubey A, Saxena RK, Penmetsa RV, Poornima KN, Kumar N, Farmer AD et al (2011) Analysis of BAC-end sequences (BESs) and development of BES-SSR markers for genetic mapping and hybrid purity assessment in pigeonpea (*Cajanus* spp.). BMC Plant Biol 11:56

Brondani RPV, Williams ER, Brondani C, Grattapaglia D (2006) A microsatellite-based consensus linkage map for species of *Eucalyptus* and a novel set of 230 microsatellite markers for the genus. BMC Plant Biol 6:20

Cappa EP, Martín Marcó D, Nikles G, Ian S (2013) Last performance of *Pinus elliottii*, *Pinus caribaea*, their F1, F2 and backcross hybrids and *Pinus taeda* to 10 years in the Mesopotamia region, Argentina. New For 44:197–218

Cupertino FB, Leal JB, Vidal PO, Gaiotto FM (2009) Parentage testing of hybrid full-sib families of Eucalyptus with microsatellites. Scan J For Res 24:2–7

Dickinson GR, Wallace HM, Lee DJ (2013) Reciprocal and advanced generation hybrids between *Corymbia citriodora* and *C. torelliana*: forestry breeding and the risk of gene flow. Ann For Sci 70:1–10

Elliott FG, Shepherd MJ, Henry RJ (2006) Verification of interspecific pine hybrids using paternally inherited chloroplast microsatellites. For Genet 12(2):81–87

Faria DA, Mamani EMC, Pappas MR, Pappas GJ Jr, Grattapaglia D (2010) A selected set of EST-derived microsatellites, polymorphic and transferable across 6 species of *Eucalyptus*. J Hered.

Grattapaglia D (2007) Marker-assisted selection in *Eucalyptus*. In: Guimarães E, Ruane J, Scherf B, Sonnino A, Dargie J (eds) Marker-assisted selection: current status and future perspectives in crops, livestock, forestry and fish. Food and Agriculture Organization of the United Nations, Rome, p 252

Grattapaglia D, Kirst M (2008) *Eucalyptus* applied genomics: from gene sequences to breeding tools. New Phytol 179:911–929

Griffin AR, Tran Duc V, Harbard JL, Wong CY, Brooker C, Vaillancourt RE (2010) Improving controlled pollination methodology for breeding *Acacia mangium* Willd. New For 40:31–142

Guichoux E, Lagache L, Wagner S, Chaumeil P, Leger P, Lepais O et al (2011) Current trends in microsatellite genotyping. Mol Ecol Res 11:591–611

He X, Wang Y, Li F, Weng Q, Li M, Xu L, Shi J et al (2012) Development of 198 novel EST-derived microsatellites in *Eucalyptus* (Myrtaceae). Am J Bot 99(4):e134–e148

Mbanjo EGN, Tchoumbougnang F, Mouelle AS, Oben JE, Nyine M, Dochez C et al (2012) Development of expressed sequence tags-simple sequence repeats (EST-SSRs) for *Musa* and their applicability in authentication of a *Musa* breeding population. Afr J Biotech 11:13546–13559

Narendrula R, Nkongolo K (2012) Genetic variation in *Picea mariana* × *P. rubens* hybrid populations assessed with ISSR and RAPD markers. Am J Plant Sci 3:731–737.

Naresh V, Yamini KN, Rajendrakumar P, Dinesh Kumar V (2009) EST-SSR marker-based assay for the genetic purity assessment of safflower hybrids. Euphytica 170:347–353

Nikles DG, Griffin AR (1992) Breeding hybrids of forest trees: definitions, theory, some practical examples, and guidelines on strategy with tropical Acacias. ACIAR Proc Ser 37:101–109

Potts BM, Dungey HS (2004) Interspecific hybridization of *Eucalyptus*: key issues for breeders and geneticists. New For 27:115–138

Rajora OP, Rahman MH (2001) Microsatellite DNA markers and their usefulness in poplars, and conservation of microsatellite DNA loci in Salicaceae. genetic response of forest systems to changing environmental conditions. For Sci 70:105–115

Schuelke M (2000) An economic method for the fluorescent labelling of PCR fragments. Nat Biotech 18:233–234

Selvakumar P, Ravikesavan R, Gopikrishnan A, Thiyagu K, Preetha S, Manikanda Boopathi N (2010) Genetic purity analysis of cotton (*Gossypium* spp.) hybrids using SSR markers. Seed Sci Technol 38:358–366

Semagn K, Beyene Y, Makumbi D, Mugo S, Prasanna BM, Magorokosho C, Atlin Gary et al (2012) Quality control genotyping for assessment of genetic identity and purity in diverse tropical maize inbred lines. Theor Appl Genet 125: 1487–1501

Stanton BJ, Neale DB, Li S (2010) *Populus* breeding: from the classical to the genomic approach. Genet Genomics Populus 8:309–348

Steenkamp ET, Van der Nest MA, Wingfield BD, Wingfield MJ (2003) Detection of hybrids in commercially propagated *Eucalyptus* using 5S rDNA sequences. For Genet 10:195–206

Sundaram RM, Naveenkumar B, Biradar SK, Balachandran SM, Mishra B, Ilyasahmed M, Viraktamath BC et al (2008) Identification of informative SSR markers capable of distinguishing hybrid rice parental lines and their utilization in seed purity assessment. Euphytica 163:215–224

Thamarus KA, Groom K, Bradley A, Raymond CA, Schimleck LR, Williams ER, Moran GF (2004) Identification of quantitative trait loci for wood and fibre properties in two full-sib pedigrees of *Eucalyptus globulus*. Theor Appl Genet 109:856–864

Varghese M, Kamalakannan R, Harwood CE, Lindgren D, Mcdonald MW (2009) Changes in growth performance and fecundity of *Eucalyptus camaldulensis* and *E. tereticornis* during domestication in southern India. Tree Genet Genom 5:629–640

Venkatesh CS, Kedharnath S (1965) Genetic improvement of *Eucalyptus* in India. Silvae Genet 14:141–176

Venkatesh CS, Sharma VK (1977) Hybrid vigour in controlled interspecific crosses of *Eucalyptus tereticornis* × *E. camaldulensis*. Silvae Genet 26:121–124

Venkatesh CS, Sharma VK (1979) Comparison of a *Eucalyptlrs tereticornis* × *E. grandis* controlled hybrid with a *E. grandis* × *E. tereticornis* putative natural hybrid. Silvae Genet 28:127–131

Wu M, Jia X, Tian L, Lv B (2006) Rapid and reliable purity identification of F$_1$ hybrids of Maize (*Zea may* L.) using SSR markers. Mol Plant Breed 4:381–384

Yashitola J, Thirumurugan T, Sundaram RM, Naseerullah MK, Ramesha MS, Sarma NP, Sonti RV (2002) Assessment of purity of rice hybrids using microsatellite and STS markers. Crop Sci 42:1369–1373

Yu Q, Tigerstedt PMA, Haapanen M (2001) Growth and phenology of hybrid aspen clones (*Populus tremula* L. × *Populus tremuloides Michx.*). Silva Fenn 35:15–25

21

Plant terpenes: defense responses, phylogenetic analysis, regulation and clinical applications

Bharat Singh · Ram A. Sharma

Abstract The terpenoids constitute the largest class of natural products and many interesting products are extensively applied in the industrial sector as flavors, fragrances, spices and are also used in perfumery and cosmetics. Many terpenoids have biological activities and also used for medical purposes. In higher plants, the conventional acetate-mevalonic acid pathway operates mainly in the cytosol and mitochondria and synthesizes sterols, sesquiterpenes and ubiquinones mainly. In the plastid, the non-mevalonic acid pathway takes place and synthesizes hemi-, mono-, sesqui-, and diterpenes along with carotenoids and phytol tail of chlorophyll. In this review paper, recent developments in the biosynthesis of terpenoids, indepth description of terpene synthases and their phylogenetic analysis, regulation of terpene biosynthesis as well as updates of terpenes which have entered in the clinical studies are reviewed thoroughly.

Keywords Terpenes · Terpene synthase · Phylogenetic analysis · Clinical trials

Introduction

Plants produce various types of secondary metabolites, many of which have been subsequently exploited by humans for their beneficial roles in a diverse array of

B. Singh (✉)
AIB, Amity University Rajasthan, NH-11C, Kant Kalwar, Jaipur 303 002, India
e-mail: bharatsingh217@gmail.com; bsingh@jpr.amity.edu

R. A. Sharma
Department of Botany, University of Rajasthan, Jaipur 302 055, India

biological functions (Balandrin et al. 1985). Several terpenoids have their roles in plant defense against biotic and abiotic stresses or they are treated as signal molecules to attract the insects of pollination. Out of the investigated terpenoids, many have pharmacological and biological activities and are, therefore, interesting for medicine and biotechnology. The first step of terpenoid biosynthesis is generation of C_5 unit like as isopentenyl diphosphate (IPP) or dimethylallyl diphosphate (DMAPP). For this study, two different separate pathways have been investigated that can generate the C_5 unit: the mevalonate and methylerythritol phosphate (MEP) pathway. On the basis of C_5 units, we can classify the terpenoids as C_5 (hemiterpenes), C_{10} (monoterpenes), C_{15} (sesquiterpenes), C_{20} (diterpenes), C_{25} (sesterpenes), C_{30} (triterpenes), C_{40} (tetraterpenes), $>C_{40}$ (polyterpenes) (Ashour et al. 2010; Martin et al. 2003).

The terpene synthases are responsible for the synthesis of terpenes; they can easily acquire new catalytic properties by minor changes in the structures (Keeling et al. 2008). The synthesis of monoterpenes is initiated by dephosphorylation and ionization of geranyl diphosphate to geranyl carbocation (Huang et al. 2010) and the synthesis of sesquiterpene starts with the ionization of farnesyl diphosphate to farnesyl cation, which can also be isomerized to nerolidyl cation (Degenhardt et al. 2009). Diterpenes are synthesized by diterpene synthases in two different pathways: via the ionization of diphosphate, as catalyzed by class I enzyme and the other is via the substrate protonation at the 14, 15-double bond of geranyl geranyl diphosphate; reaction is catalyzed by class II enzymes (Tholl 2006). The nonsteroidal triterpenoids are produced by conversion of squalene into oxidosqualene and cyclization via formation of dammarenyl cation; reaction is catalyzed by oxidosqualene cyclases (Phillips et al. 2006). Many terpenoids also possess the pharmaceutical properties and currently are being used in clinical practices. Among these terpenoids, taxol (diterpene) of *Taxus buccata* and artemisinin (sesquiterpene lactone) from *Artemisia annua* are well known antineoplastic and antimalarial agents (Croteau et al. 2006; Pollier et al. 2011).

This review deals with biosynthesis of terpenoids, phylogeny of terpene synthases, regulation of terpene biosynthesis, and also about the studies of the clinical trials of terpenoids. This review highlights the current approaches of the phylogenetic analysis of terpene synthases and regulation of terpenoids.

Biosynthesis of terpenoids

Terpenoids are important for plant survival and also possess biological and pharmacological properties that are beneficial to humans. In plants, isopentenyl diphosphate (IPP) and

dimethylallyl diphosphate (DMAPP) can be synthesized via two compartmentalized pathways. The mevalonic acid pathway of terpenoid biosynthesis operates in cytosol, the endoplasmic reticulum and peroxisomes (Carrie et al. 2007; Hemmerlin et al. 2003; Dudareva et al. 2006; Leivar et al. 2005; Merret et al. 2007; Sapir-Mir et al. 2008; Simkin et al. 2011; Lange and Ahkami 2013) (Fig. 1). The condensation of acetyl CoA three units leads to the synthesis of 3-hydroxy-3-methylglutaryl CoA, which later on produces mevalonic acid. The mevalonic acid converted to isopentenyl diphosphate through the process of the phosphorylation and decarboxylation. 3-hydroxy-3-methylglutaryl CoA reductase catalyzes the reduction of 3-hydroxy-3-methylglutaryl CoA to mevalonic acid (Luskey and Stevens 1985; Basson et al. 1988; Igual et al. 1992; Rodwell et al. 2000). In *Arabidopsis thaliana*, mevalonate-5-diphosphate is produced from mevalonic acid by the phosphorylation and the whole reaction is catalyzed by mevalonate kinase and

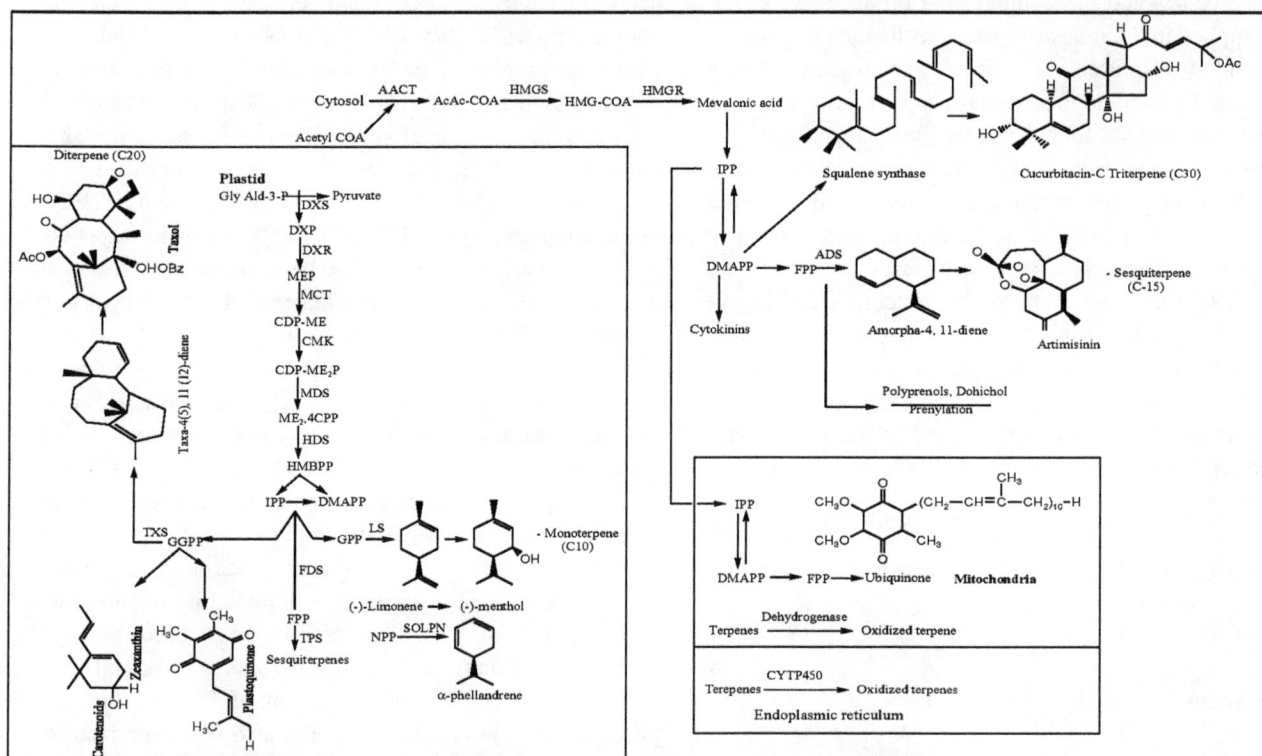

Fig. 1 Schematic overview of monoterpenoid, sesquiterpenoid, diterpenoid and triterpenoid biosynthetic pathways. *AACT* acetoacetyl-CoA thiolase, *AcAc-CoA* acetoacetyl-CoA, *HMGS* HMG-CoA synthase, *HMG-CoA* 3-hydroxy-3-methylglutaryl, *HMGR* HMG-CoA-reductase, *IPP* isopentenyl diphosphate, *DMAPP* dimethylallyl diphosphate, *FPP* farnesyl pyrophosphate, *ADS* amorpha-4,11-diene synthase, *CYT450* cytochrome P$_{450}$ hydroxylase, *GlyAld-3P* glyceraldehyde-3-phosphate, *DXP* deoxyxylulose-5-phosphate, *DXS* DXP synthase, *MEP* methylerythritol-4-phosphate, *DXR* DXP reductoisomerase, *CDP-OME* 4-(cytidine-5′-diphospho)-2-C-methyl-D-erythritol, *MCT* 2-C-methyl-D-erythritol-4-phosphate-cytidylyl transferase,

CDP-ME2P 4-(cytidine-5′-diphospho)-2-C-methyl-D-erythritol phosphate, *CMK* CDP-ME Kinase, *ME2, 4cPP* 2-C-methyl-D-erythritol, 2,4-cyclodiphosphate, *MDS* 2-C-methyl-D-erythritol-2,4-cyclodiphosphate synthase, *HMBPP* (*E*)-4-hydroxy-3-methylbut-2-enyl diphosphate, *HDS* (*E*)-4-hydroxy-3-methylbut-2-enyl diphosphate synthase, *GPP* geranyl diphosphate, *LS* limonene synthase, *NPP* neryl diphosphate, *SOLPN* α-phellandrene synthase, *FDS* farnesyl diphosphate synthase. Similarly chemical structures of (−)-methanol, α-phellandrene; taxol, artemisinin and cucurbitacin C are shown as representative examples of terpenoids

phosphomevalonate kinase (Tsay and Robinson 1991; Lluch et al. 2000). Later on, the mevalonate-5-diphosphate decarboxylase catalyzes the conversion of mevalonate-5-diphosphate to isopentenyl diphosphate, which is the end product of mevalonic acid pathway of terpenoid biosynthesis (Dhe-Paganon et al. 1994) (Fig. 1).

Another part of terpenoid biosynthetic pathway starts in plastid by the condensation of pyruvic acid and glyceraldehydes-3-phosphate, which leads to the synthesis of 1-deoxy-D-xylulose 5-phosphate. The reaction is catalyzed by the enzyme as 1-deoxy-D-xylulose 5-phosphate synthase (Sprenger et al. 1997). The 1-deoxy-D-xylulose 5-phosphate reduced to 2-C-methyl-D-erythritol 4-phosphate by 1-deoxy-D-xylulose 5-phosphate reductoisomerase (Takahashi et al. 1998). The conjugation of 2-C-methyl-D-erythritol 4-phosphate and 4-cytidine 5-phosphate leads to the formation of 4-cytidine 5-phospho-2-C-methyl erythritol and the reaction catalyzed by the enzyme 2-C-methyl-D-erythritol 4-phosphate cytidyltransferase. The 4-cytidine 5-phospho-2C-methyl erythritol converted to 2-C-methyl erythritol 2,4-cyclodiphosphate by the enzyme 2-C-methyl erythritol 2,4-cyclodiphosphate synthase (Rohdich et al. 2000; Steinbacher et al. 2003; Herz et al. 2000; Calisto et al. 2007) (Fig. 1). All the enzymes of the 2C-methyl-D-erythritol-4-phosphate pathway are localized in plastids (Hseih et al. 2008; Surie et al. 2000). In the 1-deoxy-D-xylulose 5-phosphate pathway, the synthesis of hydroxymethylbutenyl 4-diphosphate took place from 2-C-methyl erythritol 2,4-cyclodiphosphate and the reaction was catalyzed by hydroxymethylbutenyl 4-diphosphate synthase. The hydroxymethylbutenyl 4-diphosphate directly converted into the isopentenyl diphosphate and dimethylallyl diphosphate mixture by the enzyme isopentenyl diphosphate and dimethylallyl diphosphate synthase (Baker et al. 1992; Cunningham et al. 2000).

In the steps of downstream process, mevalonate converted into IPP, which involves phosphorylations and decarboxylation events. The carried out reactions catalyzed by the following enzymes: mevalonate kinase, phosphomevalonate kinase and mevalonate diphosphate decarboxylase. Although the enzymes involved in these steps are thoroughly studied in yeast and various animal systems, very little information was reported in terms of their biochemical characterization in plants (Gershenzon and Kreish 1999). Recently, a cis-prenyl transferase, neryl phosphate synthase, was reported to provide precursor for monoterpene biosynthesis in several species of Solanum (Schilmiller et al. 2009; Lange and Ahkami 2013). A condensation of one molecule of DMAPP with two molecules of IPP generates farnesyl diphosphate (C_{15}), the direct precursor of most sesquiterpenes, which is catalyzed by farnesyl diphosphate synthase. Plant genomes appear to encode various farnesyl diphosphate synthase isoforms that localize to the cytosol, plastids, mitochondria or peroxisomes (Cunillera et al. 1997; Thabet et al. 2011). In tomato, a cis-prenyl transferase, farnesyl diphosphate synthase is localized to plastids of the glandular trichomes, where it is involved in the biosynthesis of sesquiterpene volatiles (Salland et al. 2009). Diterpenes are formed from geranyl geranyl diphosphate which itself is synthesized by the catalysis of geranyl geranyl diphosphate synthase from DMAPP and three molecules of IPP. Isoforms of this enzyme have been reported to occur in plastids, the endoplasmic reticulum and mitochondria (Thabet et al. 2012; Sitthithaworn et al. 2001; Okada et al. 2000; Cheniclet et al. 1992). Terpene synthases often catalyze the formation of multiple products from a prenyl diphosphate substrate, resulting from a catalytic mechanism that involves highly reactive carbocation intermediates (Degenhardt et al. 2009). In general, monoterpene synthases are localized on plastids, whereas sesquiterpene synthases are found in the cytosol (Chen et al. 2011; Aharoni et al. 2003). A mitochondrial localization was determined for a terpene synthase in tomato (Solanum lycopersicum), but the in vivo substrate is currently unknown (Falara et al. 2011). The terpenoid skeletons are further functionized through redox, conjugation and other related reactions (Fig. 1).

The class of triterpenes includes sterols and triterpenoids, which can be synthesized as saponins and sapogenins sufficient amount in plants (Sparge et al. 2004). The linear triterpene squalene is derived from the reductive coupling of the two molecules of farnesyl pyrophosphate (FPP) by squalene synthase. Squalene is later on oxidized biosynthetically by the other enzyme squalene epoxidase to generate 2,3-oxidosqualene. 2,3-oxidosqualene converted to triterpene alcohols or aldehydes by oxidosqualene cyclases (Phillips et al. 2006; Jenner et al. 2005; Haralampidis et al. 2002). In plants, triterpenoid biosynthetic diversity has been developed and their diverse genomes encode multiple oxidosqualene cyclase enzymes to form these triterpene skeletons (Fig. 2). The level at which the structural diversity of triterpenes is generated, depends on the cyclization of 2,3-oxidosqualene by different oxidosqualene cyclases such as lupeol synthase (LS) and α/β-amyrin synthase (Mangus et al. 2006; Sawai and Saito 2011). All triterpene synthases appear to have diverged from cycloartenol synthase gene (Zhang et al. 2003), but an independent origin for β-AS in dicots and monocots has also been reported (Phillips et al. 2003; Moses et al. 2013).

The ursolic acid, oleanolic acid and betulinic acid are likely to be derived from α-amyrin, β-amyrin and lupeol, respectively, followed by successive oxidation (Augustin et al. 2011) at the C_{28} position. It has been shown that the triterpene skeletons (α-amyrin, β-amyrin and lupeol) are cyclized from 2,3-oxidosqualene, a common precursor of

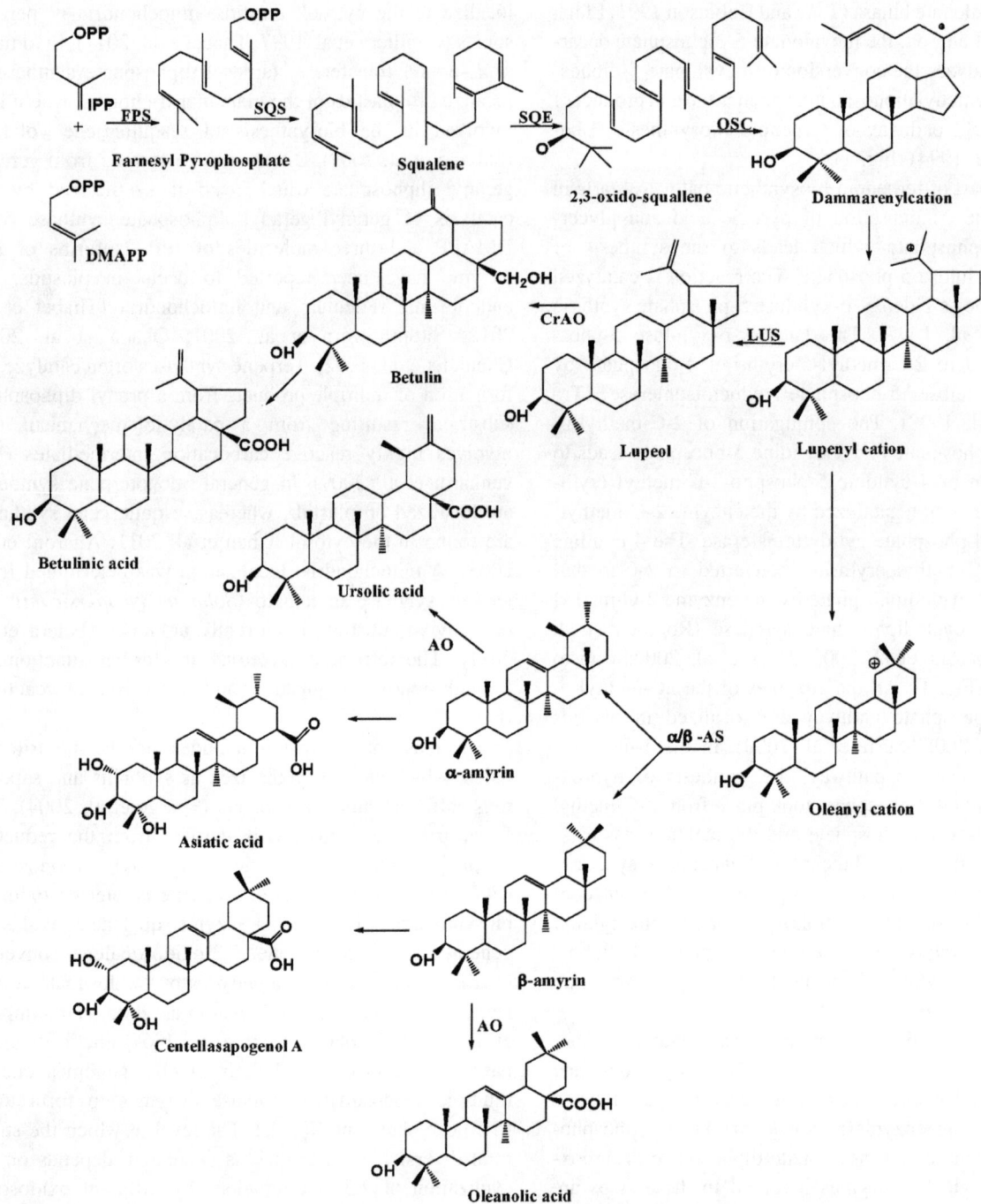

Fig. 2 Schematic overview of triterpenoid biosynthesis. Farnesyl diphosphate synthase (FPS) isomerizes isopentenyl diphosphate and dimethylallyl diphosphate (DMAPP) to farnesyl diphosphate, while squalene synthase converts to squalene. Squalene epoxide oxidizes the squalene to 2,3-oxidosqualene. Oxidosqualene cyclase (OSC) catalyzes 2,3-oxidosqualene through cationic intermediates to one or more cyclic triterpene skeletons. The other enzymes involved in the biosynthesis include α/β amyrin synthase (α/β AS) which can also form the lupenyl cation but further ring expansion and rearrangements are required before the deprotonation to α/β amyrin, the precursors of sapogenins. α-Amyrin oxidase involved in biosynthesis of ursolic acid and oleanolic acid

phytosterols and triterpenoids (Abe et al. 1993). Oxidosqualene cyclases yielded one specific product, such as lupeol synthase (Shibuya et al. 1999; Guhling et al. 2006; Moses et al. 2013), β-amyrin synthase (Kushiro et al. 1998; Kirby et al. 2008; Shibuya et al. 2009) and α-amyrin synthase (Muffler et al. 2011), cycloartenol synthase (Hayashi et al. 2000) and cucurbitadienol synthase (Shibuya 2004). Following the formation of the carbon skeletons, the triterpene alcohols are modified by various cytochrome P450s, dehydrogenases, reductases and other modification enzymes, while some triterpenoids are synthesized in all plant cells. Laticifers are elongated epithelial cells that produce chemically complex latex, which can consist of a polyterpenoids such as natural rubber (Beilen and Poirier 2007). Trichomes which are glandular in nature, are generally the known as storage organs of terpenoids and/ or phenolic compounds (Lange and Turner 2013).

The synthesis of isopentenyl diphosphate and dimethylallyl diphosphate, both are intermediates of terpenoid biosynthesis, is compartmentalized. The mevalonic acid pathway operates in cytosol, which is responsible for the formation of sesquiterpenes and sterols; 1-deoxy-D-xylulose 5-phosphate (DXP/MEP) pathway operates in the plastids, involved in the synthesis of monoterpenes, diterpenes and some sesquiterpenes as well as plastoquinones (Laule et al. 2003). In plants, mevalonic acid pathway's enzyme localization is also fragmented. The 3-hydroxy-3-methylglutaryl CoA reductase and squalene synthase are localized in the endoplasmic reticulum (Leivar et al. 2005; Sapir-Mir et al. 2008; Busquets et al. 2008) while the acetoacetyl CoA is also important enzyme of mevalonic acid pathway, localized in peroxisomes (Reumann et al. 2007). In contrast, the DXP pathway enzymes are localized in plastids of cyanobacteria (Ginger et al. 2010). However, after constant observations of certain uncertainties, it has been widely accepted that initial reactions of DXP pathway are catalyzed in the cytosol while remaining reactions operated in the plastid. In red algae and the *Cyanophora paradoxa*, both mevalonic acid and DXP pathway run concurrently (Grauvogel and Peterson 2007). However, it has been proved that both the pathways are not separated spatially. The taxol (sesquiterpenes) is synthesized by the both mevalonic and DXP pathways (Adam and Zapp 1998; Wang et al. 2003). The unidirectional proton symport system of export of terpenoid intermediates and their involvement between cytosol and plastid pathway have been proved by the Ca^{2+} gated channel (Bick and Lange 2003). Laule et al. (2003) have suggested in their experiment that some limiting plastidial membrane transporters must be operated in the exchange of terpenoid intermediate's exchange in between cytosol and plastid (Liao et al. 2006).

Defense responses of terpenoids in plants

Plant kingdom has direct and indirect defense responses when they come in contact of microbial pathogens. The direct mode of defense mechanism includes physical structures like as trichomes, thorns as well as accumulation of phytochemicals that have antibiotic activities. The compounds such as phytoalexins are low-molecular-weight compounds that are produced as part of plant defense mechanisms. In few plant species the diterpenes and sesquiterpenes act as phytoalexins, e.g., 14 diterpene phytoalexins have been investigated from *Oryza sativa*. These phytoalexins can be grouped into four types—monilactones A and B (Hwang and Sung 1989), oryzalexins A–F (Peters 2006) and oryzalexin S (Tamongani and Mitani 1993). Polycyclic diterpenoids are synthesized from geranyl geranyl diphosphate via the intermediate hydrocarbon precursors (e.g., 9-β-pimara-7, 15-diene, stemar-13-ene, ent-sandaracopimaradiene and ent-cassa-12, 15-diene). All these natural products are accumulated in leaves in response to inoculation with the pathogenic blast fungus, *Magnaporthe grisea,* or ultraviolet irradiation and exhibit antimicrobial properties (Prisic et al. 2004).

Indirect mode of defenses indicates that the plants have characteristics to defend against herbivores indirectly by enhancing the effectiveness of natural enemies of the herbivores. One of the most amazing examples of the plant indirect defense is the release of the blend of specific volatiles, which attract the carnivores of herbivores, after herbivore attacking. More attention has been paid in case of corn, lima bean, poplar and cotton that are well studied with genetics, biochemical, physiological and ecological approaches (Rodriguez-Saona et al. 2003; Arimura et al. 2004a; Mithofer et al. 2005; Schnee et al. 2006). In an olfactometer assay, the transgenic *Arabidopsis* plants used as odor sources, females of the parasitoid *Cotesia marginiventris* learned to exploit the TPS10 sesquiterpenes to locate their lepidopteran host (Schnee et al. 2006). When a strawberry nerolidol synthase gene was expressed in *Arabidopsis* transgenic plants emitted two new terpenoids (3,S)-(*E*)-neridol and its derivative (*E*)-4,8-dimethyl-1,3-7-nonatriene [(*E*)-DMNT] and attracted more carnivorous predatory mites (Kappers et al. 2005). The capacity to produce deterrents to insects from plant-derived terpenoids is typical of some *Chrysolina* species. Because feeding of herbivores alters the aromatic profile of essential oil-producing plants like *Mentha aquatica*, the issue is both economically and ecologically relevant (Burse et al. 2009; Atsbaha Zebelo et al. 2011).

Airborne terpenoids are also critical components of plant defense responses to abiotic and biotic stresses (Unsicker et al. 2009; Vickers et al. 2009). From agronomic perspective, crop losses due to insect infestation are a significant issue (El-Wakeil et al. 2010). Insecticide applications are most common and effective strategy for control of insects, but some of the agrochemicals have undesirable side effects on useful insects and can pose long-term risks to the environment (Dedryver et al. 2010; Zulak and Bohlmann 2010).

When patchoulol synthase (PTS), a sesquiterpene synthase from *Pogostemon cabli* L., was targeted to the cytosol in transgenic tobacco (*Nicotiana tabacum* L. cv. Xanthi), only small amounts of the expected product, patchoulol, were detected. When the same gene was expressed coordinately with an additional copy of farnesyl diphosphate synthase, the patchoulol accumulation in transgenic tobacco also remained very low (Wu et al. 2006), while both gene products patchoulol synthase and farnesyl diphosphate synthase were targeted to plastids, a patchoulol accumulation was increased, which appeared to be volatilized. It was also observed that volatile emitted from these transgenic plants significantly deterred tobacco hornworms and pine beetles from feeding on leaves (Wu et al. 2006; Bohlmann 2012).

Plant terpene synthases

The investigations of terpene synthases have been an interesting and active area of plant metabolic engineering research and may genes have been isolated from various plant species (gymnosperms and angiosperms), including *Picea abies* (Martin et al. 2004), *Taxus media* (Wildung and Croteau 1996), *Arabidopsis thaliana* (Chen et al. 2003; Degenhardt et al. 2009), *Cucumis sativus* (Mercke et al. 2004), and *Nicotiana attenuata* (Facchini and Chappel 1992). The total numbers of terpene synthases reported from *Thapsia laciniata* are 8 monoterpene and 5 sesquiterpene (Drew et al. 2013) which are slightly larger than the number reported in *Arabidopsis thaliana* and *Artemisia annua* (Tholl and Lee 2011). Several sesquiterpene synthases have also been cloned and characterized from maize (Kölner et al. 2004). In general, the lengths of monoterpene synthases are between 600 and 650 amino acid residues and are 50–70 amino acids are larger than sesquiterpene synthases (Martin et al. 2004). Most diterpene synthases are approximately 210 amino acids longer than monoterpene synthases because of an additional internal element that is conserved in both sequence and position (Prisic et al. 2004). All the terpene synthases contain the aspartate-rich DDxxD motif involved in the coordination of divalent metal ions for substrate binding (Lesburg et al. 1997). The terpene synthases are further sub-classified into four subfamilies—TPSa, TPSb, TPSd and TPSg. The TPSa family consists of angiosperm terpene synthases (Bohlmann et al. 1998; Dudareva et al. 2003), the TPSb family contains angiosperm monoterpene synthases (Bohlmann et al. 1998; Dudareva et al. 2003), the TPSb includes angiosperm monoterpene synthases (Bohlmann et al. 1998; Dudareva et al. 2003), TPSd of gymnosperm monoterpene synthases (Bohlmann et al. 1998; Dudareva et al. 2003) and TPSg *Antirrhinum majus* monoterpene synthases (Bohlmann et al. 1998; Chen et al. 2003) also share common evolutionary origin.

The triterpene synthases lead to synthesis of tricyclic, tetracyclic and pentacyclic molecules in complex of concerted reaction steps catalyzed by single enzyme. The sterol and triterpenoid biosynthetic pathways diverged at some point, depending on the involvement of the type of the oxidosqualene synthases. Cyclization of 2,3-oxidosqualene in the chair–boat–chair conformation leads to protosteryl cation intermediate, sterol precursor, via the synthesis of cycloartenol or lanosterol in plants (Kolesnikova et al. 2006; Suzuki et al. 2006), while in contrast to 2,3-oxidosqualene in the chair–chair–chair conformation is cyclized into a dammarenyl carbocation intermediate, which subsequently gives rise to diverse triterpenoid skeletons after further re-arrangements. Many different types of oxidosqualene synthases have been isolated from various plant species including lanosterol synthase (Baker et al. 1995; Sung et al. 1995), cycloartenol synthase (Bach 1995; Kawano et al. 2002), lupeol synthase (Hayashi et al. 2004; Segura et al. 2000) and β-amyrin synthase (Hayashi et al. 2001; Iturbe-Ormaetxe et al. 2003). Besides these synthases, some multifunctional triterpene synthase have also been characterized from other different plant species (Basyuni et al. 2006; Shibuya et al. 2007).

Phylogenetic analysis of terpenoid synthases

On the basis of the phylogeny, the gymnosperm terpene synthases have been subdivided into three distinct clades—TPS-d1 to TPS-d3. The TPS-d1 subclade are (−)-α/β-pinene synthases, (−)-linalool synthases and (*E*)-α-farnesene synthases; in TPS-d2 clade are longifoline synthase and in TPS-d3 clade are levopimaradiene/abietadiene synthases and isopimaradiene synthase (Martin et al. 2004). The functional identification of spruce terpene synthase genes account for several terpenoid compounds of the oleoresin and volatile emissions. Many terpene synthase genes (*TPSd*) of terpenoid metabolism, especially *ent*-copalyl diphosphate synthase and *ent*-kaurene synthase gene appear to be expressed as single copy genes (Bohlmann et al. 1999). These primary metabolism terpene synthase genes are basal to the specialized metabolism genes and are the descendants of an ancestral plant diterpene synthase similar to the non-vascular plant as *Physcomitrella patens* (Hayashi et al. 2006; Keeling et al. 2010, 2011).

Sesquiterpene synthases sequence of roots of *Cycus* species was investigated and it was found that gymnosperms form a distinct group from the angiosperms, which displayed a pattern that seemed to be influenced by the types of products of different plant species. By the phylogenetic analysis, it can be predicted that α-copaene synthase gene is more homologous to germacrene B, germacrene D and valencene synthase gene (Hiltpold and Turlings 2008; Wen et al. 2012). A sesquiterpene synthase gene that produces α-copaene as its sole reaction product has been reported. This enzyme is highly expressed in potato and correspondence to the difference in tuber flavor between two cultivars of potato (Ducreux et al. 2008; Zapta and Fine 2013).

The terpenoid synthases of primary metabolism, (−)-CDP synthase (Sun and Kamiya 1994; Ait-Ali et al. 1994; Bensen et al. 1995) and kaurene synthase B (Yamaguchi et al. 1996), are only distantly related to those of secondary metabolism, including members of sub-families TPSa, TPSb and TPSd. However, all plant terpene synthases share a common evolutionary origin and it appears that the bifurcation of terpenoid synthases of primary and secondary metabolism occurred before separation of angiosperms and gymnosperms. Terpene synthases of secondary metabolism constitute the most extensively studied TPS sub-families including TPSa, TPSb, TPSd and the distant and possibly ancient TPSf branch containing linalool synthase (Bohlmann et al. 1998; Chen et al. 1996). Valencene synthase (Sheron-Asa et al. 2003) and 5-epi aristolochene synthase (Back and Chappel 1996) are related to one another based on the biosynthesis of sesquiterpenes within the eremophilene class of compounds. The *Magnolia* possesses a single intron positioned near the 5′ region of the gene, similar to the first intron in all other three classes of terpene synthases from plants. The intron found in fungal trichodiene synthase gene is inserted into the middle of the trichodiene synthase gene and is not spatially oriented similar to the insertion site of the first intron in any of the plant genes including *Mg25* (Trapp and Croteau 2001; Lee and Chappel 2008). The dendrogram analysis was conducted to determine the evolutionary relatedness of chamomile terpene synthases to those of others Asteraceae. MrTPS1, MrTPS2, MrTPS3 and MrTPS5 were found to belong to the TPSa sub-family covering angiosperms, whereas MrTPS4 fell into the TPSb sub-family covering angiosperm monoterpene synthases (Irmisch et al. 2012). The monoterpenes are formed in plastids and the nucleus-encoded monoterpene synthases are targeted by N-terminal transit peptides of approximately 40–70 amino acids which reside upstream of the conserved RRx8W motif and are cleaved during import from the nucleus. (Williams et al. 1998; Turner et al. 1999). In contrast to the multiple closely related AtTPS of the TPSa and TPSb groups, only one AtTPS member is found in each of three sub-families TPSc, TPSe and TPSf (*Arabidopsis thaliana* terpene synthases). The copalyl diphosphate synthases show between 45 and 55 % identity. The AtTPS GA2 enzyme (Yamaguchi et al. 1998) is a diterpene synthase of the TPSe sub-family of kaurene synthases. Finally AtTPS04 has a TPSf type primary structure reminiscent of that of linalool synthase from *Clarkia breweri* (Dudareva et al. 1996). It has been suggested that AtTPS04 is an orthologue of this linalool synthase in *Arabidopsis thaliana* (Cseke et al. 1998; Aubourg et al. 2002). The comparison of Grtps (grape fruit terpene synthases) amino acid sequence with the sequences of other terpene synthase indicated that this *Grtps* cDNA is truncated at the 5′ terminus and that the truncation represents 18 amino acid residues of the presumptive transit peptide region at the N-terminus of the deduced protein. The peptide is found in both monoterpene and diterpene synthases and supposedly facilitates the import of these nuclear encoded gene products into plastids, a process that involves cleavage of the preproteins to the mature active enzymes (Mau and West 1994; Vogel et al. 1996; Jia et al. 2005). The phylogenetic analysis of large number of *Vitis vinifera* terpene synthase genes resolved a bifurcation of TPSb and TPSg sub-families at a juncture that was previously ambiguous and had misclassified some TPSg genes as TPSb members. Later on it was concluded that grapevine geraniol and linalool synthase matches with basil geraniol and linalool synthase and showed same proximity which indicates that these TPS functions already existed have evolved from same ancestor (Martin et al. 2010).

Azadirachta indica and *Citrus* belong to the order Rutales and the phylogenetic studies reaffirmed their taxonomic closeness. Additionally, phylogenetic studies grouped *A. indica* with *Melia* species, one that is also known to harbor bioactive compounds suggesting a common evolutionary process with regard to synthesis of these compounds in Meliaceae. The repeat analysis showed low repeat content in *A. indica* genome compared with other sequenced angiosperms. This could have been due to presence of xenobiotic terpenoids specific to the plant, which might have been a major impediment for horizontal gene transfer (Richardson and Palmer 2007; Krishnan et al. 2012). The phylogenetic analysis of cycloartenol synthase, lupeol synthases and the dicot β-amyrin synthase reported them as multifunctional enzymes. These enzymes have same specificity clusters by which authors have suggested a molecular evolution mechanism for lupeol synthase and β-amyrin synthase arising from a common ancestral cycloartenol synthase (Shibuya et al. 1999; Zhang et al. 2003).

The increasing diversification of the cyclization reaction sequence from the dammarenyl to the oleanyl cation via the lupenyl cation is consistent with this evolutionary scheme. MdOSC1, MdOSC2 and MdOSC3 are located with in the group of enzymes that produce a dammarenyl cation intermediate in *Malus domestica*. In MdOSC1 and MdOSC3 cluster, lupeol synthases are more related to β-amyrin synthases than to lupeol synthases (Basyuni et al. 2007; Guhling et al. 2006; Brendolise et al. 2011). This new class of lupeol synthase includes BgLUS, RcLUS and multifunctional triterpene synthase KcMS and another putative OSC (EtOSC—*Euphorbia terucalli* triterpene synthase) for which no triterpene synthase activity has been detected when it is expressed in yeast (Kajikawa et al. 2005).

The phylogenetic analysis using neighbor-joining methods showed that SlTTS1 and SlTTS2 (*Solanum lycopersicum*) terpene synthases are more closely related to each other than to any other oxidosqualene cyclases and they together with the *Panax ginseng* β-amyrin synthases (Kushiro et al. 1998) form the subclad within a group of oxidosqualene cyclase enzyme that were all characterized as β-amyrin synthases from different plant species. The intron patterns and exon lengths of the two *SlTTS1* genes are very similar to those of the other oxidosqualene cyclases, while *SlTTS2* gene organization most closely resembles OSC3 of *Lotus japonicus* (Sawai et al. 2006; Wang et al. 2011). The close match in localization of transcripts and metabolites makes it very likely that *SlTTS1* and *SlTTS2* genes dedicated entirely making the triterpenoids destined for the cuticular wax of the fruit surface. This major biological function can be assigned to the OSCs; the cuticular triterpenoids contribute significantly to the chemical composition and to the ecophysiological properties of the fruit cuticle (Vogg et al. 2004; Isaacson et al. 2009). It should be noted that similar biological functions had previously been attributed to a few other OSCs, for example, a glutinol synthase and a friedelin synthase from *Kalanchoe daigremontiana* (Wang et al. 2010).

It has been reported that 13 *Arabidopsis thaliana OSC* genes and the 11 triterpene synthase genes are grouped into one functional group. Furthermore, 20 out of 36 Poaceae *OSC* genes were also assigned either to the pentacyclic triterpene synthase-like group based on the characterized β-amyrin synthase from *Avena* species (Haralampidis et al. 2001; Qi et al. 2004) or to the rice isoarborinol synthase group (Xue et al. 2012). In *Arabidopsis thaliana*, a tandem cluster on chromosome 1 containing four homologous *OSC* genes, *At1g78950*, *At1g78955/CAMS1*, *At1g78960/LUP2* and *At1g78970/LUP1*, is likely to have arisen by three tandem duplication events. Another tandem duplicate gene pair *At4g15340* and *At4g15370*, encoding arabidiol synthase and baruol synthase, respectively (Xiang et al. 2006; Lodeiro et al. 2007), is located on *A. thaliana* chromosome 4. Indeed, most triterpene synthase genes in the Poaceae family appear to have arisen from *CS* genes by the D3 gene duplication event, which caused the divergence of the 20 triterpene synthase genes (D3-2) from 12 *CS* genes and other closely related genes form group D3-1. The D3 duplication event is highly likely to have been a tandem duplication that occurred during the ancient Poaceae genome before the ρ whole genome duplication, which was estimated to have occurred between 117 and 50 mya (Gaut 2002; Yu et al. 2005; Lescot et al. 2008; Jaio et al. 2011).

In pairwise comparison of all predicted *Citrus* terpene synthases with all *Arabidopsis* AtTPS proteins, it was found that the overall sequence identity varies widely from 18 to 91 %. *Citrus* terpene synthase EST contigs were long enough to allow the complete encoded protein sequences to be deduced (ranged from 547 to 617 amino acids), which corresponds to the size of known monoterpene synthases, sesquiterpene synthases and diterpene synthases of secondary metabolism (Bohlmann et al. 1998; Aubourg et al. 2002). Most terpene synthases encoded by class-III genes contain variations of a conserved motif RR(x)8W, close to the N-terminus (Dornelas and Mazzafera 2007).

2-Methyl-3-buten-2-ol (MBO-hemiterpene) is a five-carbon alcohol produced and emitted by plant species of pine in large quantities. The gene most closely related to MBO synthase is a linalool synthase from *Picea abies* with which MBO shares 82 % amino acid identity. Also closely related to MBO synthase are farnesene synthase from *P. abies* and *P. taeda*. These enzymes form a strongly supported clade of related enzymes producing MBO, linalool and E-α-farnesene nested within what is otherwise a clade dominated by enzymes producing cyclic monoterpene (Gray et al. 2011). MBO synthase and isoprene synthase comparison clearly demonstrate that hemiterpene synthase evolved independently in gymnosperms and angiosperms. The MBO synthase clusters with gymnosperm monoterpene synthase, isoprene clusters with angiosperm monoterpene synthases and these gene families diverged between 250–290 million years ago (Martin et al. 2004).

Regulation of terpenoid biosynthesis

The role of light and temperature in modulating a range of terpenoids and the corresponding transcripts has been reported, but there is no universal behaviour and it varies depending upon the type of metabolites as well as plant species. The 3-hydroxy-3-methylglutaryl CoA reductase is stimulated by light in *Triticum aestivum* (Aoyogi et al. 1993), pea (Wong et al. 1982) and potato (Korth et al. 2000), but down-regulated by light in *Lithospermum erythrorhizon* (Lange et al. 1998). The effects of light as promoter on 3-hydroxy-3-methylglutaryl CoA reductase activity has been documented and also the light-mediated alteration in 3-hydroxy-3-methylglutaryl CoA reductase transcripts (Learned and Connolly 1997; Kawoosa et al. 2010). Rodrìguez-Concepción (2006) suggested about the light-dependent regulation of terpenoid biosynthesis during the early stages of development in *Arabidopsis thaliana*. As per their model, the seedlings which were grown in the dark obtain lower level of precursors for the synthesis of sterols from the mevalonic acid pathway. Some of the prenyl diphosphates of the mevalonic acid pathway might be translocated to the plastid for the synthesis of carotenoids and gibbrellins; while those seedlings were grown in the light, the activity of mevalonic acid pathway increased and the isoprenoid precursors are not required by the plastid (Vranova et al. 2012). In *Artemisia annua*, the discharging of β-pinene fluctuates as per the rhythm of day as well as night and it is higher in the day light than night (Lu et al. 2002). The all terpenoid compounds of *Arabidopsis thaliana* flowers showed clear diurnal emission patterns (Aharoni et al. 2003).

Jasmonate was reported to be general inducer of biosynthesis of plant secondary metabolite (Memelink et al. 2001; van der Fits and Memelink 2000). Jasmonate and its derivative methyl jasmonate were shown to induce the terpenoid indole biosynthesis in suspension cell cultured with auxin and to enhance the terpenoid indole alkaloid production when cells were cultured in an auxin-free medium (Gantet et al. 1998). Genes involved in secologanin biosynthesis (*Crdxs*, *Crcpr*) were upregulated by methyl jasmonate as well as most of the known other genes of terpenoid indole alkaloid biosynthesis pathway (Hedhili et al. 2007; Siamaru et al. 2007; Zhao et al. 2004). Downstream of the conserved jasmonate hormone perception and initial signaling cascade, species-specific transcriptional machineries exist that regulate the transcriptional activity of specific biosynthetic genes (Pauwels et al. 2009; Pauwels and Goosens 2011). A few transcriptional factors regulated by the jasmonate hormone signaling cascade that activate the transcription of sesquiterpenoid biosynthetic genes have already been reported (De-Geyter et al. 2012). The triterpenoid contents in *Ocimum basilicum* was produced higher in quantity after the treatment with methyl jasmonate rather than control plants. The exogenous treatment of methyl jasmonate affected the production of terpenoids by regulating the terpene synthase genes (Li et al. 2007; Prins et al. 2010).

It is well known that the monoterpene synthase gene *SIMTS1* activity is induced by jasmonic acid (van Schie et al. 2007). The jasmonic acid has been shown to increase trichome density on newly formed leaves of *Arabidopsis* and tomato (Boughton et al. 2005; Traw and Bergelson 2003). The production of acyl sugars on the leaf surface of *Datura wrightii* plants increased without affecting trichome density (Hare and Walling 2006). The expression levels of most of the genes of the mevalonic acid pathway and monoterpene or sesquiterpene synthesis follow the same profile during development, suggesting coordinated regulation of terpenoid biosynthesis at the gene level. However, although expression of these genes is relatively trichome specific, the expression profiles do not alter metabolic accumulation during development, suggesting that terpenoid synthesis is not regulated at the transcript level in tomato trichomes but rather involves other (post-transcriptional) regulatory mechanisms. Similarly, only a loose correlation between terpenoid pathway gene expression and enzyme activity has been found in *Ocimum basilicum* terpenoid metabolism (Iijima et al. 2004; Besser et al. 2008). Jasmonic acid is essential for induction of defenses in glandular trichomes. Production of many trichome metabolites is also regulated tightly by transcriptional control, thereby allowing for temporally regulated emission of plant volatiles (Dudareva et al. 2006; Glas et al. 2012).

Methyl jasmonate is a cyclopentanone ring bearing lipophilic hormone synthesized in plants from octadecanoid pathway and they play role in development of responses to biotic stress (Creelman and Mullet 1997). The changes in accumulation of terpene synthase transcripts were also observed in methyl jasmonate-treated Norway spruce; this supports the view that the transcription of terpene synthase genes also regulated by this hormone (Fäldt et al. 2003). The accumulation of taxoid (diterpene taxadiene) was enhanced by supplementation of methyl jasmonate in cell cultures of *Taxus* (Ketchum et al. 1999; Phillips et al. 2006). Traumatic ducts are specialized anatomical structures for the accumulation of resin terpenes, which are formed in Norway spruce and other conifer plants. Like other resin ducts, traumatic ducts are lined with epithelial cells thought to be the site of terpene biosynthesis. The development of traumatic ducts in xylem of Norawy spruce, induced by the treatment of methyl jasmonate was similar in fashion, which caused by the attack of fungal elicitation and mechanical wounding (Krokene et al. 2008; Herrera et al. 2005; Martin et al. 2002). The morphological changes are accompanied by an increase in monoterpene and diterpene synthase activity peaking at the highest rate at 10–15 days after treatment of methyl jasmonate (Martin et al. 2002). The significant increase in the resin terpenoid quantity in bark and wood of Norway spruce followed by methyl jasmonate treatment has also been reported (Martin et al. 2004; Miller et al. 2005).

The level of mRNAs of squalene synthase and β-amyrin synthase was upregulated by adding methyl jasmonate to *Glycyrrhiza glabra* cell cultures and it was observed that the level was higher 3 days after the treatment and lasted for 7 days. The mRNA levels of cycloartenol synthase and oxidosqualene cyclase, which are involved in the biosynthesis of phytosterols, were relatively constant (Hayashi et al. 2003). *Artemisia annua* plants treated with methyl jasmonate showed only slight change in the transcription levels of the control plants. 3-hydroxy-3-methylglutaryl CoA reductase gene expression decreased 1.5-fold at 24 h and then increased threefold by 48 h (Mehjerdi et al. 2013).

When the zeatin and ethylene were added together to the culture medium of *Catharanthus roseus* cell cultures, the mRNA level of mevalonic acid pathway genes coordinately increased in suspension culture cells (Papon et al. 2005). In addition, the zeatin stimulates the bioconversion of exogenic secologanin to the terpenoid indole alkaloid ajmalicine, suggesting that cytokinin may also act on other downstream enzymatic steps of the terpenoid indole alkaloid biosynthesis pathway (Decendit et al. 1992). Similarly, the ethylene treatment itself induces the formation of traumatic ducts in *Pseudotsuga menziesii* and *Sequoiadendron giganteum* (Hudgins and Franceschi 2004) and stimulates the accumulation of β-thujaplicin in *C. lusitanica* at low levels (Phillips et al. 2006).

Dudareva et al. (2003) reported that the biosynthesis and emission of the monoterpenes (E)-β-ocimene and myrcene in *Antirrhinum majus* flowers correlate with specific expression patterns of the lobes of flower petals during floral development, with the highest transcripts levels detected at day four post anthesis. In *Arabidopsis* flowers, monoterpene and sesquiterpene synthases are not expressed in flower petals; instead their expression is limited to the stigma, anthers and sepals (Tholl et al. 2005). Many monoterpene and sesquiterpene synthase genes have been reported from terpene accumulating cells and tissues such as leaf glandular trichomes of *Citrus* and grapes (Picaud et al. 2005; Lücker et al. 2004; Shimada et al. 2004; Kai et al. 2006). Most of the terpene synthase genes belonging to TPSa and TPSb sub-families reached the highest expression in accordance with the peak of accumulation of the respective compounds, while in TPSg sub-family, only one gene for linalool synthase showed major transcript in the ripening of berries. The geraniol synthase had a peak of expression that started to increase and overcome the linalool concentration (Matarese et al. 2013; Chen et al. 2011; Falara et al. 2011).

Ginsenoside backbones are synthesized via the isoprenoid pathways where squalene acts as precursor. The squalene is synthesized by a series of several reactions with geranyl diphosphate synthase, farnesyl pyrophosphate synthase and squalene synthase through mevalonate pathway (Kuzuyama 2002) and subsequent reactions with squalene epoxidase yielded 2,3-oxidosqualene (Fig. 3). The cyclization of 2,3-oxidosqualene into dammarenediol and β-amyrin is catalyzed by oxidosqualene cyclases including dammarenediol-II synthase and β-amyrin synthase (Han et al. 2006, 2010; Tansakul et al. 2006; Kim et al. 2010). Both glycyrrhizin and soyasaponins share a common biosynthetic intermediate, β-amyrin (Fig. 3), which is synthesized by β-amyrin synthase, an oxidosqualene cyclase (OSCs). OSCs catalyze the cyclization of 2,3-oxidosqualene, a common intermediate of both triterpene and phytosterol biosynthesis (Abe et al. 1993; Haralampidis et al. 2002). In *Glycyrrhiza glabra* three OSCs: β-amyrin synthase, lupeol synthase and cycloartenol synthase are situated at the branching step for biosynthesis of

oleanane-type triterpene saponins, lupane-type triterpene (betulinic acid) and phytosterol, respectively (Fig. 4). cDNAs of β-amyrin synthase (Hayashi et al. 2001) and cycloartenol synthase (Hayashi et al. 2000) have already been isolated from cultured cells of licorice (Kölner et al. 2004; Hayashi et al. 2003).

Terpenoid biosynthesis occurs within specific tissues or at specific stages of development in plants (Nagegowda 2010; Vranova et al. 2012). In many plant species those have glandular trichomes, specialized structures for secreted terpenoid natural products (Lange and Turner 2013). The glycyrrhizin accumulates only in underground organs, stolons and roots of licorice plants (Seki et al. 2008). Avenacins, the bioactive saponins in *Avena sativa* accumulate only in epidermis of roots, where they develop resistance to plant pathogenic fungal organisms (Haralampidis et al. 2001). The biosynthesis of avenacin genes are co-regulated and exclusively expressed in the epidermis of roots in which the avenacins are accumulated (Qi et al. 2006; Field and Osbourn 2008).

Sometimes the regulation of terpenoid biosynthesis is induced by herbivore feeding, attack by pathogen or abiotic stresses (Nagegowda 2010; Vranova et al. 2012). The enhancement of concentration of terpenoids in response to various abiotic stresses is often is mediated by an increase in transcriptional activity of the specific terpenoid biosynthetic genes (Tholl 2006; Nagegowda et al. 2004, 2010; Xi et al. 2012). This type of transcriptional response is controlled by complex signaling cascade in which jasmonate hormone play important role. The pathogen attack causes transcriptional and metabolic changes in plant cell cultures of *Medicago sativa* (Suzuki et al. 2005). The defence mechanism-related synthesis of terpenoids has been studied by various authors (Van Poecke et al. 2001; Rodriguez-Saona et al. 2003). The synthesis of several terpenoids in poplar is induced and emitted from the *Malacosoma disstria* infested leaves (Arimura et al. 2004a). *Tetranychus urticae* infests on the *Lotus japonicas* and after this it induces the emission of the (E)-β-ocimene and also accumulation of (E)-β-ocimene synthase gene transcript (Arimura et al. 2004b).

In addition to the transcriptional, developmental and spatiotemporal modulation of terpenoid biosynthetic genes, the post-translational regulation mechanisms were also reported in the biosynthesis of terpenoids. The HMGR activity, the specific enzyme that catalyzes crucial regulatory steps of the mevalonic acid pathway, is controlled at the protein level through the activity of protein phosphatase A or by the E3 ubiquitin ligase *SUD1* (Leivar et al. 2011; Doblas et al. 2013). The role of jasmonic acid in the induction of trichome-specific terpene synthases has been well reported (van Schie et al. 2007). Upon perception of pathogens or herbivores, signal transduction pathways are activated, which lead to induced defense responses.

Fig. 3 Biosynthetic pathways of ginsenosides from squalene in *P. ginseng*

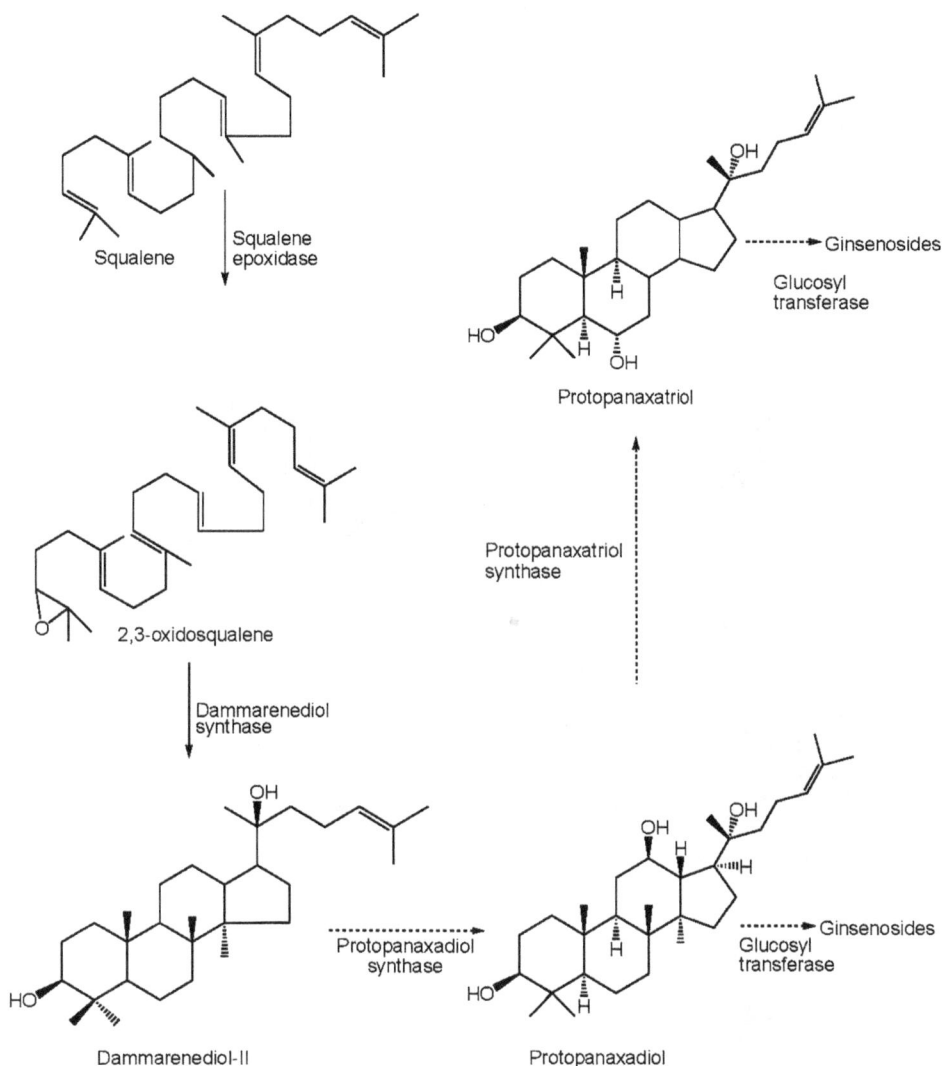

Terpenoids under clinical trials

Terpenes are the largest group of natural bioactive compounds including monoterpenes, sesquiterpenes, diterpenes, hemiterpenes and triterpenes. Out of these natural compounds, several terpenes are under studies of clinical trials, which are as follows:

D-Limonene

Monoterpenes such as D-limonene and peryl alcohol prevent mammary, live and other types of cancers. The monoterpenes have several cellular and molecular activities that could potentially underlie their positive therapeutic index. The monoterpenes inhibit the isoprenylation of small G proteins. Such inhibitions could alter signal transduction and result in altered gene expression. When mammary cancers were initiated in rats by either the direct acting carcinogen N-methyl-N-nitrosourea or indirectly

acting carcinogen DMBA, they could be prevented from developing if the carcinogen-exposed rats were fed D-limonene (Elson et al. 1988; Yoon et al. 2010).

D-Limonene is considered to have fairy low toxicity. It has been tested for carcinogenicity in mice and rats. Although initial results showed D-limonene increased the incidence of renal tubular tumours in male rats, female rats and mice in both genders showed no evidence of any tumor. Subsequent studies have determined how these tumors occur and established that D-limonene doses pose mutagenic or nephrotoxic risks to humans as well as human prostate cancers. In humans, D-limonene has demonstrated toxicity after single and repeated dosing for up to 1 year. Being a solvent of cholesterol D-limonene has been used to clinically to dissolve cholesterol containing gallstones (Igimi et al. 1991; Rabi and Bishayee 2009a, b). Because of its gastric acid neutralizing effect and its support of normal peristalsis, it has been also used for relief of heartburn and gastroesophageal reflux (Kodama et al. 1976; Sun 2007). In

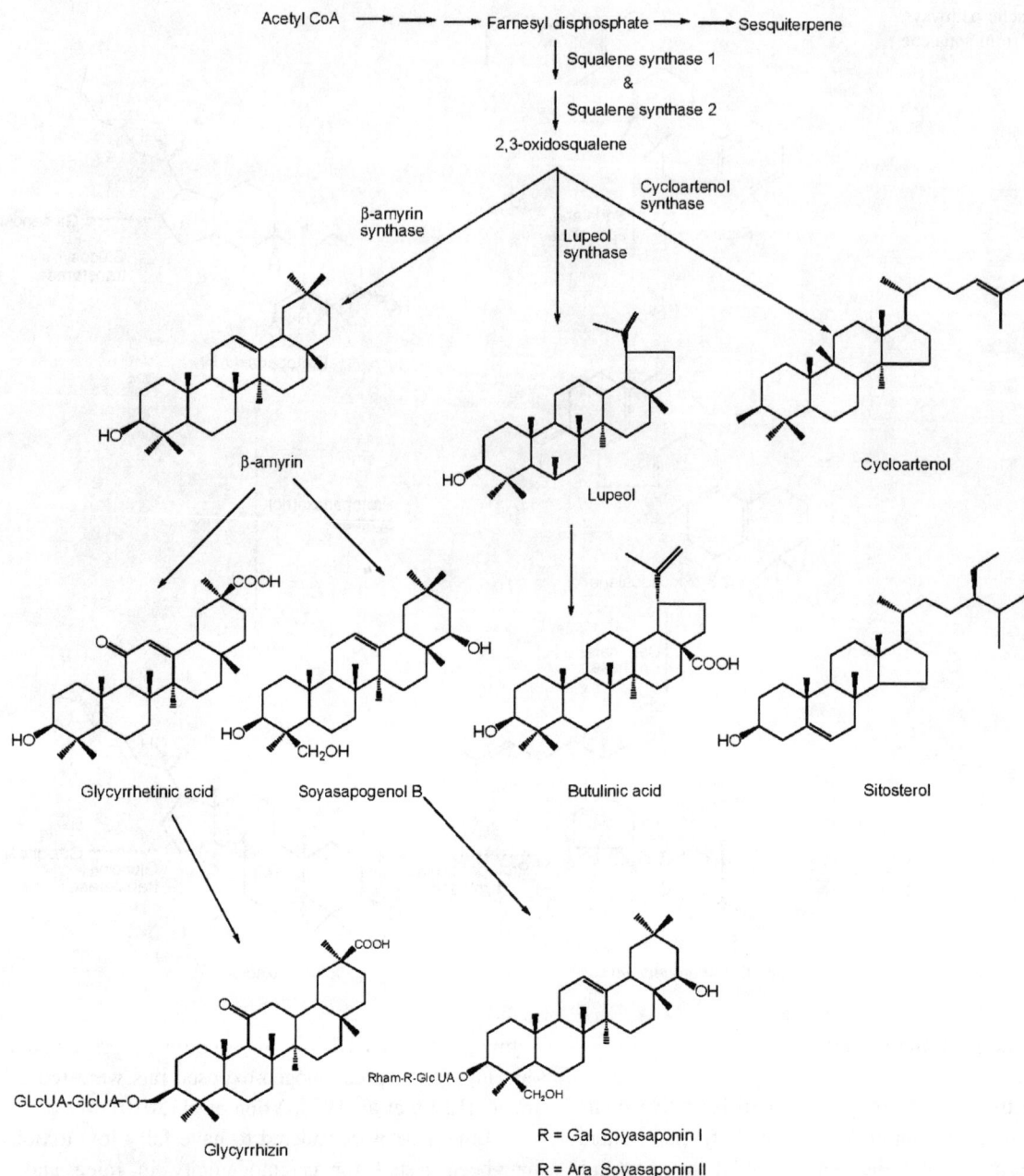

Fig. 4 Biosynthetic pathways of glycyrrhizin in *Glycyrrhiza glabra*

phase I clinical trial of orally administered D-limonene, 17 women and 15 men aged 35 to 78 with advanced metastatic solid tumors received an average of three treatment cycles of 21 days at dose ranging from 0.5 to 12.0 g/m^2 body surface area. D-Limonene was slowly absorbed, the maximal plasma concentration being attained at 1–6 h. The mean peak plasma concentrations of D-limonene were 11–20 μmol/L and the predominant metabolites were perillic acid (21–71 μmol/L), dihydroperillic acid (17–28 μmol/L) and isomers of perillic acid. After reaching these peaks, the plasma concentration decreased according to first-order kinetics (Vigushin et al. 1998;

Saldanha and Tollefsbol 2012). Carcinomas regress when D-limonene is added to the diet either when the tumour is small or still capable of spontaneously regressing. D-Limonene appears to act in cytostatic fashion. It is predicted that D-limonene inhibits the isoprenylation of small G proteins (Hogg et al. 1992; Gould et al. 1994; Miller et al. 2011).

1,8-Cineole

In humans, 1,8-cineole inhibits sensory irritations caused by octanol and methanol with sensitive volunteers. Both methanol and octanol are well-known chemicals causing skin irritation. The result that 1,8-cineole, whose ability to activate TRMP8 is lower than methanol, inhibited methanol-evoked skin irritation clearly suggests that inhibitory effects of 1,8-cineole are probably due to inhibition of TRPA1, but not activation of TRMP8 (Takashi et al. 2012; Bastos et al. 2011). 1,8-cineole not only reduces exacerbation rate but also provides clinical benefits as manifested by improved airflow obstruction, reduced severity of dyspnea and improvement of health status (Juergens et al. 1998, 2003). Therefore, it can provide a useful treatment option for symptomatic patients with COPD in addition to treatment according to the guidelines. The results have to be seen in context with socio-economic aspects. As COPD is an extremely costly disease and cause of major financial and social burden concomitant therapy with 1,8-cineole can be recommended. These finding correspond to the interpretation of the efficacy study with carbocysteine but not with acetylcysteine because this medication did not show a significant reduction of exacerbations (Zhang et al. 2008; Decramer et al. 2005; Worth et al. 2009).

Boswellic acid

More recently extracts of resin enriched in pentacyclic triterpenoid known as boswellic acid have been employed as anti-inflammatory drugs (Anthoni et al. 2006). Pilot clinical studies do indeed suggest that boswellic acid promotes pain control and dampens inflammation in osteoarthritis and colitis and helps to control the brain oedema associated with radiotherapy of cerebral tumours; anti-inflammatory effects of rodent models have also been demonstrated (Gupta et al. 2001; Kimmatkar et al. 2003). Initial attempts to clarify the molecular target of boswellic acid in inflammatory disorders determined that keto-boswellic acid can inhibit 5-lipoxygenase in low molecular concentrations (Bhushan et al. 2007). This suggested that boswellic acid preparation might dampen inflammation by blocking leukotriene synthesis (Joos et al. 2006). In vitro studies reveals boswellic acid in a dose-dependent manner blocks the synthesis of pro-inflammatory 5-lipoxygenase product including 5-hydroxyeicosatetraenoic acid and leukotriene B4, which cause bronchoconstriction chemotaxis and increase vascular permeability (Shao et al. 1998). Boswellic acid from *Boswellia serrata* also have inhibitory and apoptotic effect against the cellular growth of leukemia HL-60 cells (Huang et al. 2000). Clinical trials have demonstrated promising benefits from boswellic acids in rheumatoid arthritis, chronic colitis, ulcerative colitis, Crohn's disease and bronchial asthma in addition to benefits for brain tumour patients. The effects of boswellic acid on central signaling pathways in human platelets and on various platelet functions have been investigated. It also caused a pronounced mobilization of Ca^{2+} from internal stores and induced the phosphorylation of p38 MAPK and elicits functional platelet responses (Poeckel et al. 2005). Boswellic acids have also been observed to inhibit human leukotriene elastase which may be involved in pathogenesis of emphysema. Human leukotriene elastase also stimulates mucus secretion and thus may play a role in cystic fibrosis, chronic bronchitis and acute respiratory distress syndrome (Rall et al. 1996; Safayhi et al. 1997). However, the clinical trials of gum-resin of *Boswellia serrata* have shown to improve symptoms in patients with osteoarthritis and rheumatoid arthritis (Poeckel and Werz 2006; Poeckel et al. 2006).

Betulinic acid

Betulinic acid is a naturally occurring pentacyclic triterpene that exhibits a variety of biological activities including potent antiviral and anticancer effects (Alakurtti et al. 2006; Hsu et al. 2012). Mitochondria from cells, which were treated with betulinic acid, induced the cleavage of both caspase-3 and caspase-8 in cytosolic extracts. Cleavage of caspase-3 and 8 was preceded by disturbance of mitochondrial membrane potential and by generation of reactive oxygen species. Activation of caspase cascade was required for betulinic acid-triggered apoptosis. Interestingly, neuroblastoma cells resistant to doxorubicin-mediated apoptosis were still responsive to treatment with betulinic acid (Fulda et al. 1998; Fulda 2008). This revealed that betulinic acid inhibits the catalytic activity of topoisomerase I (Choudhary et al. 2002). Furthermore, betulinic acid exerts context dependent effects on cell cycle, it also reduces the expression of p21 protein in glioblastoma cells (Rieber and Strasberg Rieber 1998).

β-Sitosterol

It is used to prevent and relieve prostate symptoms and has been tested for thousands of years in Asia and Mediterranean where the incidence of prostate problems—including prostate cancer—is considerably lower than that in the United States and Canada (Wilt et al. 1999; Richelle et al. 2004). Taking β-sitosterol at the dose of 60–110 mg/day significantly improve urinary symptoms. It increases the maximum urinary flow and decreases the volume of the urine left in the bladder. Like saw palmetto, β-sitosterol does not affect prostate size (Awad et al. 2000). β-Sitosterol is also used to lower cholesterol. It is an ingredient in the cholesterol-lowering margarine which is used to be known as take control (Berges et al. 2000). Aging is the main cause of enlarged prostate glands (Glynn et al. 1985). Testosterone is converted into another more powerful male hormone dihydrotestosterone in prostate cells. Dihydrotestosterone is responsible for triggering the division of prostate cells so their numbers increase by cell division process. Levels of dihydrotestosterone are known to be five times higher in enlarged prostate glands in those of normal sized prostate. If the conversion of testosterone to dihydrotestosterone is prevented, the BPH may not occur and may even be reversed once it has developed. β-Sitosterol works for enlarged prostate by inhibiting an enzyme called 5-α-reductase, blocking the conversion of testosterone into dihydrotestosterone, thereby decreasing the amount of dihydrotestosterone. Since, dihydrotestosterone is considered to be responsible for the enlargement of prostate, β-sitosterol helps to support normal prostate size. β-Sitosterol has been recommended by physicians for over 20 years as natural supplement to promote prostate health (Bent and Kane 2006). Based on the highly preliminary evidence, it has been suggested that β-sitosterol may also help strengthen the immune system (Pegel 1997). One study suggests that β-sitosterol can help prevent the temporary immune weakness that typically occurs during recovery from endurance exercise and can lead to post-race infections (Bouic et al. 1999). A randomized controlled trial of 47 patients with pulmonary tuberculosis investigated adjuvant β-sitosterol therapy vs placebo (Silveira e Sá et al. 2013). The β-sitosterol treatment group (average dose 60 mg/day) demonstrated increased weight gain, higher lymphocyte and eosinophil count and a generally faster clinical recovery (Donald et al. 1997).

Ursolic acid

Mice fed with ursolic acid diet for 8 weeks delayed formation of prostate intraepithelial neoplasia. Similarly, mice fed with ursolic acid for 6 weeks inhibited progression of prostate intraepithelial neoplasia to adenocarcinoma as determined by hematoxylin and eosin staining. With respect to the molecular mechanism, it was observed that ursolic acid down regulated the activation of various pro-inflammatory mediators including NF-κB, TNF-α and IL-6 (Shanmugum et al. 2012).

Future prospects

The terpenoids are synthesized from two five-carbon building blocks, which are known as isoprenoid units. Based on the number of building blocks, terpenoids are grouped into several classes, such as monoterpenes (e.g. carvone, geraniol, D-limonene and peril alcohol), diterpenes (e.g. retinol and retinoic acid), triterpenes (e.g. betulinic acid, lupeol, oleanolic acid and ursolic acid) and tetraterpenes (e.g. α-carotene, β-carotene, lutein, and lycopene) (Thoppil and Bishayee 2011; Rabi and Bishayee 2009a; Withers and Keasling 2007). Terpenoids have been found to be useful for the treatment of various types of diseases and disorders viz, antimicrobial, antifungal, anti-parasitic, antiviral, antihyperglycemic, antihypoglycemic, anti-inflammatory and immunomodulatory properties (Wagner and Elmadfa 2003; Shah et al. 2009; Sultana and Ata 2008).

The development of transgenic plants against biotic stress like as insects has been a major successful scientific approach, mirrored by practical success of a limited number of pest-resistant transgenic crops in various countries. In some well-developed countries, this has been a result of vocal opposition to plant transformation technology itself; but in many examples, in both developed and developing countries, it is more a case of great potential economic benefits not being sufficient to make the introduction of transformed varieties of crops commercially viable.

The development of zero-cannabinoid cannabis chemotype has provided crude drugs that will facilitate discernment of the pharmacological effects and contributions of different fractions. Breeding work has already resulted in chemotypes that produce 98 % of monoterpenes as myrcene or 77 % as D-limonene. Through selective breeding of high terpenoid and phytocannabinoid-specific chemotype, has thus become rational target that may lead to novel approaches in treatment of different types of diseases and disorders.

The terpenes play important roles in plant interactions, plant defenses and the other environmental stresses (Chen et al. 2011). To better understand the physiological and ecological roles of specific terpene synthase genes and enzymes research in various areas is required. Roles of specific terpenes or general roles of classes of terpene synthases, regulation of terpene synthesis, phylogeny of terpene synthases, must be examined thoroughly in plants and ideally in the natural environments of the plants that produce these terpenes. The terpene synthase gene's manipulation and their expression in model and non-model plants will be critical at this end. Despite many discoveries about the functions of terpene synthase genes, the ongoing and future structural and biochemical investigations of terpene synthases will continue to be field of exciting new discoveries. At present, functional characterization and regulation of terpene synthases have been completed only for subsets of terpene synthase families, including *Arabidopsis*, grapevine and spruce. Similar to other plant species, the biochemical functions and phylogeny of various number of terpene synthase proteins have already been established and known; there are only few experimentally determined three-dimensional structures of terpene synthases of plant origin (Gennadios et al. 2009; Kampranis et al. 2007). At the same time, large transcriptome sequencing projects targeted at plants species that produce interesting medicinally important metabolites will enhance the identification of comprehensive sets of terpene synthase genes in a large variety of non-model systems. An improved and updated knowledge on regulation of terpene metabolism and phylogeny of terpene synthases will facilitate the manipulation of terpene biosynthetic pathways for improvement of agronomic traits, biotransformation of medicinally important terpenes, floral scents (Lücker et al. 2001), plant defense against pests and pathogens (Schnee et al. 2006) and production of known and novel phytocompounds (Bohlmann and Keeling 2008).

Acknowledgments Authors are very much grateful to Professor (Dr.) A. N. Pathak, Director, Institute of Biotechnology, Amity University Rajasthan, for proof reading and moral support in preparing of this review paper.

Conflict of interest Authors hereby declare no conflict of interest.

References

Abe I, Rohmer M, Prestwich GD (1993) Enzymatic cyclization of squalene and oxidosqualene to sterols and triterpenes. Chem Rev 93:2189–2206

Adam KP, Zapp J (1998) Biosynthesis of the isoprene units of chamomile sesquiterpenes. Phytochemistry 48:953–959

Aharoni A, Giri AP, Deuerlein S, Griepink F, de-Kogel WJ, Verstappen FWA, Verhoeven HA, Jongsma MA, Schwab W, Bouwmeester HJ (2003) Terpenoid metabolism in wild type and transgenic *Arabidopsis* plants. Plant Cell 15:2866–2884

Ait-Ali T, Swain SM, Reid JB, Sun TP, Kamiya Y (1994) The *LS* locus of pea encodes the gibberellin biosynthesis enzyme *ent*-kaurene synthase A. Plant J 11:443–454

Alakurtti S, Mäkelä T, Koskimies S, Yli-Kauhaluoma J (2006) Pharmacological properties of the ubiquitous natural product betulinic acid. Eur J Pharm Sci 29:1–13

Anthoni C, Laukoetter MG, Rizken E (2006) Mechanism underlying the anti-inflammation actions of boswellic acid derivatives in experimental colitis. Am J Physiol Gastrointest Liver Physiol 290:G1131–G1137

Aoyogi K, Beyou A, Moon K, Fang L, Ulrich T (1993) Isolation and characterization of cDNAs encoding wheat 3-hydroxy-3-methylglutaryl CoA reductase. Plant Physiol 102:623–628

Arimura G, Huber DPW, Bohlmann J (2004a) Forest tent caterpillar (*Malacosoma disstria*) induce local and systemic diurnal emissions of terpenoid volatiles in hybrid poplar (*Populus trichocarpa* × *deltoids*: cDNA cloning functional characterization and patterns of gene expression of (−)-germacrene D-synthase, PtdTPS1. Plant J 37:603–616

Arimura G, Ozawa R, Kugimiya S, Takabayashi J, Bohlmann J (2004b) Herbivore-induced defence response in a model legume: two spotted spider mites *Tetranychus urticae*, induce emission of (*E*)-β-ocimene synthase in *Lotus japonicas*. Plant Physiol 135:1976–1983

Ashour M, Wink M, Gershenzon J (2010) Biochemistry of terpenoids: monoterpenes, sesquiterpenes and diterpenes. In: Wink M (ed) Annual plant reviews: biochemistry of plant secondary metabolism, vol 40, 2nd edn. Wiley, New York

Atsbaha Zebelo S, Bertea CM, Bossi S, Occhipinti A, Gnavi G, Maffei ME (2011) *Chrysolina herbacea* modulate terpenoid biosynthesis of *Mentha aquatica* L. PLoS One 6:e17195

Aubourg S, Lecharny A, Bohlmann J (2002) Genomic analysis of the terpenoid synthase gene family of *Arabidopsis thaliana*. Mol Genet Genomics 267:730–745

Augustin JM, Kuzina V, Andersen SB, Bark S (2011) Molecular activities, biosynthesis and evolution of triterpenoid saponin. Phytochemistry 72:435–457

Awad AB, Chan KC, Downie AC, Fink KC (2000) Peanuts as source of β-sitosterol, a sterol with anticancer properties. Nutr Cancer 36:238–241

Bach TJ (1995) Some aspects of isoprenoid biosynthesis in plants? a review. Lipids 30:191–202

Back K, Chappel J (1996) Identifying functional domains within terpene cyclases using a domain-swapping strategy. Proc Natl Acad Sci USA 93:6841–6845

Baker J, Franklin DB, Parker J (1992) Sequence and characterization of the gcpE gene of *Escherichia coli*. FEMS Microbiol Lett 73:175–180

Baker CH, Matsuda SPT, Liu DR, Corey EJ (1995) Molecular cloning of the human gene encoding lanosterol synthase from a liver cDNA library. Biochem Biophys Res Commun 213:154–160

Balandrin MF, Klocke JA, Wurtele ES, Bollinger WH (1985) Natural plant chemicals: sources of industrial and medicinal materials. Science 228:1154–1160

Basson ME, Thorsness M, Finer-Moore J, Stroud RM, Rine J (1988) Structural and functional conservation between yeast and human 3-hydroxy-3-methylglutaryl CoA reductase, the rate limiting enzyme of sterol biosynthesis. Mol Cell Biol 8:3797–3808

Bastos VPD, Gomes AS, Lima FJB, Brito JB, Soares PMG, Pinho JPM, Silva CS, Santos AA, Souza MHLP, Magalhâes PJC (2011) Inhaled 1, 8-cineole reduces inflammatory parameters in airways of ovalbumin-challenged guinea pigs. Basic Clin Pharmacol Toxicol 108:34–39

Basyuni M, Oku H, Inafuka M, Baba S, Iwasaki H, Oshiro T, Okabe T, Shibuya M, Ebizuka Y (2006) Molecular cloning and functional expression of a multifunctional triterpene synthase cDNA from a mangrove species *Kandelia candel* (L.) Druce. Phytochemistry 67:2517–2524

Basyuni M, Oku H, Tsuijimoto E, Kinjo K, Baba S, Takara K (2007) Triterpene synthases from the Okinawan mangrove tribe, Rhizophoraceae. FEBS J 274:5028–5042

Beilen JB, Poirier Y (2007) Establishment of new crops for the production of natural rubber. Trends Biotechnol 25:522–529

Bensen RJ, Gohal GS, Crane VC, Tossberg JT, Schnable PS, Meeley RB, Briggs SP (1995) Cloning and characterization of the maize An1 gene. Plant Cell 7:75–84

Bent S, Kane C (2006) Saw palmetto for benign prostatic hyperplasia. N Engl J Med 354:557–566

Berges RR, Kassen A, Seng T (2000) Treatment of symptomatic benign prostate hyperplasia with β-sitosterol: an 18 month follow-up. BJU Int 85:842–846

Besser K, Harpel A, Welsby N, Schauvinhold I, Slocombe S, Li Y, Dixon RA, Broun P (2008) Divergent regulation of terpenoid metabolism in the trichomes of wild and cultivated tomato species. Plant Physiol 149:499–514

Bhushan S, Kumar A, Malik F (2007) A triterpenediol from *Boswellia serrata* induces apoptosis through both the intrinsic and extrinsic apoptotic pathways in human leukemia HL-60 cells. Apoptosis 12:1911–1926

Bick JA, Lange BM (2003) Metabolic crosstalk between cytosolic and plastidial pathways of isoprenoid biosynthesis:

unidirectional transport of intermediates across the chloroplast envelope membrane. Arch Biochem Biophys 415:146–154

Bohlmann J (2012) Pine terpenoid defences in the mountain pine beetle epidemic and in other conifer pest interactions: specialized enemies are eating holes into a diverse, dynamic and durable defence system. Tree Physiol 32:943–945

Bohlmann J, Keeling CI (2008) Terpenoid biomaterials. Plant J 54:656–669

Bohlmann J, Meyer-Gauen G, Croteau R (1998) Plant terpenoid synthases: molecular biology and phylogenetic analysis. Proc Natl Acad Sci USA 95:4126–4133

Bohlmann J, Phillips M, Ramachandiran V, Katoh S, Croteau R (1999) cDNA cloning, characterization and functional expression of four new monoterpene synthase members of the Tpsd gene family from grand fir (Abies grandis). Arch Biochem Biophys 368:232–243

Boughton AJ, Hoover K, Felton GW (2005) Methyl jasmonate application induces increased densities of glandular trichomes on tomato, Lycopersicon esculentum. J Chem Ecol 31:2211–2216

Bouic PJD, Clark A, Lamprecht J (1999) The effects of β-sitosterol glucoside mixture on selected immune parameters of marathon runners: inhibition of post marathon immune suppression and inflammation. Int J Sports Med 20:258–262

Brendolise C, Yauk YK, Eberhard ED, Wang M, Chagne D, Andre DR, Beuning LL (2011) An unusual plant triterpene synthase with predominant α-amyrin producing activity identified by characterizing oxidosqualene cyclases from Malus domestica. FEBS J 278:2485–2499

Burse A, Frick S, Discher S, Tolzin-Banasch K, Kirsch R, Strauß A, Kunert M, Boland W (2009) Always being well prepared for defense: the production of deterrents by juvenile Chrysomelina beetles (Chrysomelidae). Phytochemistry 70:1899–1909

Busquets A, Keim V, Closa M, del Arco A, Boronat A, Arro B, Ferrer A (2008) Arabidopsis thaliana contains single gene encoding squalene synthase. Plant Mol Biol 67:25–36

Calisto BM, Perez-Gil J, Bergua M, Querol-Audi J, Pita I, Imperial S (2007) Biosynthesis of isoprenoids in plants: structure of 2C-methyl-D-erythritol 2,4-cyclodiphosphate synthase from Arabidopsis thaliana. Comparison with bacterial enzymes. Protein Sci 16:2082–2088

Carrie C, Murcha MW, Millar AH, Smith SM, Whelam J (2007) Nine 3-ketoacyl-CoA thiolases (KATs) and acetoacetyl-CoA thiolases (ACATs) encoded by five genes in Arabidopsis thaliana are targeted either to peroxisomes or cytosol but not to mitochondria. Plant Mol Biol 63:97–108

Chen XY, Wang M, Chen Y, Davisson VJ, Heinstein P (1996) Cloning and heterologous expression of a second (+)-δ-cadinene synthase from Gossypium arboreum. J Nat Prod 59:944–951

Chen F, Tholl D, D'Auria JC, Farooq A, Pichersky E, Gershenzon J (2003) Biosynthesis and emission of terpenoid volatiles from Arabidopsis flowers. Plant Cell 15:481–494

Chen F, Tholl D, Bohlmann J, Pichersky E (2011) The family of terpene synthase in plants: a mid size family of genes for specialized metabolism that is highly diversified throughout the kingdom. Plant J 66:212–229

Cheniclet C, Rafia F, Salnt-Guily A, Verna A, Cadre JR (1992) Localization of the enzyme geranyl geranyl pyrophosphate synthase in Capsicum fruits by immunogold cytochemistry after conventional chemical fixation or quick freezing followed by freeze substitution: labeling evolution during fruit ripening. Biol Cell 75:145–154

Choudhary AR, Mandal S, Mitra B, Sharma S, Mukhopadhyay S, Majumdar HK (2002) Betulinic acid, a potent inhibitor of eukaryotic topoisomerase I: identification of the inhibitory step, the major functional group responsible and development of more potent derivatives. Med Sci Monit 8:BR254–BR265

Creelman RA, Mullet JE (1997) Biosynthesis and action of jasmonates in plants. Annu Rev Plant Physiol Plant Mol Biol 48:355–381

Croteau R, Ketchum RE, Long RM, Kaspera R, Wildung MR (2006) Taxol biosynthesis and molecular genetics. Phytochem Rev 5:75–97

Cseke L, Dudareva J, Pichersky E (1998) Structure and evolution of linalool synthase. Mol Biol Evol 15:1491–1498

Cunillera N, Boronat A, Ferrer A (1997) The Arabidopsis thaliana FPS1 gene generates a novel mRNA that encodes a mitochondrial farnesyl diphosphate synthase isoform. J Biol Chem 272:15381–15388

Cunningham FX, Lafend TP, Gannt E (2000) Evidence of a role for LytB in the non-mevalonate pathway of isoprenoid biosynthesis. J Bacteriol 182:5841–5848

Decendit A, Liu D, Ouelhazi L, Doireau P, Mérillon J, Rideau M (1992) Cytokinin-enhanced accumulation of indole alkaloids in Catharanthus roseus cell cultures: the factor affecting the cytokinin response. Plant Cell Rep 11:400–403

Decramer M, Rutten-van Mölken M, Dekhuijzen PN, Trooster T, van Herwaarden C, Pellegrino R, van Schayck CP, Olivieri D, Del Donno M, De Backer W, Lankhorst I, Ardia A (2005) Effects of N-acetylcysteine on outcomes in chronic obstructive disease: a randomized placebo-controlled trial. Lancet 365:1552–1560

Dedryver CA, Lee Ralee A, Fabre F (2010) The conflicting relationships between aphids and men: a review of aphid damage and control strategies. C R Biol 333:539–553

Degenhardt J, Kollner TG, Gershenzon J (2009) Monoterpene and sesquiterpene synthases and the origin of terpene skeletal diversity in plants. Phytochemistry 70:1621–1637

De-Geyter N, Gholami A, Goormachtig S, Goosens A (2012) Transcriptional machineries in jasmonate-elicited plant secondary metabolism. Trends Plant Sci 17:349–359

Dhe-Paganon S, Magrath J, Abeles RH (1994) Mechanism of mevalonate pyrophosphate decarboxylase: evidence for a carbocationic transition state. Biochemistry 33:13355–13362

Doblas VG, Amorim-Silva V, Pose D, Rosado A, Esteban A, Arro M, Azevedo H, Bombarley A, Borsani O, Valpuesta V (2013) The SUD1 gene encodes a putative E3 ubiquitin ligase and is a positive regulation of 3-hydroxy-3-glutaryl CoA reductase activity in Arabidopsis. Plant Cell 25:728–743

Donald PR, Lamprecht JH, Freestone M (1997) A randomized placebo-controlled trial of the efficacy of β-sitosterol and its glucoside as adjuvants in the treatment of pulmonary tuberculosis. Int J Tuberc Lung Dis 1:518–522

Dornelas MC, Mazzafera P (2007) A genomic approach to characterization of the Citrus terpene synthase gene family. Genet Mol Biol 30:832–840

Drew DP, Dueholm B, Weitzel C, Zhang Y, Sensen CW, Simonsen HT (2013) Transcriptome analysis of Thapsia laciniata Rouy provides insights into terpenoid biosynthesis and diversity in Apiaceae. Int J Mol Sci 14:9080–9098

Ducreux LJM, Morris WL, Prosser IM, Morris JA, Beale MH, Wright F, Shepherd T, Bryan JM, Hedley PE, Taylor MA (2008) Expression profiling of potato germplasm differentiated in quality traits leads to the identification of candidate flavor and texture genes. J Exp Bot 59:4219–4231

Dudareva J, Cseke L, Blanc VM, Pichersky E (1996) Evolution of floral scent in Clarkia: novel patterns of S-linalool synthase gene expression in the C. breweri flower. Plant Cell 8:1137–1148

Dudareva N, Martin D, Kish CM, Kolosova N, Gorenstein N, Fäldt J, Miller B, Bohlmann J (2003) (E)-β-ocimene and myrcene synthase genes of floral scent biosynthesis in snapdragon: function and expression of three terpene synthase genes of a new terpene synthase sub-family. Plant Cell 15:1227–1241

Dudareva N, Negre F, Nagegowda DA, Orlova I (2006) Plant volatiles: recent advances and future perspectives. Crit Rev Plant Sci 25:417–440

Elson CE, Maltzman TH, Boston JL (1988) Anticarcinogenic activity of D-limonene during the initiation and promotion/progression stage of DMBA-induced rat mammary carcinogenesis. Carcinogenesis 9:331–332

El-Wakeil NE, Wolkmar C, Sallam AA (2010) Jasmonic acid induces resistance to economically important insect pests in winter wheat. Pest Manag Sci 66:549–554

Facchini PJ, Chappel J (1992) Gene family for an elicitor induced sesquiterpene cyclase in tobacco. Proc Natl Acad Sci USA 89:11088–11092

Falara V, Akhtar TA, Nguyen TT, Spyropoulou EA, Blecker PM, Schauvinhold I, Matsuba Y, Bonini ME, Schimiller AL, Last RL, Schuurink RC, Richersky E (2011) The tomato terpene synthase gene family. Plant Physiol 157:770–789

Fäldt J, Martin D, Miller B, Rawat S, Bohlmann J (2003) Traumatic resin defence in Norway spruce (*Picea abies*): methyl jasmonate-induced terpene synthase gene expression and cDNA cloning and functional characterization of (+)-3-carene synthase. Plant Mol Biol 51:119–133

Field B, Osbourn AE (2008) Metabolic diversification-independent assembly of operon like gene clusters in different plants. Science 320:543–547

Fulda S (2008) Betulinic acid for cancer treatment and prevention. Int J Mol Sci 9:1096–1107

Fulda S, Susin SA, Kroemer G, Debatin KM (1998) Molecular ordering of apoptosis induced by anticancer drugs in neuroblastoma cells. Cancer Res 58:4453–4460

Gantet P, Imbault N, Thiersault M, Doireau P (1998) Necessity of a functional octadecanoic pathway for indole synthesis by *Catharanthus roseus* cell suspension cultured in an auxin-starved medium. Plant Cell Physiol 39:220–225

Gaut BS (2002) Evolutionary dynamics of grass genomes. New Phytol 154:15–28

Gennadios HA, Gonzalez V, Di Costanzo L, Li A, Yu F, Miller DJ, Allemann RK, Christianson DW (2009) Crystal structure of (+)-δ-cadinene synthase from *Gossypium arboreum* and evolutionary divergence of metal binding motifs for catalysis. Biochemistry 48:6175–6183

Gershenzon J, Kreish W (1999) Biochemistry of terpenoids: monoterpenes, sesquiterpenes, diterpenes, sterols, cardiac glycosides and steroid saponins. In: Wink M (ed) Biochemistry of plant secondary metabolism. CRC Press, Florida, pp 222–299

Ginger ML, McFadden GI, Michels PAM (2010) Rewiring and regulation of cross-compartmentalized metabolism in protists. Phil Trans R Soc Biol 365:831–845

Glas JJ, Schimmel BCJ, Alba JM, Escobar-Bravo R, Schuurink RC, Kant MR (2012) Plant glandular trichomes as targets for breeding or engineering of resistance to herbivores. Intl J Mol Sci 13:17077–17103

Glynn RJ, Campion EW, Bouchard GR, Silbert JE (1985) The development of benign of prostatic hyperplasia among volunteers in the normative aging study. Am J Epidemiol 121:78–90

Gould MN, Moore CJ, Zhang R, Wang B, Kennan WS, Hagg JD (1994) Limonene chemoprevention of mammary carcinoma induction following direct in situ transfer of v-Ha-ras. Cancer Res 54(3540):3543

Grauvogel C, Peterson J (2007) Isoprenoid biosynthesis authenticates the classification of the green alga *Mesostigma viride* as an ancient streptophyte. Gene 396:125–133

Gray DW, Breneman SR, Topper LA, Sharkey ID (2011) Biochemical characterization and homology modeling of methyl butenol synthase and implication for understanding hemiterpene synthase evolution in plants. J Biol Chem 286:20582–20590

Guhling O, Hobl B, Yeast T, Jetter R (2006) Cloning and characterization of lupeol synthase involved in the synthesis of epicuticular wax crystals on stem and hypocotyl surfaces of *Ricinus communis*. Arch Biochem Biophys 448:60–72

Gupta I, Parihar A, Malhotra P (2001) Effects of gum-resin of *Boswellia serrata* in patients with chronic colitis. Planta Med 67:391–395

Han JY, Kwon YS, Yang DC, Jung YR, Choi YE (2006) Expression and RNA interference-induced silencing of the dammarenediol synthase gene in *Panax ginseng*. Plant Physiol 47:1653–1662

Han JY, In JG, Kwon YS, Choi YE (2010) Regulation of ginsenoside and phytosterol biosynthesis by RNA interferences of squalene epoxide gene in *Panax ginseng*. Phytochemistry 71:36–46

Haralampidis K, Bryan G, Qi X, Papadopoulou K, Bakht S, Melton R, Osbourn A (2001) A new class of oxidosqualene cyclases directs synthesis of antimicrobial phytoprotectants in monocots. Proc Natl Acad Sci USA 98:13431–13436

Haralampidis K, Trojanowska M, Osbourn AE (2002) Biosynthesis of triterpenoid saponins in plants. Adv Biochem Eng Biotechnol 75:31–49

Hare JD, Walling LL (2006) Constitutive and jasmonate-inducible traits of *Datura wrightii*. J Chem Ecol 32:29–47

Hayashi H, Hiraoka N, Ikeshiro Y, Kushiro T, Morita M, Shibuya M, Ebizuka Y (2000) Molecular cloning and characterization of cDNA for *Glycyrrhiza glabra* cycloartenol synthase. Biol Pharm Bull 23:231–234

Hayashi H, Huang PY, Kirakosyan A, Inoue K, Hiraoka N, Ikeshiro Y, Kushiro T, Shibuya M, Ebizuka Y (2001) Cloning and characterization of a cDNA encoding β-amyrin synthase involved in glycyrrhizin and soyasaponin biosynthesis in licorice. Biol Pharm Bull 24:912–916

Hayashi H, Huang P, Inou K (2003) Up-regulation of soyasaponin biosynthesis by methyl jasmonate in cultured cells of *Glycyrrhiza glabra*. Plant Cell Physiol 44:404–411

Hayashi H, Hung P, Takada S, Obinata M, Inoue K, Shibuya M, Ebizuka Y (2004) Differential expression of three oxidosqualene cyclase mRNAs in *Glycyrrhiza glabra*. Biol Pharm Bull 27:1086–1092

Hayashi K, Kawaide H, Notomi M, Sakigi Y, Matsuo A, Nozaki H (2006) Identification and functional analysis of bifunctional *ent*-kaurene synthase from the moss *Physcomitrella patens*. FEBS Lett 580:6175–6181

Hedhili S, Courdavault V, Giglioli-Guivarc'h N, Gantet P (2007) Regulation of terpene moiety biosynthesis of *Catharanthus roseus* terpene indole alkaloids. Phytochem Rev 6:341–351

Hemmerlin A, Hoeffler JF, Meyer O, Tritsch D, Kagan IA, Grosdemange-Billiard C, Rohmer M, Bach TJ (2003) Cross talk between the cytosolic mevalonate and the plastidial methylerythritol phosphate pathways in tobacco bright yellow-2 cells. J Biol Chem 278:26666–26676

Herrera LP, Casas CE, Bates ML, Guest JD (2005) Ultrastructural study of the primary olfactory pathway in *Macaca fascicularis*. J Comp Neurol 488:427–441

Herz S, Wungsintaweekul J, Schulir CA, Hecht S, Luttgen H, Sagner S, Fellermeier M, Eisenreich W, Zenk MH, Bacher A, Rohdich F (2000) Biosynthesis of terpenoids: YgbB protein converts 4-diphosphocytidyl-2C-methyl-D-erythritol 2,4-cyclodiphosphate. Proc Natl Acad Sci USA 97:2486–2490

Hiltpold I, Turlings TCJ (2008) Below ground chemical signaling in maize: when simplicity rhymes with efficiency. J Chem Ecol 34:628–635

Hogg JD, Lindstorm MJ, Gould MN (1992) Limonene-induced regression of mammary carcinomas. Cancer Res 52:4021–4026

Hseih MH, Chang CY, Hsu SJ, Chen JJ (2008) Chloroplast localization of mitochondrial genes of methylerythritol-4-phosphate pathway enzymes and regulation of mitochondrial genes in ispE albino mutants in *Arabidopsis*. Plant Mol Biol 66:663–673

Hsu TI, Wang MC, Chen SY, Huang ST, Yeh YM, Su WC, Chang WC, Hung JJ (2012) Betulinic acid decreases specificity protein 1 (sp1) level via increasing the sumoylation of sp1 to inhibit lung cancer growth. Mol Pharmacol 82:1115–1128

Huang MT, Badmaev V, Ding Y (2000) Antitumor and anticarcinogenic activities of triterpenoid β-boswellic acid. Biofactors 13:225–230

Huang M, Abel C, Sohrabi R, Petri J, Haupt I, Cosmano J, Gershenzon J, Tholl D (2010) Variation of herbivore-induced volatile terpene among *Arabidopsis* ecotypes depends on allelic differences and subcellular targeting of two terpene synthases TPS02 and TPS03. Plant Physiol 153:1293–1310

Hudgins JW, Franceschi VR (2004) Methyl jasmonate-induced ethylene production is responsible for conifer phloem defense responses and reprogramming of stem cambial zone for traumatic resin duct formation. Plant Physiol 135:2134–2149

Hwang BK, Sung NK (1989) Effect of metalaxyl on capsidiol production in stems of pepper plants infected with *Phytophthora capsici*. Plant Dis 73:748–751

Igimi H, Tamura R, Toraishi K (1991) Medical dissolution of gallstones, clinical experience of D-limonene as a sample, safe and effective solvent. Dig Dis Sci 36:200–208

Igual JC, Gonzalez-Bosch C, Dopazo J, Perez-Ortin JE (1992) Phylogenetic analysis of thiolase family: implications for the evolutionary origin of peroxisomes. J Mol Evol 35:147–155

Iijima Y, Davidovich-Rikanati R, Fridman E, Garg DR, Bark K, Lewinsohn E, Pichersky E (2004) The biochemical and molecular basis for the divergent patterns in the biosynthesis of terpenes and phenyl propenes in the peltate glands of the three cultivars of basil. Plant Physiol 136:3724–3736

Irmisch S, Krause ST, Kunert G, Gershenzon J, Degenhardt J, Kollner TG (2012) The organ specific expression of terpene synthase genes contributes to the terpene hydrocarbon composition of chamomile essential oils. BMC Plant Biol 12:84

Isaacson T, Kosma DK, Matas AJ, Buda GJ, He Y, Yu B, Pravitasari A, Batteas JD, Stark RE, Jenks ME, Rose JKC (2009) Cutin deficiency in the tomato fruit cuticle consistently affects resistance to microbial infection and biomechanical properties, but not transpirational water loss. Plant J 60:363–377

Iturbe-Ormaetxe I, Haralampidis K, Papadopoulou K, Osbourn AE (2003) Molecular cloning and characterization of triterpene synthases from *Medicago truncatula* and *Lotus japonicus*. Plant Mol Biol 51:731–743

Jaio YN, Wickett NJ, Ayyampalayam S, Chanderbali AS, Landherr L, Ralph PE, Tonisho LP, Hu Y, Liang HY, Soltis PS (2011) Ancestral polyploidy in seed plants and angiosperms. Nature 473:97–100

Jenner H, Townsend B, Osbourn A (2005) Unravelling triterpene glycoside synthesis in plants: phytochemistry and functional genomics join forces. Planta 220:503–506

Jia X, Xia D, Louzada ES (2005) Molecular cloning and expression analysis of a putative terpene synthase gene from *Citrus*. J Am Soc Hortic Sci 130:454–458

Joos SS, Rosemann T, Szecsenyi J, Hahn EG, Willich SN, Brinkhaus B (2006) Use of complementary and alternative medicine in Germany—a survey of patients with inflammatory bowel disease. BMC Complement Altern Med 6:19

Juergens UR, Stober M, Vetter H (1998) Steroid like inhibition of monocyte arachidonic acid metabolism and IL-1β production by 1,8-cineole. Atemwegs Lungenkrankneitan 24:3–11

Juergens UR, Dethlefsen U, Steinkamp G, Gillisena RR, Vetter H (2003) Anti-inflammatory activity of 1,8-cineole in bronchial asthma: a double-blind placebo-controlled trial. Respir Med 97:250–256

Kai G, Zho AL, Zhang L, Li Z, Guo B, Zhang D (2006) Characterization and expression profile analysis of a new cDNA

encoding taxadiene synthase from *Taxus media*. J Biochem Mol Biol 38:668–675

Kajikawa M, Yamato KT, Fukuzawa H, Sakai Y, Uchida H, Ohyama K (2005) Cloning and characterization of a cDNA encoding β-amyrin synthase from petroleum plant *Euphorbia tirucalli* L. Phytochemistry 66:1759–1766

Kampranis SC, Loannidis D, Purvis A, Mahrez W, Ninga E, Katerelos NA, Anssour S, Dunwell JM, Degenhardt J, Makris AM, Goodenough PW, Johnson CB (2007) Rational conversion of substrate and product specificity in a *Salvia* monoterpene synthase: structural insights into the evolution of terpene synthase function. Plant Cell 19:1994–2005

Kappers IF, Aharoni A, van Harpen TWJM, Luckerhoff LLP, Dick M, Bouwmeester HJ (2005) Genetic engineering of terpenoid metabolism attracts bodyguards to *Arabidopsis*. Science 309:2070–2072

Kawano N, Ichinose K, Ebizuka Y (2002) Molecular cloning and functional expression of cDNAs encoding oxidosqualene cyclases from *Costus speciosus*. Biol Pharm Bull 27:1086–1092

Kawoosa T, Singh H, Kumar A, Sharma SK, Devi K, Dutt S, Vats SK, Sharma M, Ahuja PS, Kumar S (2010) Light and temperature regulated terpene biosynthesis: hepatoprotective monoterpene picroside accumulation in *Picrorhiza kurrooa*. Funct Integr Genomics 10:393–404

Keeling CI, Weisshar S, Lin RP, Bohlmann J (2008) Functional plasticity of paralogous diterpene synthases involved in conifer defense. Proc Natl Acad Sci USA 105:1085–1090

Keeling CI, Dullat HK, Ralph SG, Jancsik S, Bohlmann J (2010) Identification and function characterization of monofunctional *ent*-copalyl diphosphate and *ent*-kaurene synthases in white spruce (*Picea glauca*) reveal different patterns for diterpene synthase evolution for primary and secondary metabolism in gymnosperms. Plant Physiol 152:1197–1208

Keeling CI, Weisshaar S, Ralph SG, Jancsik S, Hamberger B, Dullat HK, Bohlmann J (2011) Transcriptome mining, functional characterization and phylogeny of a large terpene synthase family in spruce (*Picea* spp.). BMC Plant Biol 11:43

Ketchum RE, Gibson DM, Croteau RB, Shuler ML (1999) The kinetics of taxoid accumulation in cell suspension cultures of *Taxus* following elicitation with methyl jasmonate. Biotechnol Bioeng 62:97–105

Kim OT, Bang KH, Jung SJ, Kim YC, Kim SH, Hyun DY, Cha SW (2010) Molecular characterization of ginseng farnesyl diphosphate synthase gene and its up-regulation by methyl jasmonate. Biol Planta 54:47–53

Kimmatkar N, Thawani V, Hingorani L (2003) Efficacy and tolerability of *Boswellia serrata* extract in treatment of osteoarthritis of knee—a randomized double-blind placebo controlled trial. Phytomedicine 10:3–7

Kirby J, Romanini DW, Paradise EM, Keasling JD (2008) Engineering triterpene production in *Saccharomyces cerevisiae*-β-amyrin synthase from *Artemisia annua*. FEBS J 275:1852–1859

Kodama R, Yano T, Furukawa K, Noda K, Ide H (1976) Studies on the metabolism of D-limonene. IV. Isolation and characterization of new metabolites and species differences in metabolism. Xenobiotica 6:377–389

Kolesnikova MD, Xiong QB, Lodeiro S, Hua L, Matsuda SPT (2006) Lanosterol biosynthesis in plants. Arch Biochem Biophys 447:87–95

Kölner TG, Schnee C, Gershenzon J, Degenhardt J (2004) The variability of sesquiterpenes from two *Zea mays* cultivars is controlled by allelic variations of two terpene synthase genes encoding stereoselective multiple product enzymes. Plant Cell 16:1115–1131

Korth KL, Jaggard DAW, Dixon RA (2000) Developmental and light regulated post-translational control of 3-hydroxy-3-methylglutaryl CoA reductase level in potato. Plant J 23:507–516

Krishnan NM, Pattnaik S, Jain P, Gaur P, Chaudhary R, Vaidyanathan S, Deepak S, Hariharan AK, Bharathkrishna PG, Nair J, Varghese L, Valivarthi NK, Dhas K, Ramaswami K, Panda B (2012) A draft of the genome and four transcriptomes of a medicinal and pesticidal angiosperm *Azadirachta indica*. BMC Genomics 13:464

Krokene P, Nagy NE, Krekling T (2008) Traumatic resin duct and polyphenolic parenchyma cells in conifers. In: Schaller A (ed) Induced plant resistance to herbivory. Springer, Dordrecht, pp 147–169

Kushiro T, Shibuya M, Ebizuka Y (1998) β-Amyrin synthase: cloning of oxidosqualene cyclase that catalyses the formation of the most popular triterpene among higher plants. Eur J Biochem 256:238–244

Kuzuyama T (2002) Mevalonate and non-mevalonate pathways for isoprene units. Biosci Biotechnol Biochem 66:1619–1627

Lange BM, Ahkami A (2013) Metabolic engineering of plant monoterpene, sesquiterpene and diterpenes—current status and future opportunities. Plant Biotechnol J 11:169–196

Lange BM, Turner GW (2013) Terpenoid biosynthesis in glandular trichomes—current status and future opportunities. Plant Biotechnol J 11:2–22

Lange BM, Severin K, Bechthold A, Heide L (1998) Regulatory role of microsomal 3-hydroxy-3-methylglutaryl CoA reductase for shikonin biosynthesis in *Lithospermum erythrorhizon* cell suspension cultures. Planta 204:234–241

Laule O, Füholz A, Chang H-S, Zhu T, Wang X, Heifetz PB, Gruissem W, Lange BM (2003) Crosstalk between cytosolic and plastidial pathways of isoprenoid biosynthesis in *Arabidopsis thaliana*. Proc Natl Acad Sci USA 100:6866–6871

Learned RM, Connolly EL (1997) Light modulates the spatial patterns of 3-hydroxy-3-methylglutaryl CoA reductase gene expression in *Arabidopsis thaliana*. Plant J 11:499–511

Lee S, Chappel J (2008) Biochemical and genomic characterization of terpene synthases in *Magnolia grandiflora*. Plant Physiol 147:1017–1033

Leivar P, Gonzalez VM, Castel S, Trelease RN, Lopez-Iglesias C, Arro M, Boronat A, Campos N, Ferrer A, Fernandez-Busquets X (2005) Sub-cellular localization of *Arabidopsis* 3-hydroxy-3-methylglutaryl CoA reductase. Plant Physiol 137:57–69

Leivar P, Antolin-Llovera M, Ferroro S, Closa M, Arro M, Ferrer A, Boronat A, Campos N (2011) Multilevel control of *Arabidopsis* 3-hydroxy-3-methylglutaryl CoA reductase by protein phosphatase 2A. Plant Cell 23:1494–1511

Lesburg CA, Zhai G, Cane DE, Christianson DW (1997) Crystal structure of pentalenene synthase: mechanistic insights on terpenoid cyclization reactions in biology. Science 277:1820–1824

Lescot M, Piffnelli P, Ciampi AY, Ruiz M, Blanc G, Mack JL, Silva FR, Santos CMR, Hont AD, Garsmeur O (2008) Insights into the *Musa* genome: syntenic relationships to rice and between *Musa* species. BMC Genomics 9:58

Li Z, Wang X, Chen F, Kim H-J (2007) Chemical changes and overexpressed genes in sweet in weet basil (*Ocimum basilicum* L.) upon methyl jasmonate treatment. J Agric Food Chem 55:706–713

Liao Z-H, Chen M, Gong Y-F, Miao Z-Q, Sun X-F, Tang K-X (2006) Isoprenoid biosynthesis in plants: pathway, genes, regulation and metabolic engineering. J Biol Sci 6:209–219

Lluch MA, Masferrer A, Arró M, Boranat A, Ferrer A (2000) Molecular cloning and expression analysis of the mevalonate kinase gene from *Arabidopsis thaliana*. Plant Mol Biol 42:365–376

Lodeiro S, Xiong Q, Wilson WK, Kolesnikova MD, Onak CS, Matsuda SPT (2007) An oxidosqualene cyclase makes numerous products by diverse mechanisms: a challenge to prevailing concepts of triterpene biosynthesis. J Am Chem Soc 129:11213–11222

Lu S, Xu R, Jia JW, Pang J, Matsuda SPT, Chen XY (2002) Cloning and functional characterization of a β-pinene synthase from *Artemisia annua* that shows a circadian pattern of expression. Plant Physiol 130:477–486

Lücker J, Bouwmeester HJ, Schwab W, Blass J, van der Plas LH, Verhoeven HA (2001) Expression of *Clarkia* S-linalool synthase in transgenic *Petunia* plants results in accumulation of S-linalyl-β-D-glucopyranoside. Plant J 27:315–324

Lücker J, Bowen P, Bohlmann J (2004) *Vitis vinifera* terpenoid cyclases: functional identification of two sesquiterpene synthase cDNAs encoding (+)-valencene synthase and (−)-germacrene D synthase and expression of mono and sesquiterpene synthases in grapevine flowers and berries. Phytochemistry 65:2649–2659

Luskey KL, Stevens B (1985) Human 3-hydroxy-3-methylglutaryl CoA reductase, conserved domains responsible for catalytic activity and sterol regulated degradation. J Biol Chem 260:10271–10277

Mangus S, Bonfill M, Osuna L, Moyano E, Tortoriello J, Cusido RM, Pinot MT, Palazom J (2006) The effect of methyl jasmonate on triterpene and sterol metabolism of *Centella asiatica*, *Ruscus aculeatus* and *Galphimia glauca* cultured plants. Phytochemistry 67:2041–2049

Martin D, Tholl D, Gershenzon J, Bohlmann J (2002) Methyl jasmonate induces traumatic resin ducts, terpenoid resin biosynthesis and terpenoid accumulation in developing xylem of Norway spruce stems. Plant Physiol 129:1003–1018

Martin JJ, Pitera DJ, Withers ST, Newmann JD, Keasling JD (2003) Engineering a mevalonate pathway in *Escherichia coli* for production of terpenoids. Nat Biotechnol 21:796–802

Martin D, Fäldt J, Bohlmann J (2004) Functional characterization of nine Norway spruce terpene synthase genes and evolution of gymnosperm terpene synthases of the TPS-d sub-family. Plant Physiol 135:1908–1927

Martin DM, Aubourg S, Schouwey MB, Daviet L, Schalk M, Toub O, Lund ST, Bohlmann J (2010) Functional annotation, genomic organization and phylogeny of the grapevine terpene synthase gene family based on genome assembly, FLcDNA cloning and enzyme assay. BMC Plant Biol 10:226

Matarese F, Scalabrelli G, D'Onofrio C (2013) Analysis of the expression of terpene synthase genes in relation to aroma content in two aromatic *Vitis vinifera* varieties. Funct Plant Biol 40:552–565

Mau CJD, West CA (1994) Cloning of casbene synthase cDNA: evidence for conserved structural features among terpenoid cyclases in plants. Proc Natl Acad Sci USA 91:8497–8501

Mehjerdi MZ, Bihamata M-R, Omidi M, Naghavi M-R, Soltanloo H, Ranjbar M (2013) Effects of exogenous methyl jasmonate and 2-isopentenyladenine on artemisinin production and gene expression in *Artemisia annua*. Turk J Bot 37:499–505

Memelink J, Verpoorte R, Kijne JN (2001) ORCAnization of jasmonate-responsive gene expression in alkaloid metabolism. Trends Plant Sci 6:212–219

Mercke P, Kappers IF, Francel WAV, Oscar V, Marcel D, Harro JB (2004) Combined transcript and metabolic analysis reveals genes involved in cucumber plants. Plant Physiol 135:2012–2024

Merret R, Cirioni JR, Bach TJ, Hemmerlin A (2007) A serine involved in actin-dependent sub-cellular localization of stress induced tobacco BY-2-hydroxymethylglutaryl-CoA-reudctase isoforms. FEBS Lett 581:5295–5299

Miller B, Madilao LL, Ralph S, Bohlmann J (2005) Insect induced conifer defense. White pine weevil and methyl jasmonate induce traumatic resinosis, *de novo* formed volatile emissions and accumulation of terpenoid synthase and putative octadecanoid pathway transcripts in Sitka spruce. Plant Physiol 137:369–382

Miller JA, Thompson PA, Hakin IA, SherryChow HH, Thomson CA (2011) D-Limonene: a bioactive food component from *Citrus* and evidence for a potential role in breast cancer prevention and treatment. Oncol Rev 5:31–42

Mithofer A, Wanner G, Boland W (2005) Effect of feeding *Spodoptera littoralis* on lima bean leaves: II. Continuous mechanical wounding resembling insect feeding in sufficient to elicit herbivory-related volatile emissions. Plant Physiol 137:1160–1168

Moses T, Pollier J, Thevelein JM, Goosens A (2013) Bioengineering of plant (tri)terpenoids: from metabolic engineering of plants to synthetic biology in vivo and in vitro. New Phytol 199:1–17

Muffler K, Leipold D, Scheller MC, Haas C, Steingroewer J, Bley T, Neuhans E, Mirata MA, Schrader J, Ulber R (2011) Biotransformation of triterpenes. Process Biochem 46:1–15

Nagegowda DA (2010) Plant volatile terpenoid metabolism: biosynthetic genes, transcriptional regulation and sub-cellular compartmentation. FEBS Lett 584:2965–2973

Nagegowda DA, Bach TJ, Chye ML (2004) *Brassica juncea* HMG-CoA synthase 1: expression and characterization of recombinant wild type and mutant enzymes. Biochem J 383:517–527

Okada K, Saito T, Nakagawa T, Kawamukai M, Kamiya Y (2000) Five geranyl geranyl diphosphate synthase expressed in different organs are localized into the sub-cellular compartments in *Arabidopsis*. Plant Physiol 122:1045–1056

Papon N, Bremer J, Vansiri A, Andreu F, Rideau M, Creche J (2005) Cytokinin and ethylene control indole alkaloid production at the level of the MEP/terpenoid pathways in *Catharanthus roseus* suspension cells. Planta Med 71:572–574

Pauwels L, Goosens A (2011) The JA2 proteins: a crucial interface in the jasmonate signaling cascade. Plant Cell 23:3089–3100

Pauwels L, Inze D, Goosens A (2009) Jasmonate inducible gene: what does it mean? Trends Plant Sci 14:87–91

Pegel KH (1997) The importance of sitosterol and sitosterolin in human and animal nutrition. S Afr J Sci 93:263–268

Peters RJ (2006) Uncovering the complex metabolic network underlying diterpenoid phytoalexin biosynthesis in rice and other cereal crops. Phytochemistry 67:2307–2317

Phillips DR, Rasberry JM, Bartel B, Matsuda SPT (2003) Biosynthetic diversity in plant triterpene cyclization. Curr Opin Plant Biol 9:305–314

Phillips MA, Bohlmann J, Gershenzon J (2006) Molecular regulation of induced terpenoid biosynthesis in conifers. Phytochem Rev 5:179–189

Picaud S, Olofsson L, Brodelius M (2005) Expression, purification and characterization of recombinant amorpha-4, 11-diene synthase from *Artemisia annua* L. Arch Biochem Biophys 436:215–226

Poeckel D, Werz O (2006) Boswellic acids: biological actions and molecular targets. Curr Med Chem 13:3359–3369

Poeckel D, Tausch L, Altmann A, Feisst C, Klinkhardt U, Graff J, Harder S, Werz O (2005) Induction of central signaling pathways and select functional effects in human platelets by β boswellic acid. Br J Pharmacol 146:514–524

Poeckel D, Tausch L, George S, Jauch J, Werz O (2006) 3-O-acetyl -11-keto-boswellic acid decreases basal intracellular Ca^{2+} levels and inhibits agonist-induced Ca^{2+} mobilization and mitogen-activated protein kinase activation in human monocyte cells. J Pharmacol Exp Ther 316:224–232

Pollier J, Moses T, Goossens A (2011) Combinatorial biosynthesis in plants: a (p) review on its potential and future exploitation. Nat Prod Rep 28:1897–1916

Prins CL, Vieira IJC, Frietas SP (2010) Growth regulators and essential oil production. Braz J Plant Physiol 22:91–102

Prisic S, Xu MM, Wilderman PR, Peters RJ (2004) Rice contains two disparate *ent*-copalyl diphosphate synthases with distinct metabolic functions. Plant Physiol 136:4228–4236

Qi X, Bakht S, Leggett M, Maxwell C, Melton R, Osbourn A (2004) A gene cluster for secondary metabolism in oat: implications for the evolution of metabolic diversity in plants. Proc Natl Acad Sci USA 101:8233–8238

Qi X, Bakht S, Qin B, Leggett M, Hemmings A, Melton F, Eagles J, Wreck-Reinhardt D, Schaller H, Lesot A, Melton R, Osbourn A (2006) A different function for a member of an ancient and highly conserved cytochrome P450 family: from essential sterols to plant defense. Proc Natl Acad Sci USA 103:18848–18853

Rabi T, Bishayee A (2009a) D-Limonene sensitizes docetaxel-induced cytotoxicity in human prostate cancer cells: generation of reactive oxygen species and induction of apoptosis. J Carcinog 8:9

Rabi T, Bishayee A (2009b) Terpenoids and breast cancer chemoprevention. Breast Cancer Res Treat 115:223–239

Rall B, Ammon HP, Safayhi H (1996) Boswellic acids and protease activity. Phytomedicine 3:75–76

Reumann S, Babujee L, Ma C, Weinkoop S, Siemsen T, Antoni Celli GE, Rasche N, Luder F, Weckwerhth W, Jahn O (2007) Proteome analysis of *Arabidopsis* leaf peroxisomes reveals novel targeting plastids, metabolic pathways and defense mechanisms. Plant Cell 19:3170–3193

Richardson AQ, Palmer JD (2007) Horizontal gene transfer in plants. J Exp Bot 58:1–9

Richelle M, Enslen M, Hager C, Groux M, Tavazzi I, Godin JP, Berger A, Métairon S, Quaile S, Piquet-Welsch C, Sagalowicz L, Green H, Fay LB (2004) Both free and esterified reduce cholesterol absorption and bioavailability of β-carotene and α-tocopherol in normocholesterolemic humans. Am J Clin Nutr 80:171–177

Rieber M, Strasberg Rieber M (1998) Inhibition of p53 without increase in p21WAF1 in betulinic acid-mediated cell death is preferential for human metastatic melanoma. DNA Cell Biol 17:399–406

Rodrìguez-Concepción M (2006) Early steps in isoprenoid biosynthesis: multilevel regulation of the supply of common precursors in plant cells. Phytochem Rev 5:1–15

Rodriguez-Saona C, Crafts-Brandner SJ, Canas LA (2003) Volatile emissions triggered by multiple herbivore damage: beet armyworm and whitefly feeding on cotton plants. J Chem Ecol 29:2539–2550

Rodwell VW, Beach MJ, Bischoff KM, Bochar DA, Darnay BG, Friesen JA, Gill JF, Hedl M, Jordan-Starck T, Kennelly PJ, Kim DY, Wang Y (2000) 3-hydroxy-3-methylglutaryl CoA reductase. Methods Enzymol 324:259–280

Rohdich F, Wungsintaweekul J, Eisenreich W, Richter G, Schuhr CA, Hecht S, Zenk MH, Bacher A (2000) Biosynthesis of terpenoids: 4-diphosphocytidyl-2C-methyl-D-erythritol synthase of *Arabidopsis thaliana*. Proc Natl Acad Sci USA 97:6451–6456

Safayhi H, Rall B, Sailor AR, Ammon HP (1997) Inhibition by boswellic acids of human leucocyte elastase. J Pharmacol Exp Ther 281:460–463

Saldanha SN, Tollefsbol TO (2012) The role of neutraceuticals in chemoprevention and chemotherapy and their clinical outcomes. J Oncol 2012(Article ID 192464):1–23

Salland C, Rontein D, Onillon S, Jabesh F, Duffe P, Giacalone C, Thoraval S, Escoffier C, Herbette G, Leonhardt N, Causse M, Tissier A (2009) A novel pathway for sesquiterpene biosynthesis from farnesyl pyrophosphate in the wild tomato (*Solanum habrochaites*). Plant Cell 21:301–317

Sapir-Mir M, Mett A, Belausov E, Tal-Meshulam S, Frydman A, Gidoni D, Eyal Y (2008) Peroxisomal localization of *Arabidopsis* isopentenyl diphosphate isomerases suggests that part of the plant isoprenoid mevalonic acid pathway is compartmentalized to peroxisomes. Plant Physiol 148:1219–1228

Sawai S, Saito K (2011) Triterpenoid biosynthesis and engineering in plants. Front Plant Sci 2:1–25

Sawai S, Shondo T, Sato S, Kaneko T, Tabata S, Ayabe SI, Aoki T (2006) Functional and structural analysis of genes encoding oxidosqualene cyclases of *Lotus japonicus*. Plant Sci 170:247–257

Schilmiller AL, Schauvinhold I, Larson M, Xu R, Charbonneau AL, Schmidt A, Wilkerson C, Last RL, Pichersky E (2009) Monoterpenes in glandular trichomes in tomato are synthesized from a neryl diphosphate precursor rather than geranyl diphosphate. Proc Natl Acad Sci USA 106:10865–10870

Schnee C, Köllner TG, Held M, Turting TCJ, Gershenzon J, Degenhardt J (2006) The products of a single maize sesquiterpene synthase form a volatile defense signal that attracts natural enemies of maize herbivores. Proc Natl Acad Sci USA 103:1129–1134

Segura MJR, Meyer MM, Matsuda SPT (2000) *Arabidopsis thaliana* LUP1 converts oxidosqualene to multiple triterpene alcohols and a triterpene diol. Org Lett 2:2257–2259

Seki H, Ohyama K, Sawai S, Mizutani M, Ohnishi T, Sudo H, Akashi T, Aoki T, Saito K, Muranaka T (2008) Licorice β-amyrin 11-oxidase, a cytochrome P450 with key role in the biosynthesis of the triterpene sweetener glycyrrhizin. Proc Natl Acad Sci USA 105:14204–14209

Shah BA, Qazi GN, Taneja SC (2009) Boswellic acid: a group of medicinally important compounds. Nat Prod Rep 26:72–89

Shanmugum MK, Ong TH, Kumar AP, Lun CP, Ho PC et al (2012) Ursolic acid inhibits the initiation, progression of prostate cancer and prolongs the survival of TRAMP mice by modulating proinflammatory pathways. PLoS One 7:e32476

Shao Y, Ho CT, Chin CK, Badmaev V, Ma W, Huang M-T (1998) Inhibitory activity of boswellic acid from *Boswellia serrata* against human leukemia HL-60 cells in culture. Planta Med 64:328–331

Sheron-Asa L, Shalit M, Frydman A, Bar E, Holland D, Or E, Lavi U, Lewinsohn E, Eyal Y (2003) *Citrus* fruit flavor and aroma biosynthesis: isolation, functional characterization and developmental regulation of Cstps1, a key gene in the production of the sesquiterpene aroma compound valencene. Plant J 36:664–674

Shibuya M (2004) Cucurbitadienol synthase, the first committed enzyme for cucurbitacin biosynthesis, is a distinct enzyme from cycloartenol synthase for phytosterol biosynthesis. Tetrahedron 60:6995–7003

Shibuya M, Zhang H, Endo A, Shishikura K, Kushiro T, Ebizuka Y (1999) Two branches of the lupeol synthase gene in the molecular evolution of plant oxidosqualene cyclases. Eur J Biochem 266:302–307

Shibuya M, Xiang T, Katsube Y, Otsuka M, Zhang H, Ebizuka Y (2007) Origin of structural diversity in natural triterpenes: direct synthesis of seco-triterpene skeletons by oxidosqualene cyclase. J Am Chem Soc 129:1450–1455

Shibuya M, Katsube Y, Otsuka M, Zhang H, Tansakul P, Xiang T, Ebizuka Y (2009) Identification of a product specific β-amyrin synthase from *Arabidopsis thaliana*. Plant Physiol Biochem 47:26–30

Shimada T, Endo T, Fuji H, Hara M, Veda T, Kita M, Omura M (2004) Molecular cloning and functional characterization of four monoterpene synthase genes from *Citrus unshiu* Marc. Plant Sci 166:49–58

Siamaru H, Orihara Y, Tansakul P, Kang YH, Shibuya M, Ebizuka Y (2007) Production of triterpene acids by cell suspension cultures of *Olea europaea*. Chem Pharm Bull 55:784–788

Silveira e Sá RC, Andrade LN, de Sousa DP (2013) A review on antiinflammatory activity of monoterpenes. Molecules 18:1227–1254

Simkin AJ, Guirmand G, Papon N, Courdavault V, Thabet I, Ginis D, Bouzid S, Giglioli-Guivarch N, Clastre M (2011) Peroxisomal

localization of the final steps of the mevalonic acid pathway in plants. Planta 234:903–914

Sitthithaworn W, Kojima N, Viroonchatapan E, Suh DY, Iwanami N, Hayashi T, Noji M, Saito K, Niwa Y, Sankawa U (2001) Geranyl diphosphate synthase from *Scoparia dulcis* and *Croton sublyratus*. Plastid localization and conversion to a farnesyl diphosphate synthase by mutagenesis. Chem Pharm Bull 49:197–202

Sparge SG, Light ME, van Staden J (2004) Biological activities and distribution of plant saponins. J Ethnopharmacol 94:219–243

Sprenger GA, Schörken U, Wiegest T, Grolle S, Graff AA, Taylor SV, Begley TP, Bringer-Meyer S, Sahm H (1997) Identification of a thiamine dependent synthase in *Escherichia coli* required for the 1-deoxy-D-xylulose 5-phosphate precursor to isoprenoids, thiamine and pyridoxol. Proc Natl Acad Sci USA 94:12857–12862

Steinbacher S, Kaisar J, Eisenreich W, Huber R, Bacher A, Rohdich F (2003) Structural basis of fosmidomycin action revealed by the complex with 2-C-methyl-D-erythritol 4-phosphate synthase (IspC). Implication for the catalytic mechanisms and antimalarial drug development. J Biol Chem 278:18401–18407

Sultana N, Ata A (2008) Oleanolic acid and related derivatives as medicinally important compounds. J Enzyme Inhib Med Chem 23:739–756

Sun J (2007) D-Limonene: safety and clinical applications. Altern Med Rev 12:259–264

Sun TP, Kamiya Y (1994) The *Arabidopsis* GA1 locus encodes the cyclase *ent*-kaurene synthase A of gibberellin biosynthesis. Plant Cell 6:1509–1518

Sung CK, Shibuya M, Sankawa U, Ebizuka Y (1995) Molecular cloning of cDNA encoding human lanosterol synthase. Biol Pharm Bull 18:1459–1461

Surie C, Bouvier F, Backhaus RA, Begu D, Bonneu M, Camara B (2000) Cellular localization of isoprenoid biosynthetic enzymes in *Marchantia polymorpha*. Uncovering a new role of oil bodies. Plant Physiol 124:971–978

Suzuki H, Reddy MS, Naoumkina M, Aziz N, May GD, Huhman DR, Sumner LW, Blount JW, Mendes P, Dixon RA (2005) Methyl jasmonate and yeast elicitor induce differential transcriptional and metabolic re-programming in cell suspension cultures of the model legume *Medicago truncatula*. Planta 220:696–707

Suzuki M, Xiang T, Ohyama K, Seki H, Saito K, Muranaka T, Hayashi H, Katsube Y, Kushiro T, Shibuya M (2006) Lanosterol synthase in dicotyledonous plants. Plant Cell Physiol 47:565–571

Takahashi S, Kuzuyama T, Watanabe H, Seto H (1998) A 1-deoxy-D-xylulose 5-phosphate reductoisomerase catalyzing the formation of 2-C-methyl-D-erythritol 4-phosphate in an alternative nonmevalonate pathway for terpenoid biosynthesis. Proc Natl Acad Sci USA 95:9879–9884

Takashi M, Fujita F, Uchida K, Yamamoto S, Sawada M, Hatai C, Shimizu M, Tominaga M (2012) 1,8-cineole, a TRPM8 agonist, is a novel natural antagonist of human TRPA1. Mol Pain 8:86

Tamongani S, Mitani M (1993) Oryzalexin S structure: a new stemarane-type rice plant phytoalexin and its biosynthesis. Tetrahedron 49:2025–2032

Tansakul P, Shibuya M, Kushiro T, Ebizuka Y (2006) Dammarenediol-II synthase, the first dedicated enzyme for ginsenoside biosynthesis in *Panax ginseng*. FEBS Lett 580:5143–5149

Thabet I, Guirimand G, Courdavault V, Papon N, Godet S, Dutillent C, Bouzid S, Giglioli-Guivarch N, Clastre M, Simkin AJ (2011) The sub-cellular localization of periwinkle farnesyl diphosphate synthase provides insight into the role of peroxisomes in isoprenoid biosynthesis. J Plant Physiol 168:2110–2116

Thabet I, Guirimand G, Guihur A, Lanoue A, Courdvault V, Papon N, Bouzid S, Giglioli-Guivarich N, Simkin AJ, Clastre M (2012) Characterization and sub-cellular localization of geranyl diphosphate synthase from *Catharanthus roseus*. Mol Biol Rep 39:32–3245

Tholl D (2006) Terpene synthases and the regulation, diversity and biological roles of terpene metabolism. Curr Opin Plant Biol 9:297–304

Tholl D, Lee S (2011) Terpene specialised metabolism in *Arabidopsis thaliana*. Arabidopsis Book 9:e143

Tholl D, Chen F, Petri J, Gershenzon J, Pichersky E (2005) Two sesquiterpene synthases are responsible for the complex mixture of sesquiterpenes emitted from *Arabidopsis* flowers. Plant J 42:757–771

Thoppil RJ, Bishayee A (2011) Terpenoids as potential chemopreventive and therapeutic agents in liver cancer. World J Hepatol 3:228–249

Trapp SC, Croteau RB (2001) Genomic characterization of plant terpene synthases and molecular evolutionary implications. Genetics 158:811–832

Traw MB, Bergelson J (2003) Interactive effects of jasmonic acid, salicylic acid and gibberellin on induction of trichomes in *Arabidopsis*. Plant Physiol 133:879–886

Tsay YH, Robinson GW (1991) Cloning and characterization of ERG8, an essential gene of *Saccharomyces cerevisiae* that encodes phosphomevalonate kinase. Mol Cell Biol 11:620–631

Turner G, Gershenzon J, Neilson EE, Froehlich JE, Croteau R (1999) Limonene synthase, the enzyme responsible for monoterpene biosynthesis in peppermint, localized to leucoplasts of oil gland secretory cells. Plant Physiol 120:879–886

Unsicker SB, Kunret G, Gershenzon J (2009) Protective perfumes: the role of vegetative volatiles in plant defense against herbivores. Curr Opin Plant Biol 12:479–485

van der Fits L, Memelink J (2000) ORCA3, a jasmonate-responsive transcriptional regulator of plant primary and secondary metabolism. Science 289:295–297

Van Poecke RMP, Posthumus MA, Dicke M (2001) Herbivore-induced volatile production by *Arabidopsis thaliana* leads to attraction of the parasitoid *Cotesia rubecula*: chemical behavioral, and gene-expression analysis. J Chem Ecol 27:1911–1928

Van Schie CCN, Haring MA, Schuurink RC (2007) Tomato linalool synthase is induced in trichomes by jasmonic acid. Plant Mol Biol 64:251–263

Vickers CE, Gershenzon J, Lerdan MT, Loreto F (2009) A unified mechanism of action for volatile isoprenoids in plant abiotic stress. Nat Chem Biol 5:283–291

Vigushin DM, Poon GK, Boddy A, English J, Harbert GW, Pagonis C, Jarman M, Coombs RC (1998) Phase I and pharmacokinetic study of D-limonene in patients with advanced cancer. Cancer Chemother Pharmacol 42:111–117

Vogel BS, Wildung M, Vogel G, Croteau R (1996) Abietadiene synthase from grand fir (*Abies grandis*). J Biol Chem 271:23262–23268

Vogg G, Fischer S, Leide J, Emmanuel E, Jetter R, Levy AA, Riedrer M (2004) Tomato fruit cuticular waxes and their effects on transpirational barrier properties: functional characterization of mutant deficient in a very long chain fatty acid β-ketoacyl-CoA-synthase. J Exp Bot 55:1401–1410

Vranova E, Coman D, Gruissem W (2012) Structure and dynamics of the isoprenoid pathway network. Mol Plant 5:318–333

Wagner KH, Elmadfa I (2003) Biological relevance of terpenoids. Overview of focusing on mono-, di- and tetraterpenes. Ann Nutr Metab 47:95–106

Wang YD, Yuan YJ, Lu M, Wu JC, Jiang JL (2003) Inhibitor studies of isopentenyl pyrophosphate biosynthesis in suspension cultures of the new *Taxus chinensis* var. mairei. Biotechnol Appl Biochem 37:39–43

Wang Z, Yeast T, Han H, Jetter R (2010) Cloning and characterization of oxidosqualene cyclases from *Kalanchoe daigremontiana*:

enzymes catalyzing up to ten rearrangement steps yielding friedelin and other triterpenoids. J Biol Chem 285:29703–29712

Wang Z, Guhling O, Yao R, Li F, Yeast TH, Rose JKC, Jetter R (2011) Two oxidosqualene cyclases responsible for biosynthesis of tomato fruit cuticular triterpenoids. Plant Physiol 155:540–552

Wen CH, Tseng YH, Chu FH (2012) Identification and functional characterization of a sesquiterpene synthase gene from *Eleutherococcus trifoliatus*. Holzforschung 66:183–189

Wildung MR, Croteau R (1996) A cDNA clone for taxadiene synthase, the diterpene cyclase that catalyzes the committed step of taxol biosynthesis. J Biol Chem 271:9201–9204

Williams DC, McGarvey DJ, Katahira EJ, Croteau R (1998) Truncation of limonene synthase preprotein provides a fully active 'pseudomature' from this monoterpene cyclase and reveals the amino-terminal arginine pair. Biochemistry 37:12213–12220

Wilt TJ, Ma Donald R, Ishani A (1999) β-Sitosterol for the treatment of benign prostate hyperplasia: a systematic review. BJU Intl 83:976–983

Withers ST, Keasling JD (2007) Biosynthesis and engineering of isoprenoid small molecules. Appl Microbiol Biotechnol 73:980–990

Wong RJ, McCormack DK, Russell DW (1982) Plastid 3-hydroxy-3-methylglutaryl CoA reductase has distinctive kinetic and regulatory features: properties of the enzyme and positive phytochrome control of activity in pea seedlings. Arch Biochem Biophys 216:631–638

Worth H, Schacher C, Dethlefsen U (2009) Concomitant therapy with cineole (eucalyptole) reduces exacerbations in COPD: a placebo-controlled double-blind trial. Respir Res 10:69

Wu SQ, Schalk M, Clark A, Miles RB, Coates R, Chappel J (2006) Re-direction of cytosolic or plastidic isoprenoid precursors elevates terpene production in plants. Nat Biotechnol 24:1441–1447

Xi Z, Bradley RK, Wurdack KJ, Wong KM, Sugumaran M, Bomblies K, Rest JS, Davis CC (2012) Horizontal transfer of expressed genes in parasitic flowering plants. BMC Genom 13:227

Xiang T, Shibuya M, Katsube Y, Tsutsumi T, Otsuka M, Zhang H, Masuda K, Ebizuka Y (2006) A new triterpene synthase from *Arabidopsis thaliana* produces a tricyclic triterpene with two hydroxyl groups. Org Lett 13:2835–2838

Xue Z, Duan L, Liu D, Guo J, Ge S, Dicks J, Omaille P, Osbourn A, Qi X (2012) Divergent evolution of oxidosqualene cyclases in plants. New Phytol 193:1022–1038

Yamaguchi S, Saito T, Abe H, Yamane H, Murofushi N, Kamiya Y (1996) Molecular cloning and characterization of a cDNA encoding the gibbrellin biosynthesis enzyme *ent*-kaurene synthase B from pumpkin (*Cucurbita maxima* L.). Plant J 10:203–213

Yamaguchi S, Sun TP, Kawaide H, Kamiya Y (1998) The GA2 locus of *Arabidopsis thaliana* encodes *ent*-kaurene synthase of gibberellin biosynthesis. Plant Physiol 117:1271–1278

Yoon WJ, Lee NH, Hyun CG (2010) Limonene suppresses lipopolysaccharide-induced production of nitric oxide, prostaglandin E2 and pro-inflammatory cytokines in RAW 264-7 macrophages. J Oleo Sci 59:415–421

Yu J, Wang J, Lin W, Li SG, Li H, Zhou J, Ni PX, Dong W, Hu SN, Zeng CQ et al (2005) The genomes of *Oryza sativa*: a history of duplications. PLoS Biol 3:e38

Zapta F, Fine PA (2013) Diversification of the monoterpene synthase gene family (Tpsb) in *Protium*, a highly diverse genus of tropical tree. Mol Phylogenet Evol 68:432–442

Zhang H, Shibuya M, Yokota S, Ebizuka Y (2003) Oxidosqualene cyclases from cell suspension cultures of *Betula platyphylla* var. japonica: molecular evolution of oxidosqualene cyclases in higher plants. Biol Pharm Bull 26:642–650

Zhang JP, Kang J, Huang SG, Chen P, Yao WZ, Yang L, Bai CX, Wang CZ, Wang C, Chen BY, Shi Y, Liu CT, Chen P, Li Q, Wang ZS, Huang YJ, Luo ZY, Chen FP, Yuan JZ, Yuan BT, Qian HP, Zhi RC, Zhong NS (2008) Effect of carbocisteine on acute exacerbation of chronic obstructive pulmonary disease: a randomized placebo-controlled study. Lancet 371:2013–2018

Zhao J, Zheng SH, Fujita K, Sakai K (2004) Jasmonate and ethylene signalling pathway leading to β-thujaplicin biosynthesis in *Cupressus lusitanica* cell cultures. J Exp Bot 55:1003–1012

Zulak KG, Bohlmann J (2010) Terpenoid biosynthesis and specialized vascular cells of conifer defense. J Intgr Plant Biol 52:86–97

22

A statistical approach for the production of thermostable and alklophilic alpha-amylase from *Bacillus amyloliquefaciens* KCP2 under solid-state fermentation

Vimal S. Prajapati · Ujjval B. Trivedi ·
Kamlesh C. Patel

Abstract The bacterial strain producing thermostable, alklophilic alpha-amylase was identified as *Bacillus amyloliquefaciens* KCP2 using 16S rDNA gene sequencing data (NCBI Accession No: KF112071). Medium components were optimized through the statistical approach for the synthesis of alpha-amylase by the organism under solid-state fermentation using wheat bran as the substrate. The medium components influencing the enzyme production were identified using a two-level fractional factorial Plackett–Burman design. Among the various variables screened, starch, ammonium sulphate and calcium chloride were found to be most significant medium components. The optimum levels of these significant parameters were determined employing the response surface Central Composite design which significantly increased the enzyme production with the supplementation of starch 0.01 g, ammonium sulphate 0.2 g and 5 mM calcium chloride in the production medium. Temperature and pH stability of the alpha-amylase suggested its wide application in the food and pharmaceutical industries.

Keywords Thermostable alpha-amylase ·
Plackett–Burman design · Response surface
methodology · Solid-state fermentation ·
Bacillus amyloliquefaciens KCP2

Introduction

Amylases find potential application in number of industrial processes such as bread making, brewing, starch processing, pharmacy, textile industries. Alpha-amylase is an extracellular enzyme that randomly cleaves the 1,4 α-D-glucosidic linkages between adjacent glucose units in the linear amylose chain. It is secreted as primary metabolite of microorganisms and its production is a growth-related process (Kammoun et al. 2008). It possesses approximately 25–33 % share of the world's marketable enzymes. Microbial alpha-amylases are the most stable and produced more economically compared to plant and animal alpha-amylases (Gupta et al. 2003).

To meet the growing demands of amylase for industrial application, it is necessary to produce the highly efficient enzymes at large scale with reduced production cost (Haq et al. 2003). Submerged fermentation (SmF) has been traditionally used for the production of industrially important enzymes because of the ease of handling and greater control of environmental factors such as temperature and pH. Use of agro-industrial residues as the substrate for the fermentation has growing interests as they are inexpensive energy-rich resources and also eliminates large-scale accumulation of the biomass (Pandey et al. 2000; Ramachandran et al. 2007). Solid-state fermentation has been generally referred useful for agro-residues utilization (Pandey et al. 2000; Babitha et al. 2007; Binod et al. 2007), although most of the commercial processes are based on submerged fermentation. The growth and the enzyme production by the organisms are strongly influenced by medium composition, and hence optimization of the medium components may lead to improved enzyme productivity (Djekrif et al. 2006).

V. S. Prajapati · U. B. Trivedi · K. C. Patel (✉)
BRD School of Biosciences, Sardar Patel University, Sardar
Patel Maidan, Vadtal Road, Satellite Campus, Post Box No. 39,
Vallabh Vidyanagar 388-120, Gujarat, India
e-mail: comless@yahoo.com

The main strategy used for optimizing the medium composition is by changing one medium component as a parameter and keeping the others at a constant level (Prajapati et al. 2013). Such optimization studies do not consider the interaction effects among the variables which influence the overall process for the production of a desired metabolite (Silva et al. 2001). Single variable optimization methods are not only tedious, but also can lead to misinterpretation of results, especially because the interaction effects between different factors are overlooked (Wenstet-Botz 2000). Limitations of the single factor optimization can be eliminated by employing response surface methodology (RSM) which is used to explain the combined effects of all the factors in a fermentation process (Elibol 2004). Response surface methodology may be summarized as a collection of experimental strategies, mathematical methods and statistical inference for constructing and exploring an approximate functional relationship between a response variable and a set of design variables.

In the present study, Plackett–Burman design is used for identifying various nutrients as significant variables influencing alpha-amylase production by *Bacillus amyloliquefaciens* KCP2. The levels of the significant variables are further optimized using response surface methodology.

Materials and methods

Strain isolation and identification

Bacterial strain isolated from municipal food waste samples collected from Vallabh Vidyanagar, Gujarat, India, on Luria agar (LA) was screened for amylase production on starch agar plate. Culture was maintained at 4 °C on Bushnell Hass agar (BHA) slants containing 1 % starch. Bushnell Hass mineral salt solution has the following composition (g/l): $MgSO_4$ 0.2, $CaCl_2$ 0.02, KH_2PO_4 1.00, K_2HPO_4 1.00, NH_4NO_3 1.00 and $FeCl_3$ 0.05 (pH 7.0). Genomic DNA of the bacterial isolate was extracted and the 16S rDNA gene was amplified using the universal primers (F, 5′ AGAGTTTGATCCTGGCTCAG 3′; R, 5′ GGTTACCTTGTTACAGCTT 3′). Amplification was carried out in a thermal cycler (Applied Biosystems) with reaction profile: initial denaturation at 95 °C for 1 min, followed by 30 cycles of denaturation at 95 °C for 30 s, annealing at 55 °C for 45 s, extension at 72 °C for 45 s and finally extension at 72 °C for 5 min. The purified PCR product was sequenced and the phylogenic relationship of the isolate was determined by comparing the sequence data with the existing sequences available through the gene bank database of the National Center for Biotechnology Information (NCBI).

Preparation of inoculum

A volume of 50 ml of nutrient broth taken in a 250-ml Erlenmeyer flask was inoculated with a loop full of cells from a 24-h-old culture and kept at 37 °C in a rotary shaker. After 18 h of incubation, an appropriate aliquot of inoculum was added to get one optical density unit in all the experimental flasks.

Enzyme production, extraction and assay procedure

The experiments were performed according to the design matrix (Tables 2, 4) in 250-ml Erlenmeyer flasks containing 5 g wheat bran as solid substrate and 10 ml of the salt solution to provide the adequate moisture content. After inoculation, all the experimental flasks were incubated under static condition at 37 ± 2 °C and were harvested after 72 h interval followed by the enzyme extraction with 40 ml of 0.05 M phosphate buffer (pH 8.0) on a rotary shaker at 150 rpm for 30 min at 25 °C. The content was filtered through muslin cloth, centrifuged at 8,000 rpm for 25 min and the clear supernatant was used for determining alpha-amylase activity, which is expressed as U/gds (Units/ gram dry substrate). The reaction mixture consisted of 1.0 ml of 1 % starch, 0.9 ml 0.05 M phosphate buffer (pH 8.0), and 0.1 ml of enzyme extract. After 10 min of incubation at 65 °C, the liberated reducing sugars were estimated by the dinitrosalicylic acid (DNS) method of Miller (1959). The colour developed was read at 560 nm using a spectrophotometer (Shimadzu UV-160). TLC analysis of products of enzymatic hydrolysis of starch indicated that it is alpha-amylase since the major end product was found to be maltose.

Effect of the pH and temperature on the enzyme activity and stability

Enzyme produced by the *B. amyloliquefaciens* KCP2 was assayed at different temperatures and pH ranging 30–90 °C and 4–10, respectively. Stability of enzyme was tested by incubating in buffers of different pH (4–9) at 30 °C up to 180 min and residual activity was determined after each 10 min of incubation under standard assay condition. The thermal stability was studied by incubating enzymes at various temperatures (30–90 °C) and residual activity was measured after each 10 min of interval up to 180 min under standard assay condition.

Screening of significant variables using Plackett–Burman design

Plackett–Burman design is a powerful technique for screening and evaluating the important variables that

Table 1 Variables representing medium components used in Plackett–Burman design

Variables	Medium components	Positive values	Negative values
X1	Dextrose	1.5 g	0.05 g
X2	Starch	1.5 g	0.05 g
X3	Lactose	1.5 g	0.05 g
X4	Soy bean meal	1 g	0.05 g
X5	Yeast extract	1 g	0.05 g
X6	Ammonium sulphate	1 g	0.05 g
X7	Ammonium nitrate	1 g	0.05 g
X8	Calcium chloride	5 mM	1 mM

influences the response (Plackett and Burman 1946). This technique significantly decreases the number of experiments needed to decide the important variables. In the present study, various medium components such as dextrose, starch, lactose, soya bean meal, yeast extract, ammonium sulphate, ammonium nitrate and calcium chloride were investigated as variables using PB design to identify the components that significantly affected alpha-amylase production. In PB design each selected variable was considered at two levels, high $(+)$ and low $(-)$ as shown in Table 1. Using the selected levels for each variable and three-dummy variable setup with 12 runs of experiment was generated using the software as shown in Table 2. Each row represents a trial, and each column represents an independent (assigned) or dummy (unassigned) variables. The effect of each variable was determined by the following equation:

$$E(Xi) = 2(\Sigma Mi^+ - Mi^-),\qquad(1)$$

where $E(Xi)$ is the concentration effect of the tested variable Mi^+ and Mi^- representing the alpha-amylase

production from trials where the variable (Xi) measured was presented at high and low concentrations, respectively. N is the total number of trials, i.e. 12. Experimental error was estimated by calculating the variance among the dummy variables as:

$$V_{\text{eff}} = \Sigma (E_d)^2 / n,\qquad(2)$$

where V_{eff} is the variance of the effect of level, E_d is the effect of the level for the dummy variables and n is the number of dummy variables used in the experiment. The standard error (SE, Es) of concentration effect was the square root of variance of an effect, and the significance level (P value) of each concentration effect was determined using the Student's t test.

$$t(Xi) = E(Xi) / Es,\qquad(3)$$

where $E(Xi)$ is the effect of variable Xi.

Response surface methodology

The concentration of the medium components found as a significant variable and the interaction effects between them which may influence the alpha-amylase production significantly were analysed and optimized by response surface Central Composite design (CCD). RSM is useful for small number of variables (up to five) but is impractical for large number of variables, due to high number of experimental runs required. In the present study, concentrations of the three major medium components, starch, ammonium sulphate and calcium chloride (identified by Plackett–Burman design) were optimized, keeping temperature, pH, moisture and inoculum size constant.

According to the design, the total number of treatment combinations is $2^k + 2k +$ no, where k is the number of

Table 2 Design matrix and experimental results of Plackett–Burman design

Run no.	Components											Alpha-amylase production (U/gds)
	X1	X2	X3	X4	X5	X6	X7	X8	D1	D2	D3	
1	1	1	−1	1	1	1	−1	−1	1	1	−1	9.334
2	−1	1	1	−1	1	1	1	−1	−1	−1	1	0.502
3	1	−1	1	1	−1	1	1	1	−1	−1	−1	3.702
4	−1	1	−1	1	1	−1	1	1	−1	−1	−1	12.746
5	−1	−1	1	−1	1	1	−1	1	1	1	−1	16.474
6	−1	−1	−1	1	−1	1	1	−1	1	1	1	9.308
7	1	−1	−1	−1	1	−1	1	1	1	1	1	21.73
8	1	1	−1	−1	−1	1	−1	1	−1	−1	1	10.26
9	1	1	1	−1	−1	−1	1	−1	1	1	−1	5.50
10	−1	1	1	1	−1	−1	−1	1	1	1	1	19.17
11	1	−1	1	1	1	−1	−1	−1	−1	−1	1	27.63
12	−1	−1	−1	−1	−1	−1	−1	−1	−1	−1	−1	39.05

D1–D3 represent dummy variables

Table 3 Experimental range and levels of the independent variables of selected components used for response surface Central Composite design

Variables	Components	Range	Levels of variable studied				
			$-\alpha$	-1	0	$+1$	$+\alpha$
X1	Starch (w/w)	0.01–0.50	−0.15	0.01	0.25	0.50	0.66
X2	Ammonium sulphate (w/w)	0.01–0.20	−0.05	0.01	0.10	0.20	0.264
X3	Calcium chloride (mM)	0.50–5.00	−1.03	0.50	2.75	5.00	6.53

independent variables and no is the number of repetition of experiments at the central point. Each factor in the design was studied at five different levels ($-\alpha$, -1, 0, $+1$, $+\alpha$) as shown in Table 3. All variables were set at a central coded value of zero. The minimum and maximum ranges of variables were determined on the basis of our previous experiments. The full experimental plan with respect to their values in actual and coded form is presented in Table 4. The alpha-amylase activity was measured in triplicate for all 20 different experimental runs. The enzyme production was analysed by using a second-order polynomial equation, and the data were fitted into the

equation by multiple regression procedure. The model equation for analysis is given as:

$$Y = \beta_0 + \sum \beta i X i + \sum \beta i i X i^2 + \sum \beta i j X i X j, \qquad (4)$$

where β_0, βi, $\beta i i$ and $\beta i j$ represent the constant process effect in total, the linear, quadratic effect of Xi and the interaction effect between Xi and Xj, respectively, for the production of alpha-amylase. Later, an experiment was run using the optimum values for variables given by response optimization to confirm the predicted value and experimental value of enzyme production.

Software and data analysis

The results of the experimental design were analysed and interpreted using Design Expert Version 8.0 (Stat-Ease Inc., Minneapolis, Minnesota, USA) statistical software.

Results

Identification of the bacterial isolate

A 800-bp size 16S rDNA sequence of the isolate was obtained through PCR amplification and sequencing. The

Table 4 Full experimental Central Composite design with coded and actual level of variables and the response function

Run no.	A: starch w/w		B: ammonium sulphate w/w		C: calcium chloride (mM)		Alpha-amylase (U/gds)	
	Actual level	Coded level	Actual level	Coded level	Actual level	Coded level	Observed	Predicted
1	0.01	−1	0.01	−1	0.5	−1	49.27	48.96
2	0.5	+1	0.01	−1	0.5	−1	46.78	46.04
3	0.01	−1	0.2	+1	0.5	−1	48.55	48.78
4	0.5	+1	0.2	+1	0.5	−1	42.77	41.82
5	0.01	−1	0.01	−1	5	+1	51.08	52.78
6	0.5	+1	0.01	−1	5	+1	46.86	47.39
7	0.01	−1	0.2	+1	5	+1	60.01	61.48
8	0.5	+1	0.2	+1	5	+1	51.21	52.06
9	−0.15	$-\alpha$	0.105	0	2.75	0	54.88	53.40
10	0.66	$+\alpha$	0.105	0	2.75	0	42.60	43.02
11	0.255	0	−0.05	$-\alpha$	2.75	0	46.95	46.60
12	0.255	0	0.26	$+\alpha$	2.75	0	51.10	50.37
13	0.255	0	0.105	0	−1.03	$-\alpha$	45.38	46.80
14	0.255	0	0.105	0	6.53	$+\alpha$	61.10	58.62
15	0.255	0	0.105	0	2.75	0	48.59	47.94
16	0.255	0	0.105	0	2.75	0	47.07	47.94
17	0.255	0	0.105	0	2.75	0	48.93	47.94
18	0.255	0	0.105	0	2.75	0	46.10	47.94
19	0.255	0	0.105	0	2.75	0	48.42	47.94
20	0.255	0	0.105	0	2.75	0	48.34	47.94

Fig. 1 Phylogenetic relationship on the basis of homology index for a bacterial isolate *B. amyloliquefaciens* KCP2

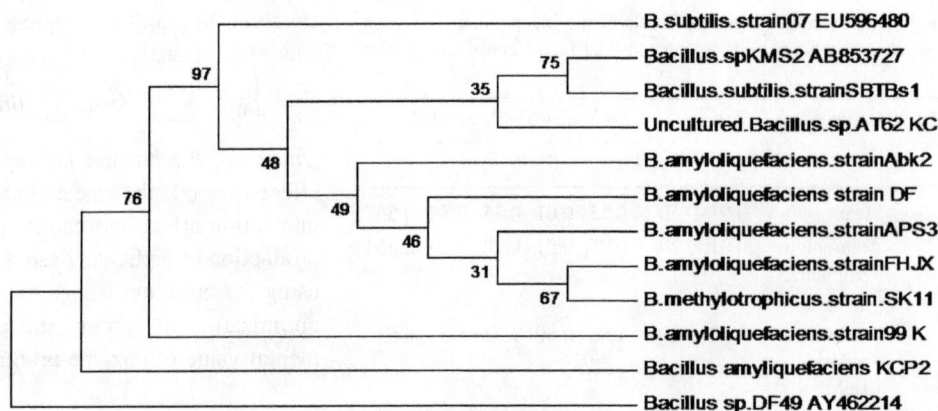

sequence was subjected to a multiple sequence alignment using the BLAST programme of NCBI. The sequence showed a homology of 99 % with *B. amyloliquefaciens*. The sequence was deposited in the gene bank of the NCBI (Accession No: KF112071). The phylogenetic tree (Fig. 1) was drawn using bioinformatics software MEGA 4.0 (Tamura et al. 2007) after alignment of the sequences with the Clustal X software.

Activity and stability of alpha-amylase from *B. amyloliquefaciens* KCP2

The alpha-amylase produced by *B. amyloliquefaciens* KCP2 was assessed at different pH and temperature ranges and it was observed that the activity was very low at acidic pH, but increase in the pH led to gradual increase in enzyme activity. The maximum activity of the enzyme was observed at pH 8.0 and was found to be highly stable in alkaline pH (Fig. 2a, b). The optimum temperature for alpha-amylase activity was found to be 65 °C and it showed good stability in the temperature range of 30–90 °C. More than 85 % of residual activity was observed in case of temperature ranging from 30 to 70 °C (Fig. 3a, b).

Screening of significant parameter for alpha-amylase production using Placket–Burman design

A statistical approach has been used to screen the most effective supplement and select their concentration to achieve highest possible alpha-amylase production by *B. amyloliquefaciens* KCP2 under solid-state fermentation using wheat bran as solid substrate. Usually, initial screening of the ingredients is done to understand the significance of their effect on the product formation and then a few better ingredients are selected for further optimization. Plackett–Burman design was used to screen eight different medium components as carbon and nitrogen sources as a 12-run experiment with two levels of

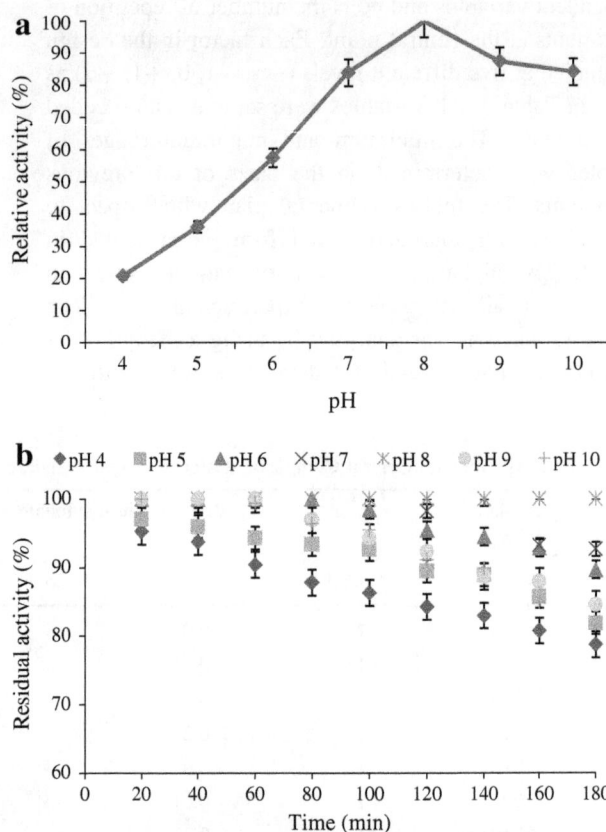

Fig. 2 Effect pH on enzyme activity (**a**) and stability (**b**)

concentration of each variable. Studies were carried out under solid-state fermentation at 37 °C for 72 h. The medium components selected as independent variables and their respective high and low concentrations used in optimization study are presented in Table 1, whereas the Plackett–Burman experimental design followed for the optimization of medium components for alpha-amylase production for 12 trials with two levels of concentration of each variable is given in Table 2. The variable X1–X8 represented the medium constituents and D1–D3 represented the dummy variable/unassigned variables. The

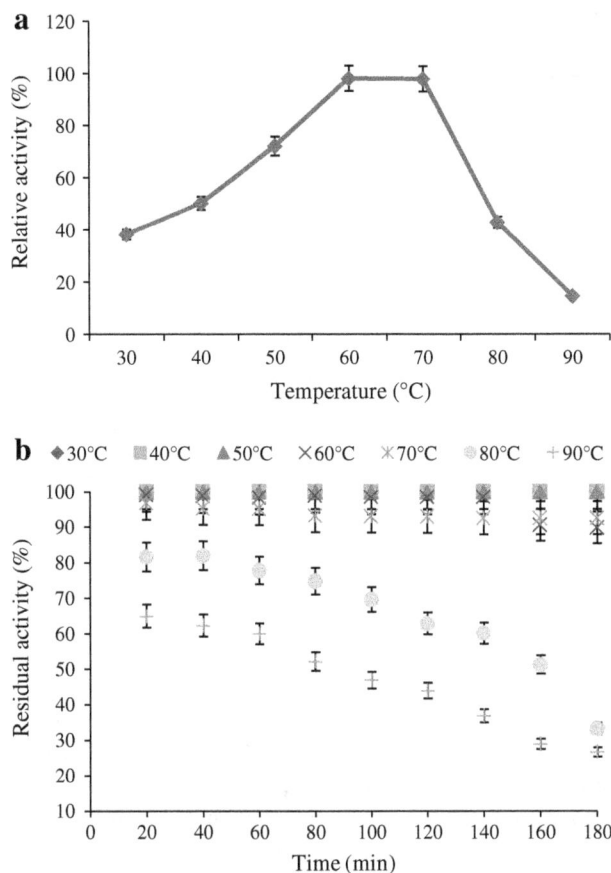

Fig. 3 Effect temperature on enzyme activity (**a**) and stability (**b**)

Table 5 Statistical analysis of components for alpha-amylase production by *B. amyloliquefaciens* KCP2

Components	Effect	Standard error	t value	P	Confidence (%)
Dextrose	−3.18	1.64	−1.98	0.140	85.91
Starch	−10.06	1.64	−6.29	0.008	99.18
Lactose	−4.90	1.64	−3.06	0.054	94.53
NH_4NO_3	−1.93	1.64	−1.21	0.312	68.77
Yeast extract	0.238	1.64	0.148	0.891	10.88
NH_4SO_4	−12.71	1.64	−7.94	0.004	99.58
$CaCl_2$	−11.40	1.64	−7.12	0.005	99.43
Soy bean meal	−1.20	1.64	−0.75	0.505	49.47

Response surface methodology

The Central Composite design was employed to study the interaction among the significant components and also determine their optimal levels. In the present work, experiments were planned to obtain a quadratic model consisting of 2^3 trials. The plan includes 20 experiments and two levels of concentration for each component. In order to study the combined effect of these medium components, experiments were performed at different combinations. Table 4 summarizes the Central Composite experimental plan along with the predicted and observed response for each individual experiment. It shows the production of alpha-amylase (U/gds) corresponding to combined effect of all three components in the specified ranges.

The optimum levels of the selected variables were obtained by solving the regression equation and by analysing the response surface contour and surface plots (Abdelhay et al. 2008). The regression equation obtained after the analysis of variance (ANOVA) provides an estimate of the level of alpha-amylase production as a function of starch, ammonium sulphate, and calcium chloride concentration.

The production of alpha-amylase may be best predicted by the following model:

$$\text{Alpha-amylase production } (Y) = (47.94) - (3.08 \times A)$$
$$+ (1.11 \times B) + (3.51 \times C) - (1.00 \times A \times B)$$
$$- (0.61 \times A \times C) + (2.22 \times B \times C) + (0.09 \times A^2)$$
$$+ (0.19 \times B^2) + (1.68 \times C^2), \tag{5}$$

where Y is alpha-amylase production (U/gds), A is starch concentration (w/w), B is ammonium sulphate concentration (w/w), and C is calcium chloride concentration (mM).

The statistical significance of the second-order model equation was evaluated by F test analysis of variance which revealed that this regression is statistically highly significant for alpha-amylase production. The model

result of Plackett–Burman experiment with respect to amylase production, the effect, standard error, t(xi), p and confidence level of each component are represented in Tables 2 and 5. The components were screened at the confidence level of 95 % on the basis of their effects. When components show significance at or above 95 % confidence level and its effect is negative, it is considered effective for production but the amount required may be lower than the indicated low (−1) concentration in Plackett–Burman experiment. If the effect is found positive, a higher concentration that the indicated high value (+) concentration is required. In our experiment, starch, ammonium sulphate and calcium chloride gave confidence level >95 % and could be considered significant and were short-listed for further optimization of their required concentration and their interaction effect leading to maximum enzyme production. Remaining components such as lactose, yeast extract, soya bean meal, ammonium nitrate and dextrose showed confidence level <95 % and were considered insignificant in the study. The methodology of Plackett–Burman was thus found to be very useful for determination of relevant variables for further optimization.

Table 6 Analysis of variance (ANOVA) for response surface quadratic model of alpha-amylase production from *B. amyloliquefaciens* KCP2

Source	Sum of squares	df	Mean square	F value	p value	
Model	407.6871	9	45.29857	17.97587	<0.0001	Significant
A-starch	130.1606	1	130.1606	51.65175	<0.0001	Significant
B-Ammonium sulphate	17.12545	1	17.12545	6.795908	0.0262	Significant
C-Calcium chloride	168.6796	1	168.6796	66.93727	<0.0001	Significant
AB	8.142854	1	8.142854	3.231337	0.1025	
AC	3.058689	1	3.058689	1.213782	0.2964	
BC	39.48642	1	39.48642	15.66943	0.0027	Significant
A^2	0.12979	1	0.12979	0.051505	0.8250	
B^2	0.54180	1	0.5418	0.215003	0.6528	
C^2	40.94109	1	40.94109	16.24669	0.0024	Significant
Residual	25.19965	10	2.519965			
Lack of fit	19.24826	5	3.849652	3.23425	0.1118	Not significant
Pure error	5.951384	5	1.190277			
Core total	432.8868	19				

CV 3.22 %, adequate precision 17.51

F value of 17.98 implies that the model is significant. There is only a 0.01 % chance that a large "Model F value" could occur due to noise. Values of "Prob > F" <0.05 indicate that the model terms are significant. In this case A, B, C, BC and C^2 are significant model terms (Table 6). The "lack of fit F value" of 3.23 implies the lack of fit is not significant relative to the pure error. Non-significant lack of fit is good for the model to fit. The R^2 value (multiple correlation coefficient) closer to 1 denotes better correlation between observed and predicted values. The coefficient of variation (CV) indicates the degree of precision with which the experiments are compared. The lower reliability of the experiment is usually indicated by high value of CV. In the present case, a low CV (3.22 %) denotes that the experiments performed are reliable. Adequate precision measures the signal-to-noise ratio. A ratio >4 is desirable. In our case, the ratio is of 17.51, which indicates an adequate signal. This model can be used to navigate the design space.

The effect of interaction of variables on enzyme (alpha-amylase production) yield was studied against any two independent variables while keeping the other independent variables at their constant level. These response surface plots or contour plots can be used to predict the optimal values for different test variables. Therefore, three response surfaces were obtained by considering all the possible combinations. Three-dimensional response plot shown in Fig. 4a describes the behaviour of alpha-amylase production, main effect, interaction effect and squared effect (nonlinear) of starch and ammonium sulphate at different concentrations. Both the components at their lower level did not show any significant effect on the alpha-amylase

production. The shape of the response surface curves showed a moderate interaction between these tested variables. Increase in the starch concentration leads to gradual decrease in the enzyme production, while increase in the ammonium sulphate concentration results in the significant increase in the alpha-amylase production. Figure 4b depicts three-dimensional curve and contour plot of the calculated response surface from the interaction between starch and calcium chloride while keeping fixed concentration of ammonium sulphate. The level of the calcium chloride in the production medium showed prominent effect on the alpha-amylase production. Increase in the calcium chloride concentration leads to concomitant increment in the alpha-amylase yield, while both components at their lower level did not result in the higher enzyme yield. Figure 4c depicts the interaction of ammonium sulphate and calcium chloride where the shape of the response surface indicates positive interaction between these two factors. The enzyme yield was found to increase with simultaneous increase in concentration of both the components. Both the components at their lower level did not show any significant effect on the enzyme yield.

Validation of the quadratic model

Validation was carried out under conditions predicted by the response surface model. The optimal concentrations estimated for each variable were 0.01 g starch, 0.2 g ammonium nitrate and 5 mM calcium chloride per 5 g of wheat bran. The predicted alpha-amylase production obtained from the model using the above optimum concentration of medium components was 61.48 U/gds. To

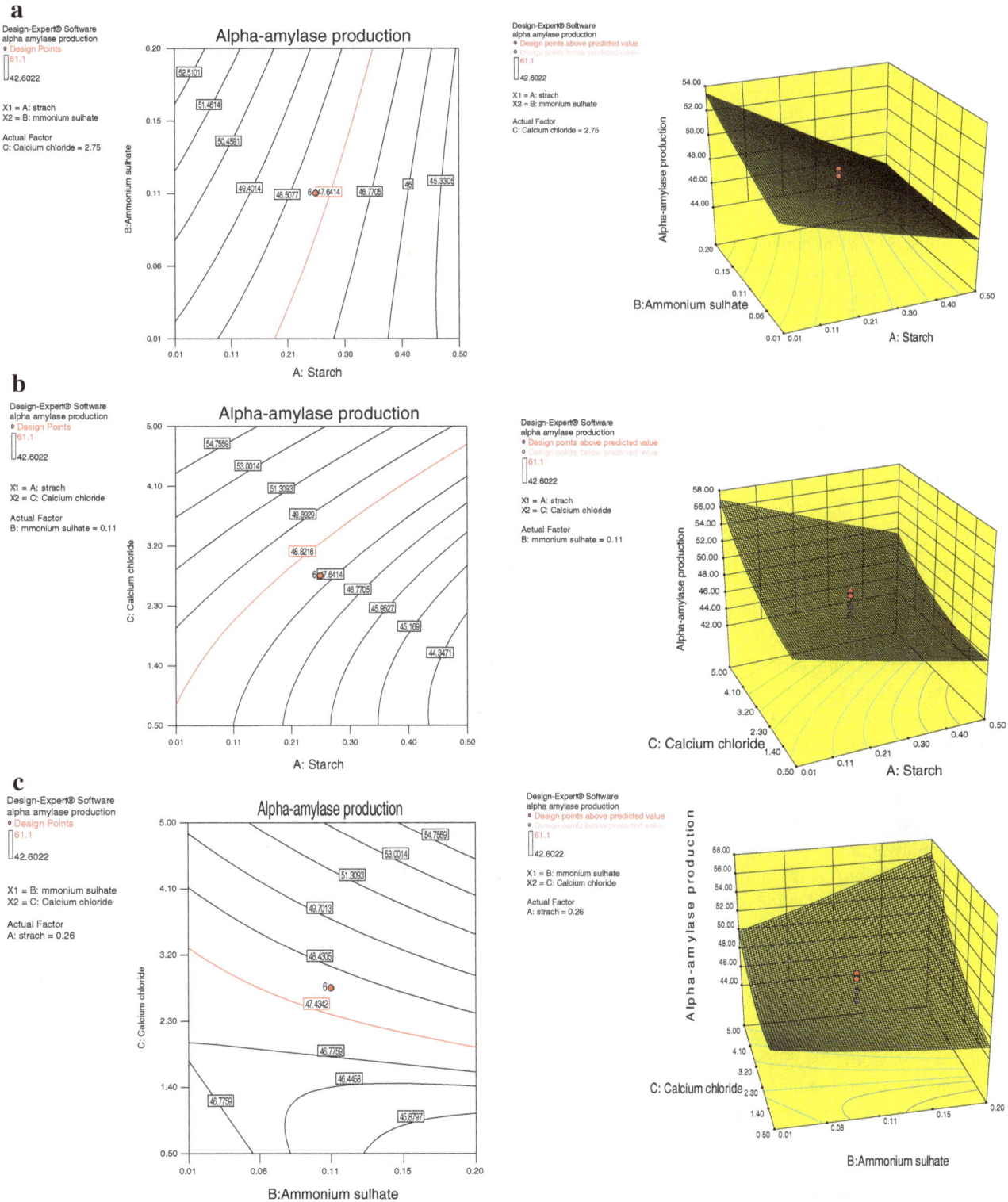

Fig. 4 Response surface graph showing interaction effects between concentration of starch and ammonium sulphate (**a**), starch and calcium chloride (**b**), and ammonium sulphate and calcium chloride (**c**)

validate the prediction of the model, additional experiments in triplicate were performed with the optimized medium. These experiments yielded the maximum amylase activity of 63.12 U/gds. Good agreements between the predicted and experimental results verified the validity of the model and the existence of the optimal points.

Discussion

Amylases have several interesting potential application in the food, detergent, pharmaceutical, leather, textile, cosmetic, and paper industries. Still their application in starch-based industries is the major market and thus demand of amylases would always be very high in this sector. There is a constant search for microorganisms producing enzyme with desired properties of pH and temperature stability considering their industrial application. In this respect the isolate KCP2 was found to have production of enzyme with such properties and was characterized further. As shown in the Fig. 1, the evolutionary history was inferred using the neighbor-joining method (Saitou and Nei 1987). The optimal tree with the sum of branch length of 1.51712318 is shown and the percentage of replicate trees in which the associated taxa clustered together in the bootstrap test (500 replicates) are shown next to the branches (Felsenstein 1985). The evolutionary distances were computed using the maximum composite likelihood method (Tamura et al. 2004) and are in the units of the number of base substitutions per site. Codon positions included were 1st + 2nd + 3rd + non-coding and all positions containing alignment gaps and missing data were eliminated only in pairwise sequence comparisons (pairwise deletion option). There were a total of 1,579 positions in the final dataset.

A marked enhancement in production of alpha-amylase by *B. amyloliquefaciens* in flask fermentation using statistical methods was reported by Zhao et al. (2011). Use of Plackett–Burman design to screen different nutrients affecting production of thermostable β-amylase and pullulanase by *Clostridium thermosulfurogenes* SV2 has been reported by Reddy et al. (1999). The substrate employed in the present investigation, i.e. wheat bran, has been reported as a potent substrate for production of alpha-amylase under SSF (Gangadharan et al. 2006; Ramachandran et al. 2004; Mulimani et al. 2000). It is well documented that wheat bran is rich source of carbon and nitrogen, thus supplementation of other nitrogen sources in the medium does not show any significant rise in the enzyme production but sometimes presence of starch as an additional carbon source was found to have inductive effect, and it also has remarkable efficiency in the production of enzyme, being an inexhaustible source of carbon compared to other carbon sources (Prajapati et al. 2013). In the present investigation, apart from starch, lactose was also found to act as an inducer for enzyme production as it gave 94.53 % confidence level during components screening study. Added nitrogen sources also have been reported for an inducing effect for the production of various enzymes including alpha-amylase in an SSF system (Pedersen and Nielsen 2000; Akher et al. 1973). Earlier reports show that among various inorganic nitrogen sources tested,

ammonium sulphate, ammonium chloride and ammonium hydrogen phosphate favour growth and enzyme secretion (Narang and Satyanarayana 2001).

Alpha-amylase is an inducible enzyme, which is generally induced in the presence of starch or its hydrolytic product maltose (Rama and Srivastav 1995). *Bacillus thermooleovorans* preferred starch, glucose, lactose, maltose and maltodextrin as favourable carbon sources for amylase secretion (Narang and Satyanarayana 2001). In some cases, hydrolyzed starch and glucose were found to repress the enzyme yield, which may be due to feedback inhibition caused by the presence of reducing sugars. Easily metabolizable carbohydrates may result in the better growth of the organism along with reduction in the enzyme formation (Rama and Srivastav 1995). The supplementation of metal ions has been reported to provide good growth and also influence higher enzyme yield (Sivaramakrishnan 2006). Most of the alpha-amylase is metalloenzymes and in most of the cases, Ca^{+2} ions are required for maintaining the spatial conformation of the enzyme, thus play an important role in the enzyme stability and its activity. Amylase from the halophilic *Bacillus* sp. Strain *TSCVKK* showed stability at a wide pH range of 6.5–10.5 with maximum activity with pH 8.0 suggesting the alkalitolerant nature (Kondepudi and Chandra 2008). Majority of the literature showed that 50 °C is the optimum temperature for alpha-amylase activity, but in our case we found that 65 °C is the optimum temperature for the activity of enzyme produced by *B. amyloliquefaciens* KCP2. Temperature and pH stability of this enzyme justify its alklophilic and thermophilic nature. Statistical approach has also been applied for the production of various enzymes, such as cyclodextrin glucanotransferase (CGTase) (Gawande and Patkar 1999; Mahat et al. 2004), chitinase (Gohel et al. 2005), pectinase (Nair and Panda 1997), vitamin riboflavin (Punjari and Chandra 2000) and glucoamylase (Prajapati et al. 2013).

Present investigation has allowed rapid screening and level optimization of a large number of nutrient parameters influencing thermophilic and alklophilic alpha-amylase production from *B. amyloliquefaciens* KCP2 using statistical methodology. The results also showed the use of cheap agro-residue as substrate for fermentation, thus contributing to the reduction in cost of production medium. The enzyme yield and the production were found to be significantly influenced by starch, ammonium sulphate and calcium chloride concentration. The data obtained after optimization has resulted in 63.12 U/gds enzyme production. Even though SSF is widely applied for enzyme production using filamentous fungi, the results of the present study proved that a bacterial isolate such as *B. amyloliquefaciens* KCP2 can be successfully used for the production of alpha-amylase employing wheat bran within a relatively shorter time interval of 3 days.

Acknowledgments The authors are grateful to the Department of Biotechnology, Ministry of Sciences and Technology, Government of India, for providing the financial assistance during the course of this investigation.

Conflict of interest We, all authors, declare that this manuscript does not have any financial/commercial conflicts of interest.

References

Abdelhay A, Magnin JP, Gondrexon N, Baup S, Willison J (2008) Optimization and modeling of phenanthrene degradation by *Mycobacterium* sp. 6PY1 in a biphasic medium using response surface methodology. Appl Microbiol Biotechnol 78:881–888

Akher M, Leithy MA, Massafy MK, Kasim SA (1973) Optimal conditions of the production of bacterial amylase. Zentralbl Bakteriol Parasitenk Infektionskr Hyg 128:483–490

Babitha S, Soccol CR, Pandey A (2007) Solid-state fermentation for the production of *Monascus* pigments from jackfruit seed. Bioresour Technol 98:1554–1560

Binod P, Sandhya C, Suma P, Szakacs G, Pandey A (2007) Fungal biosynthesis of endochitinase and chitobiase in solid-state fermentation and their application for the production of *N*-acetyl-D-glucosamine from colloidal chitin. Bioresour Technol 98:2742–2748

Djekrif DS, Gheribi AB, Meraihi Z, Bennamoun L (2006) Application of the statistical design to the optimization of the culture medium for alpha-amylase production by *Aspergillus niger* ATCC16404 grown on orange waste powder. J Food Eng 73:190–197

Elibol M (2004) Optimization of medium composition for actinorhodin production by *Streptomyces coelicolor* A3(2) with response surface methodology. Process Biochem 39:1057–1062

Felsenstein J (1985) Confidence limits on phylogenies: an approach using the bootstrap. Evolution 39:783–791

Gangadharan D, Sivaramakrishnan S, Nampoothiri KM, Pandey A (2006) Solid culturing of *Bacillus amyloliquefaciens* for alpha amylase production. Food Technol Biotechnol 44:269–274

Gawande BN, Patkar AY (1999) Application of factorial design for optimization of cyclodextrin glycosyltransferase production from *Klebsiella pneumoniae* AS-22. Biotechnol Bioeng 64:168–172

Gohel V, Jiwan D, Vyas P, Chatpar HS (2005) Statistical optimization of chitinase production by *Pantoea dispersa* to enhance degradation of crustacean chitin waste. J Microbiol Biotechnol 15:197–201

Gupta R, Gigras P, Mohapatra H, Goswami VK, Chauhan B (2003) Microbial alpha-amylases: a biotechnological perspective. Process Biochem 38:1599–1616

Haq I, Ashraf H, Iqbal J (2003) Production of alpha-amylase by *Bacillus licheniformis* using an economical medium. Bioresour Technol 87:57–61

Kammoun R, Naili B, Bejar S (2008) Application of a statistical design to the optimization of parameters and culture medium for α-amylase production by *Aspergillus oryzae* CBS 819.72 grown on gruel (wheat grinding by product). Bioresour Technol 99:5602–5609

Kondepudi KK, Chandra TS (2008) Production of surfactant and detergent stable, halophilic, and alkalitolerant alpha-amylase by moderately halophilic *Bacillus* sp. strain TSCVKK. Appl Microbiol Biotechnol 77:1023–1031

Mahat MK, Illias RM, Rahman RA, Rashid NAA, Mahmood NAN, Hassan O (2004) Production of cyclodextrin glucanotransferase (CGTase) from alkalophilic *Bacillus* sp. TSI-1: media optimization using experimental design. Enzym Microb Tech 35:467–473

Miller GL (1959) Use of dinitrosalicylic acid reagent for determination of reducing sugar. Anal Chem 31:426–429

Mulimani VH, Patil GN, Ramalingam (2000) α-Amylase production by solid state fermentation: a new practical approach to biotechnology courses. Biochem Educ 28:161–163

Nair SR, Panda T (1997) Statistical optimization of medium components for improved synthesis of pectinase by *Aspergillus niger*. Bioproc Biosyst Eng 16:169–173

Narang S, Satyanarayana T (2001) Thermostable α-amylase production by an extreme thermophile *Bacillus thermooleovorans*. Lett Appl Microbiol 32:31–35

Pandey A, Nigam P, Soccol CR, Singh D, Soccol VT, Mohan R (2000) Advances in microbial amylases. Biotechnol Appl Biochem 31:135–152

Pedersen H, Nielsen J (2000) The influence of nitrogen sources on the alpha amylase productivity of *Aspergillus oryzae* in continuous cultures. Appl Microbiol Biotechnol 53:278–281

Plackett RL, Burman JP (1946) The design of optimum multifactorial experiments. Biometrika 33:305–325

Prajapati VS, Trivedi UB, Patel KC (2013) Optimization of glucoamylase production by *Colletotrichum* sp. KCP1 using statistical methodology. Food Sci Biotechnol 22:31–38

Pujari V, Chandra TS (2000) Statistical optimization of medium components for enhanced riboflavin production by a UV mutant of *Eremothecium ashbyii*. Process Biochem 36:31–37

Rama R, Srivastav SK (1995) Effect of various carbon substrates on α-amylase production from *Bacillus* species. J Microb Biotechnol 10:76–82

Ramachandran S, Patel AK, Nampoothiri KM, Francis F, Nagy V, Szakacs G, Soccol CR, Pandey A (2004) Coconut oil cake-a potential raw material for the production of alpha amylase. Bioresour Technol 93:169–174

Ramachandran S, Singh S, Larroche C, Soccol CR, Pandey A (2007) Oil cakes and their biotechnological applications: a review. Bioresour Technol 98:2000–2009

Reddy PRM, Reddy G, Seenayya G (1999) Production of thermostable β-amylase and pullulanase by *Clostridium thermosulfurogenes* SV2 in solid-state fermentation: screening of nutrients using Plackett–Burman design. Bioproc Biosyst Eng 21:175–179

Saitou N, Nei M (1987) The neighbor-joining method: a new method for reconstructing phylogenetic trees. Mol Biol Evol 4:406–425

Silva CJSM, Roberto IC (2001) Optimization of xylitol production by *Candida guilliermondii* FTI 20037 using response surface methodology. Process Biochem 36:119–124

Sivaramakrishnan S, Gangadharan D, Nampoothiri KM, Pandey A (2006) α-Amylases from microbial sources: an overview on recent developments. Food Technol Biotechnol 44:173–184

Tamura K, Nei M, Kumar S (2004) Prospects for inferring very large phylogenies by using the neighbor-joining method. PNAS 101:11030–11035

Tamura K, Dudley J, Nei M, Kumar S (2007) MEGA4: molecular evolutionary genetics analysis (MEGA) software version 4.0. Mol Biol.

Wenster-Botz D (2000) Experimental design for fermentation media development: statistical design or global random search? J Biosci Bioeng 90:473–483

Zhao W, Zheng J, Wang YG, Zhou H (2011) A marked enhancement in production of amylase by *Bacillus amyloliquefaciens* in flask fermentation using statistical methods. J Cent South Univ Technol 18:1054–1062

23

Eggplant (*Solanum melongena* L.) polyphenol oxidase multi-gene family: a phylogenetic evaluation

Aravind Kumar Jukanti · Ramakrishna Bhatt

Abstract Polyphenol oxidases (PPOs) in different *Solanum* species including eggplant have been studied. PPOs have been implicated in undesirable enzymatic browning of eggplant fruit and also in plant defense. The main objective of this study was to identify and accelerate the further functional characterization of additional eggplant PPOs that are involved in food biochemistry and defense-related functions. Eggplant PPOs identified earlier were used in "Basic local alignment search tool (BLAST)" search against expressed sequence tag and nucleotide databases. We have identified seven additional sequences which were almost complete in length. The sequences of the PPOs were aligned and their phylogenetic and evolutionary relationships established. The sequences are quite diverse, broadly falling into two major clusters; three PPOs form a separate branch/minor cluster. The thirteen sequences had conserved copper A binding sites but copper B binding sites differed considerably in two new PPO sequences (AFJ79642 and ACR61398). A third conserved 'Histidine-rich' region has been identified at the 'C' terminus of the eggplant PPOs. In addition, all the seven new PPOs exhibited at least one glycosylated sequon in the mature PPO sequence. Identification of additional PPO genes will further help in functional and biological characterization of these PPOs.

Keywords Eggplant · Polyphenol oxidase · Phylogenetic analysis · Multi-gene family · N-glycosylation

A. K. Jukanti (✉) · R. Bhatt
Central Arid Zone Research Institute, Jodhpur 342003, Rajasthan, India
e-mail: jukanti5@yahoo.com; aravindjukanti@gmail.com

Introduction

Polyphenol oxidases (PPOs) can oxidize specific phenolic substrates in the presence of oxygen in contrast to peroxidases which oxidize phenols in presence of H_2O_2. PPOs are ubiquitously distributed in plants (Mayer and Harel 1979); they play a role in food quality and in plant defense against pest and pathogens (Thipyapong et al.1995; Thipyapong and Steffens 1997; Wang and Constabel 2004). PPOs are also involved in: time-dependent darkening and discoloration of cereal-based products (Baik et al. 1994), biosynthesis of flavonoids (Ono et al. 2006) and in oxidation of flavonoids (Pourcel et al. 2005). PPOs come into contact with phenolic substrates that are released due to tissue damage. The phenols are oxidized to highly reactive *o*-quinones which either self-polymerize or further react with nucleophiles to produce dark colored pigments that are usually undesirable in fresh or processed foods (Anderson and Morris 2001).

PPOs contain two copper (Cu) binding sites (Cu A and Cu B) and the Cu ion is bound by conserved histidine residues. PPOs interact with molecular oxygen and phenolic substrates at Cu-A and Cu-B sites (Van Gelder et al. 1997). PPOs are nuclear encoded enzymes, synthesized as precursor proteins in cytosol, processed to mature proteins and imported into plastidial thylakoid membranes (Koussevitzky et al. 1998; Sommer et al. 1994). The typical N-terminal transit peptide of PPOs is about 80–100 amino acids in length and during the chloroplast import the molecular weight of the enzyme is reduced from ∼65–70 to <60 kDa (Dry and Robinson 1994; Van Gelder et al. 1997). PPOs are basically three types based on the substrates they catalyze: cresolases (monophenol oxidases), *o*-diphenol oxidases (catecholases) and laccase-like multi-copper oxidases (Shetty et al. 2011). Among the three classes of PPOs, *o*-diphenol oxidases have been extensively characterized in plants. Catecholases mostly occur as multi-gene families, introns are absent among the dicotyledonous PPOs genes reported so far (Shetty et al. 2011). But introns have been reported in monocot species like pineapple, banana and wheat (Massa et al. 2007). Up-regulation of PPO genes upon mechanical wounding and damage due to pests has been reported in several crop species (Thipyapong et al. 1995, 1997; Wang and Constabel 2004).

The role of PPOs in plant defense and enzymatic discolouration/browning affecting quality of crop/plant products has led to extensive identification and characterization of PPO genes in several plant species. Identification and characterization of all/most PPOs in a plant species will aid in better understanding the structural and functional differences among the multi-gene family. Solanaceae crops like potato, tomato and eggplant/brinjal form an important part of the daily diet in many parts of the world. Specifically, eggplant is an important constituent of the Indian cuisine. The role of PPOs in enzymatic browning of eggplant fruit has been vastly studied and also chlorogenic acid is shown to be the most predominant phenolic in the flesh of its fruit (Whitaker and Stommel 2003; Singh et al. 2009). Studies on eggplant PPOs describing their biochemistry, enzymatic action and genes have been published (Pérez-Gilabert and García Carmona 2000; Shetty et al. 2011). But due to their role in plant defense and food quality the identification of any additional PPO genes could be critical. Therefore, the main objective of this manuscript was to identify any additional eggplant PPOs utilizing bioinformatic tools and publicly available databases.

Materials and methods

The PPO gene sequences published by Shetty et al. (2011) were utilized to analyze public databases for additional eggplant PPOs using basic local alignment search tool (BLAST; Altschul et al. 1990) against National Center for Biotechnology Information (NCBI; http://www.ncbi.nlm.nih.blast). The *S. melangena* hits were exported to San Diego Super Computer Center biology workbench (http://seqtool.sdsc.edu) for sequence analysis. Sequences were aligned using CLUSTALW tool (Thompson et al. 1994) after translation to protein sequences. Phylogenetic relationships among the identified sequences were calculated using PHYLIP (Felsenstein 1989). Sequence identity matrix of all the thirteen PPOs was computed using LALIGN program (http://workbench.sdsc.edu/). The molecular weight (statistical analysis of protein sequence (SAPS), Brendel et al. 1992) and isoelectric point (http://seqtool.sdsc.edu/CGI/BW.cgi) of different PPO sequences

were also determined. The N-glycosylation sites and their position in the mature novel PPO sequences were analyzed using NetNGlyc software (www.cbs.dtu.dk/services/NetNGlyc/). All the additional sequences obtained (except one) were almost complete with minimal gaps, therefore, were further not sequenced. All thirteen eggplant PPOs are aligned and analyzed as described above.

Results

Six PPO sequences were reported in eggplant by Shetty et al. (2011). However, large scale work on eggplant PPO ESTs/gene sequences has been reported in past few years. Therefore, a detailed analysis of PPOs in eggplant was performed; it resulted in identifying a multi-gene family of thirteen PPOs, grouped into two major clusters and a third minor group (Fig. 1). The published sequences (Shetty et al. 2011) were used as a search tool for NCBI BLAST search, seven additional eggplant PPO sequences with a sequence identity >62 % in overlaps of at least 266 amino acids were identified in this study. A phylogeny tree and evolutionary relationships of all the PPOs was constructed using PHYLIP (Fig. 1a, b). It was observed that six novel sequences identified are paired up with ADG56700 as a separate cluster and the seventh sequence (BAA85119) was closely related to ACT22523. Four previously characterized sequences (ADY184109, 18410, 18411 and 18411) form a second major cluster with ADY18409 forming a separate branch. Based on sequence comparison with the other eggplant PPOs the new sequences represent almost full length sequences except for one (BAA85119). The eggplant PPOs were compared with other plant PPOs (data not shown) including potato (*Solanum tuberosum* L.), tomato (*Lycopersicum esculentum* L.), tobacco (*Nicotiana tabacum* L.) and grape (*Vitis vinifera* L.). The tentative start site of mature PPOs is indicated based on the sequence alignment with higher plant and previously characterized eggplant PPOs. Tentative proteolytic processing sites for both stromal and thylakoid peptidases are also indicated. The length and molecular weight of mature PPOs ranged 584–601 amino acids and 65.9–67.7 kDa, respectively (Table 1). The molecular weight of mature PPOs was about ~56–58 kDa (after proteolytic processing). The isoelectric point of different PPOs was determined and it ranged from 6.062 to 7.956 (Table 1)

The percentage identity among the different PPOs was also calculated using LALIGN software. Eggplant PPO with GenBank accession number ACR61398 was identified to be an additional eggplant PPO. This sequence demonstrates ~61–87 % identity with the sequences used as search tool, the new sequence shows highest identity to an eggplant PPO with GenBank accession number,

Fig. 1 **a** Phylogenetic analysis (*unrooted tree*) of eggplant PPOs, **b** phylogenetic analysis (*rooted tree*) of different eggplant PPOs

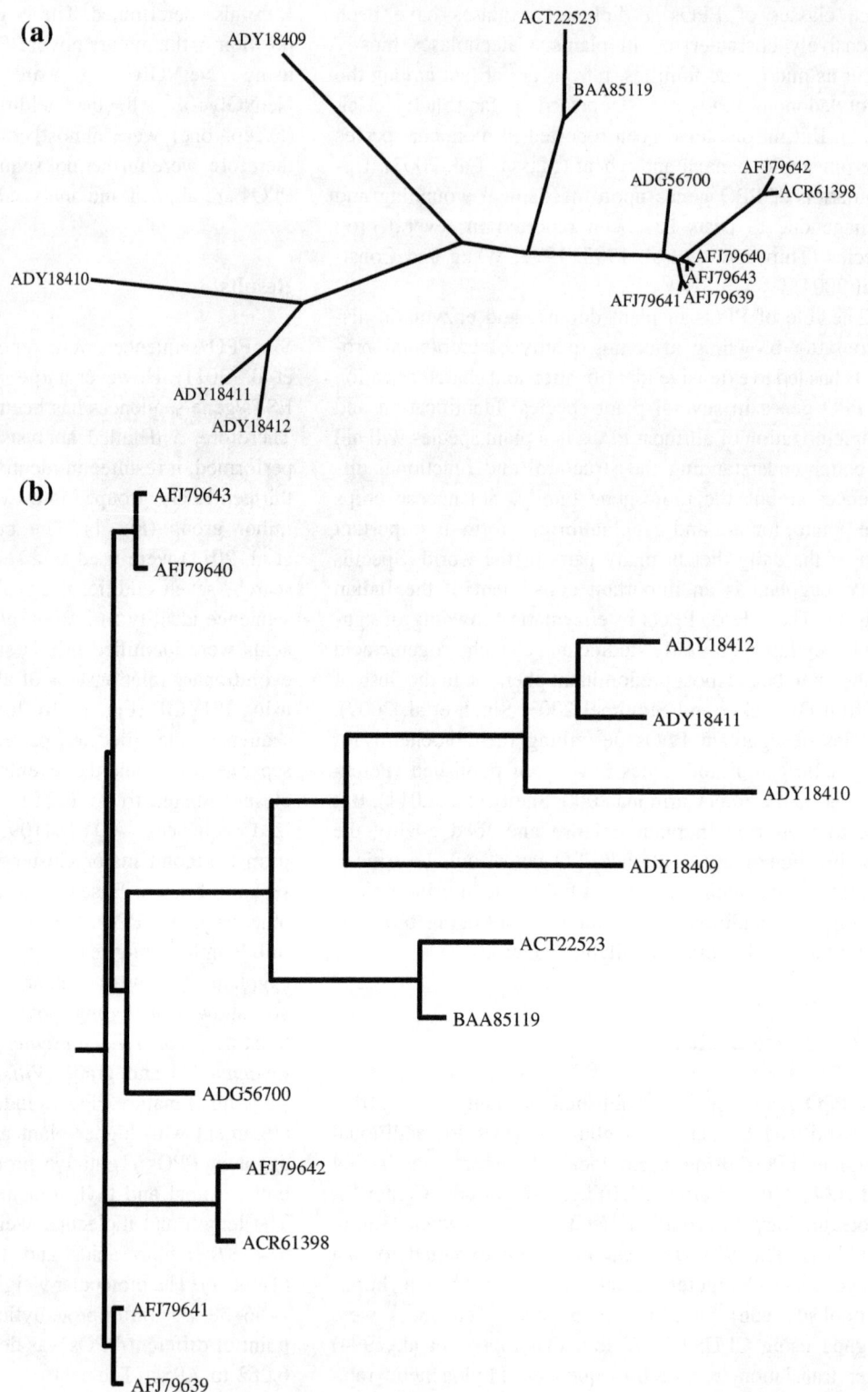

ADG56700 (Table 2). ACR61398 sequence has relatively higher sequence identity (71–99 %) to the additional five PPO sequences (AFJ79639, AFJ79640, AFJ79641, AFJ79642, and AFJ79643) identified in this study, being almost identical to AFJ79642 (Table 2). Another novel

sequence with an accession number, AFJ79639 was similar to two other novel sequences, AFJ79640 and AFJ79643 differing only at six different amino acid (Aa) sites over a length of 595 Aa (data not shown). AFJ79639 also shows a similar trend to ACR61398 in sequence identity. AFJ79641

Table 1 Characteristics of eggplant PPOs

PPO sequence	Amino acid number	Molecular weight (kDa)	Isoelectric point
ACR 61398	593	66.9	6.488
AFJ 79639	594	66.5	6.062
AFJ 79640	595	66.7	6.147
AFJ 79641	594	66.6	6.147
AFJ 79642	590	66.7	6.488
AFJ 79643	595	66.7	6.236
ADG 56700	601	67.6	6.549
ADY 18409	590	67.1	7.465
ADY 18410	584	66.3	6.294
ADY 18411	586	66.0	6.162
ADY 18412	584	65.9	6.832
ACT 22523	600	67.7	7.233
BAA 85119	266	30.4	7.956

is ~98 % identical to AFJ79639, AFJ79640 and AFJ79643. The sequence AFJ79643 follows same trend as the other new sequences. But the amino acid identity of

AFJ79642 was comparatively lower with the other novel sequences identified. Overall, the newly reported sequences vary considerably in their similarity to the search tool sequences ranging ~61–95 % (Table 2).

The amino acid sequences of the newly identified PPO sequences were analyzed for determining the most conserved features of all PPOs: two copper-binding sites and transit peptide sequence (Koussevitzky et al. 1998). The chloroplast transit peptide sequence harboring both thylakoid and stromal targeting domains extending ~80–90 amino acids is also found in the novel sequences. The transit peptide sequences of all the eggplant PPOs demonstrated conserved cleavage sites for both stromal peptidases (V, S, C, K/N) and thylakoid peptidases (L, A/T, A, S/N, A; Fig. 2). Both the copper-binding (A and B) regions of all the thirteen PPOs show considerable conservation of amino acid sequence. But among the two, 'A' site is relatively more conserved than 'B'. Two new sequences identified (AFJ79642 and ACR61398) in this study differed drastically from other PPOs especially in the copper 'B' region (Fig. 2). In these two sequences, the most conserved

Table 2 Identity matrix of eggplant PPO sequences as percentages

	ACR 61398	AFJ 79639	AFJ 79640	AFJ 79641	AFJ 79642	AFJ 79643	ADG 56700	ADY 18409	ADY 18410	ADY 18411	ADY 18412	ACT 22523	BAA 85119
ACR 61398	100	92	92	93	99	92	87	67	61	66	66	74	71
AFJ 79639	92	100	99	98	92	99	94	71	64	70	70	75	83
AFJ 79640	92	99	100	98	92	98	93	71	64	69	69	75	82
AFJ 79641	93	98	98	100	93	98	94	70	64	70	70	75	82
AFJ 79642	99	92	92	93	100	92	89	67	61	67	67	76	71
AFJ 79643	92	99	98	98	92	100	95	70	64	69	69	75	82
ADG 56700	87	94	93	94	89	95	100	69	63	68	68	77	78
ADY 18409	67	71	71	70	67	70	69	100	65	69	68	70	75
ADY 18410	61	64	64	64	61	64	63	65	100	81	80	61	62
ADY 18411	66	70	69	70	67	69	68	69	81	100	92	66	70
ADY 18412	66	70	69	70	67	69	68	68	80	92	100	65	66
ACT 22523	74	75	75	75	76	75	77	70	61	66	65	100	93
BAA 85119	71	83	82	82	71	82	78	75	62	70	66	93	100

```
                                              ▼                                                    ▽
AFJ79641  MAS------LSNSSIQPS-TPFT----SLGSTPKPSQLFLHGKRKQTFKVSCKVSNNNGDQNQNEVEKNSVDRRNVLLGLGGMYGAANFAPLAASAAPTPPPDLSSC
AFJ79639  MAS------LSNSSIQPS-NSLP----SLGSVPKPSQLFLHGKRKQTFKVSCKVSNNNGDQNQNEVEKNSVDRRNVLLGLGGMYGAANFAPLAASAAPTPPPDLSSC
AFJ79643  MAS------LSNSSIQPLKNSLP----SLGSVPKPSQLFLHGKRKQTFKVSCKVSNSNGDQNQNEVEKNSVDRRNVLLGLGGMYGAANFAPLAASAAPTPPPDLSSC
AFJ79640  MAS------LSNSSIQPTQTALP----SLGSVPKPSQLFLHGKRKQTFKVSCKVSNNNGDQNQNEVEKNSVDRRNVLLGLGGMYGAANFAPLAASAAPTPPPDLSSC
AFJ79642  MAS------LSNSSIQP----FT----SLGSTPKPSQLFLHGKRKQTFKVSCKVSNNNGDQNQNEVEKNSVDRRNVLLGLGGMYGAANFAPLAASAAPTPPPDLSSC
ACR61398  MAS------LSNSSIQPS-NSFT----SLGSTPKPSQLFLHGKRKQTFKVSCKVSNNNGDQNQNEVEKNSVDRRNVLLGLGGMYGAANFAPLAASAAPTPPPDLSSC
ADG56700  MAS------VCNTSTATLKSSFIPSPNSLGSTPKPSQLFLHGKRKQTFKVSCKVSNNNGDQNQNEVEKNSVDRRNVLLGLGGMYGAANFAPLAASAAPTPPPDLSSC
ACT22523  MAS------VCHTSTATLKSSFIPSPNSLGSTPKPSQLFLHGKRNQAFKVSCKVTNNNGDQNQNVVDTNSVDRRNVLLGLGGLYGVANAIPLAASATPIPAPNAPSC
BAA85119  --------------------------------------------------------------------------------------------------------
ADY18409  MASS----FLPLCTHPAFSNTS----ESSFLPKPSQLFLQRRHNQRFKVSCNANKHEKD------NLDVVDRRNVLLGLGGL-GAANLAPLTANAPSPPPDFKTC
ADY18412  MSSSS---TLPLCTNKSLS-SFT----NSSFLAKPSQLFLHRSRSQSFKVSCNANNVG----EHDKNLDAIDRRNVLLGLGGLYGAANLAPLAANAPIPSPDPKSC
ADY18411  MSSSSSTTTLPLCTNKSLS-SFT----NSSFLAKPSQLFLHRSRSQSFKVSCNANNVG----EHDKNLDAVDRRNVLLGLGGLYGAANLAPLAANAPIPPPDLKSC
ADY18410  --SSS---TLPLCNSKSLFFSFC----NSPFLPQPSKLFLQRTRSQRFKVSCNANNVG----EHDKNLDAVDRRNVLLGLGGLYGAANLAPLAANAPIPPPELKTC
          * ***   .           .        .:**:***:..:              : :  :*********** :* *..:* *. :  :*

AFJ79641  SIAKITE--TEEVSYSCCAP-TPDDLNKIPYYKFPSMTKLRIRQPAHAADEEYIAKYNLAISRMKHLDTTEPLNPIGFKQQANIHCAYCNGAYKIGDKVLQVHNSWL
AFJ79639  SIAKITE--TEEVSYSCCAP-TPDDLNKIPYYKFPSMTKLRIRQPAHAADEEYIAKYNLAISRMKHLDTTEPLNPIGFKQQANIHCAYCNGAYKIGDKVLQVHNSWL
AFJ79643  SIAKITE--TEEVSYSCCAP-TPDDLNKIPYYKFPSMTKLRIRQPAHAADEEYIAKYNLAISRMKHLDTTEPLNPIGFKQQANIHCAYCNGAYKIGDKVLQVHNSWL
AFJ79640  SIAKITE--TEEVSYSCCAP-TPDDLNKIPYYKFPSMTKLRIRQPAHAADEEYIAKYNLAISRMKHLDTTEPLNPIGFKQQANIHCAYCNGAYKIGDKVLQVHNSWL
AFJ79642  SIAKITE--TEEVSYSCCAP-TPDDLNKIPYYKFPSMTKLRIRQPAHAADEEYIAKYNLAISRMKHLDTTEPLNPIGFKQQANIHCAYCNGAYKIGDKVLQVHNSWL
ACR61398  SIAKITE--TEEVSYSCCAP-TPDDLNKIPYYKFPSMTKLRIRQPAHAADEEYIAKYNLAISRMKHLDTTEPLNPIGFKQQANIHCAYCNGAYKIGDKVLQVHNSWL
ADG56700  SIAKITE--TEEVSYSCCAP-TPDDLNKIPYYKFPSMTKLRIRQPAHAADEEYIAKYNLAISRMKHLDTTEPLNPIGFKQQANIHCAYCNGAYKIGDKVLQVHNSWL
ACT22523  GTATISD--GPEVPYTCCPPGMPEDIEKILYYKFPSATKLRIRQPAHAVDEELIAKYNLAISKMRELDTTDHFSPLAFKQQANIHCAYCNGAYKIGGKELQVHNSWL
BAA85119  --------------------------------------------------------------------------------------------------------
ADY18409  GIATITAD-GPPVPYTCCHLPMPSNVNTIPYYKLPSMTKVRIRQPAHTVDEEFIAKYNLAISRMKELDEKEPLNPLGFKQQANIHCAYCNGAYKIGEKVLQVHQSWL
ADY18412  SKAHIKPNK--EVPYSCCPP-PPQDIDSVPYYKFPPMTKLRIRPPAHAVDEEYIAKYQLATSRMRELD-KDPFDPLGFKQQANIHCSYCCGAYKVGGKVLQVHSSWL
ADY18411  SKAHINPDK--EVTYSCRPP-IPQDIDSVPYYKFPPMTKLRIRPPAHAVDEEYIAKYQLATSRMRELG-KDPFTPLGFKQQANIHCAYCNGAYKVGEKELQVHFSWL
ADY18410  GRAVVNDTTGELVKYSCCPP-IPDDIDSVPYYKFPPMTKLRIRPPAHAVDEEYIAKYQLATSQMRELD-KDPFGPIGFKQQANIHCAYCNGAYKAGGKELQVHFSWL
          * :.     * *:*      *.::*.: ***:*. **:*** ***:.*** ****:** *:*:.*. .:  :  *:.*********:** **** * ***** ***

AFJ79641  FFPFHRWYLYFYERILGSIIDDPTFALPYWNWDHPKGMRMPAMFDGEGTALYDQVRNQSHR-NGRVMDLG-SFGDEVQTTELQLMSNNLTLMYRQMVTNAPCPRMFF
AFJ79639  FFPFHRWYLYFYERILGSIIDDPTFALPYWNWDHPKGMRMPAMFDGEGTALYDQVRNQSHR-NGRVMDLG-SFGDEVQTTELQLMSNNLTLMYRQMVTNAPCPRMFF
AFJ79643  FFPFHRWYLYFYERILGSIIDDPTFALPYWNWDHPKGMRMPAMFDGEGTALYDQVRNQSHR-NGRVMDLG-SFGDEVQTTELQLMSNNLTLMYRQMVTNAPCPRMFF
AFJ79640  FFPFHRWYLYFYERILGSIIDDPTFALPYWNWDHPKGMRMPAMFDGEGTALYDQVRNQSHR-NGRVMDLG-SFGDEVQTTELQLMSNNLTLMYRQMVTNAPCPRMFF
AFJ79642  FFPFHRWYLYFYERILGSIIDDPTFALPYWNWDHPKGMRMPAMFDREGTALYDQVRNQSHR-NGRVMDLG-SFGDEVQTTELQLMSNNLTLMYRQWYY-APCPRMFL
ACR61398  FFPFHRWYLYFYERILGSIIDDPTFALPYWNWDHPKGMRMPAMFDGEGTALYDQVRNQSHR-NGRVMDLG-SFGDEVQTTELQLMSNNLTLMYRQWYY-APCPRMFL
ADG56700  FFPFHRWYLYFYERILGSIIDDPTFALPYWNWDHPKGMRMPAMFDREGTALYDQVRNQSHRPREGVMDFWFLLVMKVQTTKLQLMSNTLTLMYRQMVTNAPCPRMFF
ACT22523  FFPFHRWYLYFYERILGKLIDDPTFALPYWNWDHPKGMRLPPIFDRQGTALYDERRSTQVR-NGTVMDLG-SFGDKVQTTQLQLMSNNLTLMYRQMVTNAPCPLFIL
BAA85119  -----------------------------------------------NGTVMDLG-SFGDQVQTTQLQLMSNTLTLMYRQMVTNAPCPLLFF
ADY18409  FFPFHRWYLYFYERMLGKLIDDPTFALPYWNWDLPKGMRLPPMFDREGSPLYDERRNPQVR-NGTVMDLG-SFGDQVQTTELQLMSNNLTLMYRHMVTNASCPLLFF
ADY18412  FFPFHRWFLYFYERILGSLINDPTFALPYWNWDHPKGMRIPPMFDHEGSSLYDEKRNQNHR-NGKIINLG-FSCKETQTTELQTMTNNLTLMYRQMVTNAPCPLLFF
ADY18411  FFPFHMVFV-FYERILGSLINDPTFALPYWNWDHPKGMRIPPMFDREGSSLYDEKRNQNHR-NGKIIDLG-FFGTETQTTELQTMTNNLTYMYRQMVTNAPCPLLFF
ADY18410  FFPFHKAGGKELQVHFSWLFYERILGSLINDPTFALPYWNWDHPKGMRLPPHFDVEGSSLYDAKRNQSHR-NGKIIDLG-FFGQETETELQTMTNNLTLMYRQMVTNAPCPLLFF
          *****  ::  ****:**.:*:************* *****:*:.** *:.*** *. .  .  ::::  . ::**:** *:*.** ***:  *.** :::

AFJ79641  G-APYVLGNNVEA-PGTIEVIPHGPVHVWTGTVPGTTLPNGRTSHGENMGHFYSAGLDPVFFCHHSNVDRMWSEWKAIGGKRRDISHKDWLNSEFFFYDENGDPFRV
AFJ79639  G-APYVLGNNVEA-PGTIEVIPHGPVHVWTGTVPGTTLPNGRTSHGENMGHFYSAGLDPVFFCHHSNVDRMWSEWKAIGGKRRDISHKDWLNSEFFFYDENGDPFRV
```

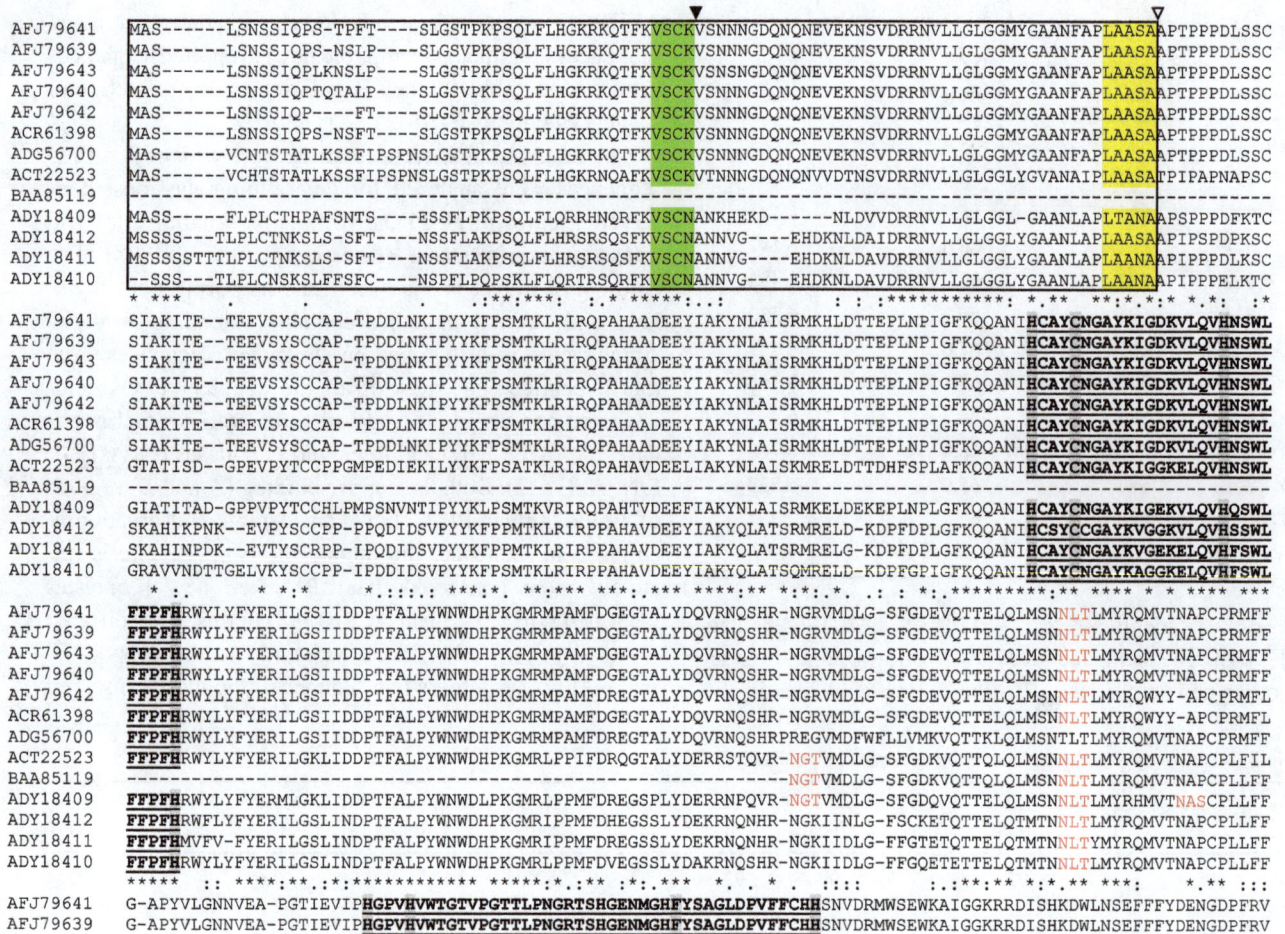

Fig. 2 Sequence alignment of predicted amino acid sequence of thirteen eggplant PPOs. Transit peptide is boxed and proteolytic processing sites are indicated by (stromal peptidase, green) and (thylakoid peptidase, *yellow*). Conserved copper binding regions (A & B) are in bold font with underline, conserved histidine and cysteine amino acid (histidine, 'H') was absent. In addition to the two known copper-binding domains of all PPOs, it was noted that the eggplant PPOs consists of a third conserved 'Histidine (His)-rich' region at the 'C' terminus (Fig. 2). NetNGlyc, N-glycosylation software identified four glycosylated sites in the different mature PPO sequences. The 'NLT' sequon was the most widely conserved among the different eggplant PPOs (except ADG56700). The other glycosylated sites identified are: NGT/NTS—ACT22523 and BAA85119; NGT/NAS—ADY18409.

(thio-ether linkage) are highlighted in light *grey*. Third 'His-rich' region is highlighted in dark *grey* at the 'C' terminus. N-glycosylated (Asn-Xaa-Thr/Ser) sequons are highlighted in *red*. (*) - single, fully conserved residue; (:) - conservation of strong groups; (.) - conservation of weak groups

Discussion

During the past few years higher plant PPOs have been extensively studied due to their potential role in food biochemistry (Feillet et al. 2000) and plant defense (Constabel et al. 1995). Specifically, eggplant PPOs have been investigated due to their perceived role in browning of eggplant fruit which is rich in phenols (Shetty et al. 2011). Chlorogenic acid is the major phenolic compound present in the flesh of eggplant fruit, accounting for ~70–95 % of the total phenolics (Whitaker and Stommel 2003; Singh et al. 2009). The biological importance of PPOs demands a comprehensive study of the number and role of different PPOs present in eggplant. Analysis of publicly available data has identified seven additional eggplant PPO sequences, constituting the eggplant PPO multi-gene family of thirteen genes. The phylogenetic and identity matrix analysis of the thirteen eggplant PPOs indicate the presence of two major clusters. In addition, it was also observed that three sequences (ACT22523 and BAA85119; ADY18409) probably represent different branch/clusters, these could include a few more yet to be identified PPO sequences.

The N-terminal region of the eggplant PPOs contained chloroplast transit peptide, these regions consists of both stromal and thylakoid targeting domains. Stromal and thylakoid targeting domains help in importing the mature

```
AFJ79643  G-APYVLGNNVEA-PGTIEVIPHGPVHVWTGTVPGTTLPNGRTSHGENMGHFYSAGLDPVFFCHHSNVDRMWSEWKAIGGKRRDISHKDWLNSEFFFYDENGDPFRV
AFJ79640  G-ALYVLGNNVEA-PGTIEVIPHGPVHVWTGTVPGTTLPNGRTSHGENMGHFYSAGLDPVFFCHHSNVDRMWSEWKAIGGKRRDISHKDWLNSEFFFYDENGDPFRV
AFJ79642  A-RLTFLGITLKP-QEPLKSSLTVLSTFGLVQCQVQPCLNGRTSHGENMGHFYSAGLDPVFFCHHSNVDRMWSEWKAIGGKRRDISHKDWLNSEFFFYDENGDPFRV
ACR61398  A-RLTFLGITLKP-QEPLKSSLTVLSTFGLVQCQVQPCLNGRTSHGENMGHFYSAGLDPVFFCHHSNVDRMWSEWKAIGGKRRDISHKDWLNSEFFFYDENGDPFRV
ADG56700  G-APYVLGNNVEA-PGTIEVIPHGPVHVWTGTVPGTTLPNGRTSHGENMGHFYSAGLDPVFFCHHSNVDRMWSEWKAIGGKRRDISHKDWLNSEFFFYDENGDPFRV
ACT22523  WRTLRSWEITLKP-QETVEVIPHIPVHIWVGTARGSKFPDGSTSYGEDMGNFYSAGLDPVFYCHHSNVDRMWNEWKQIGGKRRDISQRDWLNSEFFFFDENKNPYRV
BAA85119  G-APYVLGNNVEA-PGTVEVIPHIPVHIWVGTARGSKFPDGSTSYGEDMGNFYSAGLDPVFYCHHSNVDRMWNEWKQIGGKRRDISQRDWLNSEFFFFDENKNPYRV
ADY18409  G-GRYVLGSTQGV-QGTIEKIPHTPVHIWVGTKKDSILPNGKKSYGEDMGNFYSAALDPVFYCHHSNVDRMWNEWKQIGGKRRDLSQKDWLDSEFFFYDENKNPYLV
ADY18412  G-NPYPLGTDPKPGMGTIENIPHNAVHNWTGDQP-------RQPNGEHMGTFYSAGLDPVFYSHHANVDRMWNEWKAIGGKRRDLADKDWLNSEFFFYDENRNPFKV
ADY18411  G-NPYPLGTDPSPGMGTIENIPHNPVHIWTGDSP-------RQPNGEDMGNFYSAGLDPVFYCHHANVDRMWNEWKAIGGKRRDLADKDWLNSESFFYDENRNPFKV
ADY18410  G-NPYPLGTDPKPGMGTIENIPHTPVHIWTGDSP-------RQPNGEDMGNFYSAGLDPVFYCHHANVDRMWNEWKAIGGKRRDLADKDWLNSEFFFYDENRNPFKV
                              .:         . **.** ****.:****:.**:******.*** ******.::.:***:** **:*** :*: *

AFJ79641  KVRDCLDTKKMGYDYAPMPRPWRNFKPITKASVGKVDTSSLPPVSQVFPLAKLDK-AISFSINRPASSRTQQEKNEQEEMLTFNNIKYDNRNYVRFDVFLNVDSNVN
AFJ79639  KVRDCLDTKKMGYDYAPMPTPWRNFKPITKASVGKVDTSSLPPVSQVFPLAKLDK-AISFSINRPASSRTQQEKNEQEEMLTFNNIKYDNRNYVRFDVFLNVDSNVN
AFJ79643  KVRDCLDTKKMGYDYAPMPRPWRNFKPITKASVGKVDTSSLPPVSQVFPLAKLDK-AISFSINRPASSRTQQEKNEQEEMLTFNNIKYDNRNYVRFDVFLNVDSNVN
AFJ79640  KVRDCLDTKKMGYDYAPMPTPWRNFKPITKASVGKVDTSSLPPVSQVFPLAKLDK-AISFSINRPASSRTQQEKNEQEEMLTFNNIKYDNRNYVRFDVFLNVDSNVN
AFJ79642  KVRDCLDTKKMGYDYAPMPTPWRNFKPITKASVGKVDTSSLPPVSQVFPLAKLDK-AISFSINRPASSRTQQEKNEQEEMLTFNNIKYDNRNYVRFDVFLNVDSNVN
ACR61398  KVRDCLDTKKMGYDYAPMPRPWRNFKPITKASVGKVDTSSLPPVSQVFPLAKLDK-AISFSINRPASSRTQQEKNEQEEMLTFNNIKYDNRNYVRFDVFLNVDSNVN
ADG56700  KVRDCLDTKKMGYDYAPMPTPWRNFKPITKASVGKVDTSSLPPVSQVFPLAKLDK-AISFSINRPASSRTQQEKNEQEEMLTFNNIKYDNRNYVRFDVFLNVDSNVN
ACT22523  RVRDCLDTKTMGYDYAPMPTPWRNFKPKTKASSGKANTSAFPPASQVFPLAKMDK-VITFSIKRPASSRTQQEKNEKEEMLTFNNIKYDNREYVRFDVFLNVDNNVN
BAA85119  RVRDCLDTKTMGYDYAPMPTPWRNFKPKTKASSGKANTSAFPPASQVFPLAKMDK-VITFSIKRPASSRTQQEKNEQEEMLTFNNIKYDNREYVRFDVFLNVDNNVN
ADY18409  KVRDCLDTKKMGYDYAPSSTVWRNFKPNKKNTDGKVNTGSLPSATKIFPIFKLDK-AISFSINRPASSRTQQEKNEQEELLTFSYIKYDNREYIRFDVFVNVDKNVK
ADY18412  KVRDCLDSKKMGFDYAPMPTPWRNFKPIRKTTSGKANIGSIPPASKVFPIAKLDR-AISFSINRSASSRTQAEKNEQEEILTFNKVQYDDSQCVRFDVFLNVDKTVN
ADY18411  KVRDCLDSKKMGFDYAPMPTPWRNFKPIRKTTSGKANIGSIPPASKVFPIAKLDR-AISFSINRSASSRTQAEKNEQEEILTFNKVKYDDSQYVRFDVFLNVDKTVN
ADY18410  KVRDCLDSKKMGFDYAPMPTPWRNFKPVRRTTSGKANTRSIPPASKVFPTCETRQSDFIFHRQTSFVKDSKAEKNEQEGDTNIRQIQYDDSQYVRFDVFLNVDKTVK
             :*******:*.**:**.:.   ***** :: **.:  ::*..::** : : * : .. :: ****:*   .:   ::**: :  :*****:***..*:

AFJ79641  ADELDKAEFAGSYTNLPHVHRVGENTDHVATATLQLAITELLEDIGLEDEDTIAVTLVPKKGGEGISIEGATISLADC
AFJ79639  ADELDKAEFAGSYTNLPHVHRVGENTDHVATATLQLAITELLEDIGLEDEDTIAVTLVPKKGGEGISIEGATISLADC
AFJ79643  ADELDKAEFAGSYTNLPHVHRVGENTDHVATATLQLAITELLEDIGLEDEDTIAVTLVPKKGGEGISIEGATISLADC
AFJ79640  ADELDKAEFAGSYTNLPHVHRVGENTDHVATATLQLAITELLEDIGLEDEDTIAVTLVPKKGGEGISIEGATISLADC
AFJ79642  ADELDKAEFAGSYTNLPHVHRVGENTDHVATATLQLAITELLEDIGLEDEDTIAVTLVPKKGGEGISIEGATISLADC
ACR61398  ADELDKAEFAGSYTNLPHVHRVGENTDHVATATLQLAITELLEDIGLEDEDTIAVTLVPKKGGEGISIEGATISLADC
ADG56700  ADELDKAEFAGSYTNLPHVHRVGENTDHVATATLQLAITELLEDIGLEDEDTIAVTLVPKKGGEGISIEGVEISLADC
ACT22523  ANELDKAEFAGSYTSLPHVHRASQ-TDHVATATLQLAITELLEDIGLEDEDTIAVTLVPKKGGEGISIEGVEISLADC
BAA85119  ANELDKAEFAG----------------------------------------------------------------
ADY18409  ADELDKIEYAGSYTSLPHVHKDGD-KDHIATATLQLALTELLEDIGLENEETIAVTLVPKKGGEGLSIGCVEIKLEDC
ADY18412  ADELDKPEFAGSYTSLPHVH--GDNNTHVTSVTFKLVITELLEDIGLEDEDTIAVTLVPKEGGEGISIENAEIVLMDC
ADY18411  ADELDKAEFAGSYTSLPHVH--GDNNTHVTSVTFNLAITELLEDIGLEDEDTIALTLVPKQGGEGISIDNAEIVLVDC
ADY18410  ALELDQPEFAGSYTSLPHVH--GDKDR-APVTFKLAITELLEDNNLEDEESIVITLIPKAGGDGISIQNAVIDLVDC
             * ***: *:*****.*****  .:   :..*::*.:****** .**:*::*.:**:** **:*:** . * * **
```

Fig. 2 continued

PPOs into chloroplast stroma and thylakoid lumen, respectively. The molecular weight of plant PPOs varies considerably probably owing to the presence of multiple genes coding for plant PPOs and also due to partial proteolysis of these enzymes. The mature plant PPO proteins are usually ~52–62 kDa (Chevalier et al. 1999). The seven novel sequences analyzed possessed the widely conserved signature motifs of all PPOs: copper A/B binding sites (Koussevitzky et al. 1998). Though the two sites were mostly conserved there were some noticeable amino acid substitutions and deletions especially in copper 'B' region. Further, two PPOs in particular (AFJ79642 and ACR61398) exhibited considerable deviation in the conserved copper 'B' region compared with other sequences. This could indicate the difference in structural classes or function of different PPOs in eggplant. In addition, the presence of a third 'His-rich' region has been reported in potato (Hunt et al. 1993), tomato (Shahar et al. 1992) and pokeweed (Joy et al. 1995). However, the biological significance of the third 'His-rich' in the plant species is unknown.

The previously characterized eggplant PPOs (Shetty et al. 2011) were present in several tissues including root, leaves (young and mature), flowers (pre- and post-anthesis) and fruit. Considering the tissue from which these sequences have been reported, it appears that two genes, ACR61398 and BAA81159 are expressed in fruit. Further, based on sequence comparison and identity matrix (Table 2) data we could speculate the expression of the other novel PPOs identified in this study. Even though ACR61398 (from fruit) and AFJ79642 are highly identical, AFJ79642 is not expressed in mature fruit tissue (data not shown). This is surprising especially in view of its close proximity to ACR61398, but AFJ79642 could still be expressed at early stages of fruit development. Semi-quantitative PCR data has shown high levels of expression of ADG56700 in root and young leaves, with reduced expression levels in pre-anthesis flowers and fruit (Shetty et al. 2011). Therefore, the remaining four eggplant PPOs (AFJ79639, AFJ79640, AFJ79641 and AFJ79643) for which the tissue specificity is unknown, it is possible that these might be expressed in any of the following tissues: root, young leaves, pre-anthesis flowers and fruits. Data reported in this manuscript presents a systematic characterization of several additional eggplant PPOs. Based on sequence information some of these additional eggplant PPOs identified could be expressed in fruit. This information will aid in identifying different PPO genes that are primarily involved in browning of eggplant fruit. In addition, the data presented in this manuscript could help in better understanding the implicated role of eggplant PPOs in defense against pests and pathogens.

Acknowledgments We would like to thank Dr. R Sai Kumar, Director (Retd), Directorate of Maize, ICAR, India, for helping us in preparing this manuscript.

References

Anderson JA,Morris CF (2001) An improved whole-seed assay for screening wheat germplasm for polyphenol oxidase activity. Crop Sci 41:1697–1705

Baik BK, Czuchajowska Z, Pomeranz Y (1994) Comparison of polyphenol oxidase activities in wheats and flours from Australian and US cultivars. J Cereal Sci 19:291–296

Brendel V, Bucher P, Nourbakhsh IR, Blaisdell BE, Karlin S (1992) Methods and algorithms for statistical analysis of protein sequences. Proc Natl Acad Sci USA 89:2002–2006

Chevalier T, de Rigal D, Mbeguie AMD, Gauillard F, Richard-Forget F, Fils-Lycaon BR (1999) Molecular cloning and characterization of apricot fruit polyphenol oxidase. Plant Physiol 119:1261–1270

Constabel CP, Bergey D, Ryan CA (1995) Systemin activates synthesis of wound-inducible tomato leaf polyphenol oxidase via the octadecanoid defense signaling pathway. Proc Natl Acad Sci USA 92:407–411

Dry IB, Robinson SP (1994) Molecular cloning and characterization of grape berry polyphenol oxidase. Plant Mol Biol 26:495–502

Feillet P, Autran J-C, Icard-Verniere C (2000) Pasta brownness: an assessment. J Cereal Sci 32:215–233

Felsenstein J (1989) PHYLIP—Phylogeny Inference Package (Version 3.2). Cladistics 5:164–166

Hunt MD, Eannetta NT, Yu H, Newman SM, Steffens JC (1993) cDNA cloning and expression of potato polyphenol oxidase. Plant Mol Biol 21:59–68

Joy RW, Sugiyama M, Fukuda H, Komamine A (1995) Cloning and characterization of polyphenol oxidase cDNAs of *Phytolacca americana*. Plant Physiol 107:1083–1089

Koussevitzky S, Ne'eman E, Sommer A, Steffens JC, Harel E (1998) Purification and properties of a novel chloroplast stromal peptidase. Processing of polyphenol oxidase and other imported precursors. J Biol Chem 273:27064–27069

Massa AN, Beecher B, Morris CF (2007) Polyphenol oxidase (PPO) in wheat and wild relatives: molecular evidence for a multigene family. Theor Appl Genet 114:1239–1247

Mayer AM, Harel E (1979) Polyphenol oxidases in plants. Phytochemistry 18:193–215

Ono E, Hatayama M, Isono Y, Sato T, Watanabe R, Yonekura-Sakakibara K, Fukuchi-Mizutani M, Tanaka Y, Kusumi T, Nishino T, Nakayama T (2006) Localization of a flavonoid biosynthetic polyphenol oxidase in vacuoles. Plant J 45:133–143

Pérez-Gilabert M, García Carmona F (2000) Characterization of catecholase and cresolase activities of eggplant polyphenol oxidase. J Agric Food Chem 48:695–700

Pourcel L, Routaboul JM, Kerhoas L, Caboche M, Lepiniec L, Debeaujon I (2005) TRANSPARENT TESTA10 encodes a laccase-like enzyme involved in oxidative polymerization of flavonoids in Arabidopsis seed coat. Plant Cell 17:2966–2980

Shahar T, Hennig N, Gutfinger T, Hareven D, Lifschitz E (1992) The tomato 66.3-kD polyphenoloxidase gene: molecular identification and developmental expression. Plant Cell 4:135–147

Shetty SM, Chandrashekar A, Venkatesh YP (2011) Eggplant polyphenol oxidase multigene family: cloning, phylogeny, expression analyses and immunolocalization in response to wounding. Phytochemistry 72:2275–2287

Singh AP, Luthria D, Wilson T, Vorsa N, Singh V, Banuelos GS, Pasakdee S (2009) Polyphenols content and antioxidant capacity of eggplant pulp. Food Chem 114:955–961

Sommer A, Ne'eman E, Steffens JC, Mayer AM, Harel E (1994) Import, targeting, and processing of a plant polyphenol oxidase. Plant Physiol 105:1301–1311

Thipyapong P, Steffens JC (1997) Tomato polyphenol oxidase: differential response of the polyphenol oxidase F promoter to injuries and wound signals. Plant Physiol 115:409–418

Thipyapong P, Hunt MD, Steffens JC (1995) Systemic wound induction of potato (*Solanum tuberosum*) polyphenol oxidase. Phytochemistry 40:673–676

Thipyapong P, Joel DM, Steffens JC (1997) Differential expression and turnover of the tomato polyphenol oxidase gene family during vegetative and reproductive development. Plant Physiol 113:707–718

Thompson JD, Higgins DG, Gibson TJ (1994) CLUSTAL W: improving the sensitivity of progressive multiple sequence alignment through sequence weighting, position-specific gap penalties and weight matrix choice. Nucleic Acids Res 22:4673–4680

Van Gelder CWG, Flurkey WH, Wichers HJ (1997) Sequence and structural features of plant and fungal tyrosinases. Phytochemistry 45:1309–1323

Wang J, Constabel CP (2004) Polyphenol oxidase overexpression in transgenic populus enhances resistance to herbivory by forest tent caterpillar (*Malacosoma disstria*). Planta 220:87–96

Whitaker BD, Stommel JR (2003) Distribution of hydroxycinnamic acid conjugates in fruit of commercial eggplant (*Solanum melongena* L.) cultivars. J Agric Food Chem 51:3448–3454

24

Characterization of parasporin gene harboring Indian isolates of *Bacillus thuringiensis*

N. K. Lenina · A. Naveenkumar · A. E. Sozhavendan ·
N. Balakrishnan · V. Balasubramani ·
V. Udayasuriyan

Abstract *Bacillus thuringiensis* (Bt) is popularly known as insecticidal bacterium. However, non-insecticidal Bt strains are more extensively available in natural environment than the insecticidal ones. Parasporin (PS) is a collection of genealogically heterogeneous Cry proteins synthesized in non-insecticidal isolates of Bt. An important character generally related with PS proteins is their strong cytocidal activity preferentially on human cancer cells of various origins. Identification and characterization of novel parasporin protein which are non-hemolytic and non-insecticidal but having selective anticancer activity raise the possibility of a novel application of Bt in medical field. In the present study, seven new indigenous isolates (T6, T37, T68, T98, T165, T186, and T461) of Bt showed variation in colony morphology, crystal characters and protein profiles with each other. Out of the seven new isolates screened for parasporin (*ps*) and *cry* genes, two of the new indigenous isolates (T98 and T186) of Bt showed the presence of *ps4* gene. Partial *ps4* gene was cloned from the two new isolates and the sequence of partial *ps4* gene showed high homology with its holotype *ps4Aa1*. These two isolates were characterized based on the proteolytic processing of the inclusion proteins and the proteolytic products were found to be comparable to the PS4 reference strain A1470. The two isolates of Bt did not show toxicity toward *Spodoptera litura* and *Helicoverpa armigera*. Based on the results of this study, it can be concluded that the isolates T98 and T186 are parasporin producers.

Keywords *Bacillus thuringiensis* · Parasporin ·
δ-endotoxin · Non-insecticidal inclusions ·
Cytocidal protein

Introduction

Bacillus thuringiensis (Bt) is an aerobic gram-positive and endospore-forming bacterium, first isolated in Japan from diseased larvae of the silkworm, *Bombyx mori*, as an entomopathogenic bacterium (Ishiwata 1901). It produces large crystalline parasporal inclusions in sporangia during sporulation (stationary phase of its growth cycle). This character is used to discriminate two taxonomically closely related species, *B. thuringiensis* and *B. cereus* (Logan 2005; Ohba et al. 2009). The parasporal inclusions often contain δ-endotoxin proteins that are specifically toxic to agriculturally and medically important insect pests of several orders, including Lepidoptera, Diptera, and Coleoptera (Beegle and Yamamoto 1992) and to even nematodes, mites, and protozoa (de Maagd et al. 2001), but are not pathogenic to mammals, birds, amphibians, or reptiles (http://www.lifesci.sussex.ac.uk/home/Neil_Crickmore/Bt/) (Schnepf et al. 1998). This makes *B. thuringiensis*, a promising microbial agent in the control of insect pests in agriculture, forestry, veterinary, and public health management (Schnepf et al. 1998).

Meanwhile, non-insecticidal *B. thuringiensis* strains are ubiquitous in natural environments and are more widely distributed than insecticidal ones (Ohba 1996). It is remarkable that the non-insecticidal isolates frequently account for more than 90 % of the natural populations from soils (Ohba et al. 2002; Yasutake et al. 2007; Mizuki et al. 1999a, b). This raises the query whether non-insecticidal

N. K. Lenina · A. Naveenkumar · A. E. Sozhavendan ·
N. Balakrishnan · V. Balasubramani · V. Udayasuriyan (✉)
Department of Plant Biotechnology, Centre for Plant Molecular
Biology and Biotechnology, Tamil Nadu Agricultural
University, Coimbatore 641003, India
e-mail: udayvar@yahoo.com

inclusions have any biological activity which is yet to be undiscovered (Ohba et al. 1988). An extensive effort to screen Cry proteins for biological activity other than insecticidal toxicity was initiated in 1996. This led to the discovery of a unique activity, which is preferential for certain human cancer cells (Mizuki et al. 1999a). The protein was first categorized and defined as bacterial parasporal proteins and these proteins are non-hemolytic but cytocidal to human cancer cells (Mizuki et al. 1999a, 2000). Globally, six different parasporin types, PS1–PS6 have been identified in countries, viz. Japan, Vietnam, India, Canada, and Caribbean Islands (Gonzalez et al. 2011) and classified by the Committee of Parasporin Classification and Nomenclature (http://parasporin.fitc. pref.fukuoka.jp/list.html). In view of potential application of these proteins, this study was undertaken to characterize new isolates of Bt collected from Western Ghats, India, based on colony and crystal morphology, protein profile, screening for presence of *cry* or *parasporin* genes by PCR and insect bioassay.

Materials and methods

Bacterial strains and plasmids

The bacterial strains used in this study were *B. thuringiensis* soil isolates (T6, T37, T68, T98, T165, T186, and T461) from Western Ghats of Tamil Nadu State, India (Ramalakshmi and Udayasuriyan 2010), and maintained in the Department of Plant Biotechnology, CPMB&B, Tamil Nadu Agricultural University, Coimbatore. The reference strains for parasporin (*ps*) genes, A1190 (*ps1*), A1547 (*ps2*), A1462 (*ps3*), and A1470 (*ps4*), provided by Dr. Natsuko Kurata, Biotechnology and Food Research Institute, Fukuoka Industrial Technology Centre, Japan, were used in this study. Bt strains, HD1 (*cry1* and *cry2*) and 4Q7 (acrystalliferous) were used as reference strains. *Escherichia coli* (DH5α) was used as a host for cloning the gene. The vector, pTZ57R/T (Fermentas Inc., Canada) was used to clone parasporin gene fragments amplified from new isolates of Bt. The antibiotic concentration used for selection of *E. coli* transformants was 100 µg/ml of ampicillin.

Culture conditions

Bacillus thuringiensis culture was grown on T3 medium (Martin and Travers 1989) at 30 °C at 200 rpm for 2–8 days and the bacterial sporulation was monitored through phase contrast microscope for 2–8 days. *E. coli* was grown on LB medium for 24 h at 37 °C at 200 rpm.

Characterization of isolates for colony and crystal morphology

The *B. thuringiensis* isolates streaked on T3 agar plates were incubated at 30 °C for 2–8 days. Colony morphology was studied on single colonies developed on T3 agar plates. The Bt isolates inoculated in 5 ml of T3 broth were incubated at 30 °C at 200 rpm for 2–8 days, and the bacterial sporulation was monitored through phase contrast microscope at 100×. After about 90 % of cell lysis, a smear of 10 µl lysed culture was made on glass slide and heat fixed. After heat fixing, drops of the Coomassie Brilliant Blue stain (0.133 % Coomassie Brilliant Blue G250 in 50 % acetic acid) were added and kept as such for 1 min. Then, the smear was washed gently in running tap water. After blot drying with blotting paper, the stained cultures were observed through bright field microscopy for presence of crystalline inclusions (Ramalakshmi and Udayasuriyan 2010).

Preparation of inclusion proteins

The spore–crystal mixture was isolated from seven new isolates of Bt and reference strain A1470, as described by Lenin et al. (2001). Single colony of Bt strains was inoculated into 5 ml T3 broth and incubated in a rotary shaker, maintained at 30 °C at 200 rpm for 2–8 days, and the bacterial sporulation was monitored through phase contrast microscope. When more than 90 % of cells were lysed, the sporulated broth culture was transferred to 4 °C, at least half-an-hour before harvesting. The T3 broth containing spore–crystal mixture was centrifuged for 10 min at 10,000 rpm at 4 °C. The pellet was washed once with 5 ml of ice-cold 1× Tris–EDTA buffer [Tris 10 mM, EDTA 1 mM, pH 8.0 with 1 mM phenyl methyl sulphonyl fluoride (PMSF)], once with 5 ml of ice-cold 0.5 M NaCl followed by two more washes with 5 ml of Tris–EDTA buffer containing 1 mM PMSF by centrifuging at the same speed and time. Finally, the spore–crystal pellet was suspended in 100 µl of sterile distilled water containing 1 mM PMSF and stored at −20 °C.

Screening of *parasporin* and *cry* genes

Screening of the test isolates for *ps* and *cry* genes was carried out in a 25-µl PCR reaction. Total genomic DNA isolated from Bt strains using Genei pure bacterial DNA purification kit (Genei, Bangalore, India) was used as template for PCR screening. The PCR was accomplished using an Eppendorf thermal cycler with a reaction mixture containing 50–100 ng of total genomic DNA of Bt, 1× PCR buffer (10 mM Tris–HCl; pH 9.0, 50 mM KCl, 1.5 mM $MgCl_2$), 75 µM each of dNTPs, 50 ng each of forward and reverse primers (Table 1) and 1.5 U of *Taq* DNA polymerase.

Table 1 Primers used for screening of Bt isolates for different *cry* and *ps* genes

Primer sequences	Annealing °C	Gene	Amplicon size (bp)	Primer position in ORF		Reference
				FP	RP	
F: CATGATTCATGCGGCAGATAAAC	62	*cry1*	278	2,783	3,060	Ben-Dov et al. (1997)
R: TTGTGACACTTCTGCTTCCCATT						
F: GTTATTCTTAATGCAGATGAATGGG	64	*cry2*	702	570	1,271	Ben-Dov et al. (1997)
R: CGGATAAAATAATCTGGGAAATAGT						
F: ATCAAGAATTTTCCGATAATC	50	*ps1*	1,136	154	1,289	Yasutake et al. (2007)
R: CCAAAAGTGCCAGAATG						
F: TGTTGGGACTGTTCAGTACGT	56	*ps2*	503	341	843	*
R: CGTCACGGTACCTCTTAGTGT						
F: GGAATCCAGGTGCACTGCT	67	*ps3*	701	264	964	*
R: GTCCCGGATCATACGTTGGA						
F: AGTGGTCTCCAGGCTCATACTGG	59	*ps4*	681	81	761	*
R: TGATATTCCCGAACCTGCCCT						

* Designed in this study using Fast PCR 6.0

Template DNA was preheated at 94 °C for 2 min. Then it was denatured at 94 °C for 1 min, annealed to primers for 45 s and extensions of PCR products were achieved at 72 °C for 1 min. The PCR was performed for 30 cycles. The PCR products were analyzed on a 1.2 % agarose gel. Amplified product was ligated in pTZ57R/T PCR cloning kit and transformed into *E. coli* DH5α.

Proteolytic processing of inclusion proteins

Spore–crystal mixture isolated from parasporin producing isolate was washed thrice with 1 M NaCl and resuspended in sterile water and transferred to a microfuge tube. After centrifugation at 13,000 rpm for 5 min at 4 °C, the pellet containing purified inclusions was solubilized in 50 mM Na_2CO_3 (pH 10.0) containing 1 mM EDTA and 10 mM dithiothreitol for 1 h at 37 °C (200 μl/25 mg pellet). After centrifugation at 13,000 rpm for 5 min at 4 °C, the supernatant was passed through 0.2-μm filter to remove unsolublized materials. The pH of the filtrate was adjusted to 8.0 and split into two equal aliquots. One of the aliquots of solubilized proteins was treated with proteinase K (final conc. 60 μg/ml), in 50 mM Na_2CO_3 (pH 10.0) for 90 min at 37 °C. After proteinase K treatment, 1 mM PMSF was added to the mixture to stop the proteolytic reaction. Both the aliquots (solubilized and proteinase K-treated inclusions) were subjected to sodium dodecyl sulfate-polyacrylamide gel electrophoresis (SDS-PAGE) analysis (Okumura et al. 2006; Saitoh et al. 2006).

Toxicity analysis of new isolates

The laboratory cultures of *S. litura* and *H. armigera* reared on a semi-synthetic diet (Patel et al. 1968) were used to determine the insecticidal activity of the isolates T98 and T186 using diet surface contamination method. Approximately 1 ml of the

semi-synthetic diet was dispensed into 1.8 ml cryovials (Tarson®; 1 cm dia.) and allowed to cool for an hour. After solidification of the diet, 10 μl spore–crystal mixture was coated on the diet surface and allowed to air dry for 30 min. Neonate larvae of *S. litura* and *H. armigera* were released using a soft hairbrush and the tube closed with a screw cap. All the above steps were carried out in a laminar airflow chamber. Vials without crystal mixture served as a control. Each treatment was replicated four times and ten vials were maintained for each replication. Larval mortality was recorded for 7 days and subjected to ANOVA. All the experiments were carried out in a room with a photoperiod of 14:10 (L:D) at an average temperature of 27 °C and 60 % relative humidity.

Results and discussion

Bacillus thuringiensis formed white rough colonies which spread out and expanded over the plate quickly. Seven new isolates of Bt and six reference strains were observed for colony morphology on T3 plates. All the seven isolates produced creamy white colonies after 24 h of inoculation on T3 agar plates. The colony characteristics of test isolates showed slight variation with each other (Table 2). Chaterjee et al. (2006) also found similar variation in the morphological characteristics of Bt isolates on nutrient agar medium and reported circular, white, flat, and undulate colonies of the Bt isolates of West Bengal, India. The time taken for 90 % cell lysis in T3 broth was also observed for the new isolates along with the reference strains. The reference strains, HD1 and 4Q7 took 2–3 days, while the parasporin reference strains and the seven new isolates took 6–8 days for 90 % of cell lysis.

Morphology of the parasporal inclusion bodies of Bt was reported to be heterogeneous (Ohba et al. 2001). Crystal

Table 2 Morphological characteristics of new isolates

Isolate	Color of colonies	Shape of colony	Margin of colony	Elevation of colony	Shape of inclusion
T6	Creamy white	Irregular	Undulated	Raised	Spherical
T37	Creamy white	Circular	Entire	Raised	Irregular
T68	Creamy white	Circular	Entire	Raised	Spherical
T98	Creamy white	Irregular	Undulated	Flat	Irregular
T165	Creamy white	Irregular	Undulated	Flat	Spherical
T186	Creamy white	Circular	Entire	Raised	Irregular
T461	Creamy white	Irregular	Undulated	Raised	Spherical

morphology of Bt isolates are of cuboidal, spherical, rhomboidal, and irregular shapes (Bernhard et al. 1997). However, four distinct crystal morphologies are apparent; the bipyramidal crystals are related to Cry1 proteins (Aronson and Fritz-James 1976), cuboidal inclusions related to Cry2 proteins and usually associated with bipyramidal crystals (Ohba and Aizawa 1986); square crystals related to Cry3 proteins (Herrnstand et al. 1986; Lopez-Meza and Ibarra 1996); amorphous and composite crystals related to Cry4 and Cyt proteins (Federici et al. 1990). There is a striking correlation between shape of crystal and spectrum of toxicity (Chambers et al. 1991; Ramalakshmi and Udayasuriyan 2010). Recent reports show parasporin protein inclusions which do not have insecticidal properties also exhibit variation in crystal morphology. The crystal morphology of parasporin producers varies from spherical, bipyramidal to irregular (Kitada et al. 2006). In the present study, four new isolates (T6, T68, T165, and T461) showed spherical inclusions. The isolates T37, T98, and T186 showed irregular-shaped inclusions (Table 2). The crystal shape of the isolates T98 and T186 is similar to the reference strain of *ps4*, A1470 (Fig. 1). Variation in crystal morphology may indicate the diversity of crystal proteins in isolates (Schnepf et al. 1998; Rampersad and Ammons 2005; Ibarra et al. 2003).

Grouping of Bt isolates according to crystal protein(s) profile analyzed through SDS-PAGE will give a

prelude for the presence of diversity in *cry* and *ps* genes. Crystal protein profile of the seven new isolates, reference strains for Cry1, Cry2 (HD1), reference strains for PS1–PS4 (A1190, A1547, A1462, and A1470) and the acrystalliferous reference strain (4Q7) were compared. The reference strain HD1 showed a prominent 135-kDa protein of *cry1* gene and 65-kDa protein of *cry2*. The Bt strain HD1 was included as one of the reference strains, even though it is known to be toxic to lepidopteran insects (Hofte and Whiteley 1989), to observe whether the new isolates produced distinct protein similar to that of HD1 or not. The acrystalliferous reference strain of Bt 4Q7 did not show prominent protein bands as reported earlier (Schnepf et al. 1985; Adang et al. 1985; Widner and Whiteley 1989). The reference strains of *ps1, ps2, ps3,* and *ps4* showed various sized proteins ranging from 29 to 140 kDa as reported earlier workers (Mizuki et al. 2000; Kim et al. 2000; Yamashita et al. 2005; Okumura et al. 2004). All the new isolates had different protein profile when compared to reference strains (Fig. 2). Protein profile of the test isolates T68, T461, and T98 showed prominent multiple bands, whereas the isolates T37, T165, and T186 showed faint multiple bands. The new isolate T6 did not show any prominent protein band. All the new isolates and reference strains of Bt differed from each other suggesting that the parasporin and crystal proteins of the new isolates could be the novel one.

Among several methods available for characterization of Bt strains, such as PCR, RFLP, Southern blot analysis, and bioassay (Kronstad and Whiteley 1986), PCR is rapid and highly sensitive method for detecting and identifying novel Bt genes. Carrozi et al. (1991) proposed PCR as an accurate and rapid method for identification of novel strains with unknown crystal producing genes. The efficacy of PCR for *cry* genes and *ps* genes identification relies on the alternation of conserved and variable nucleotide regions. All the seven new isolates of Bt were screened for the presence of *cry* genes (*cry1* and *cry2*) and *ps* genes (*ps1, ps2, ps3,* and *ps4*) by PCR. Primers specific for *cry1* and *cry2* family genes gave

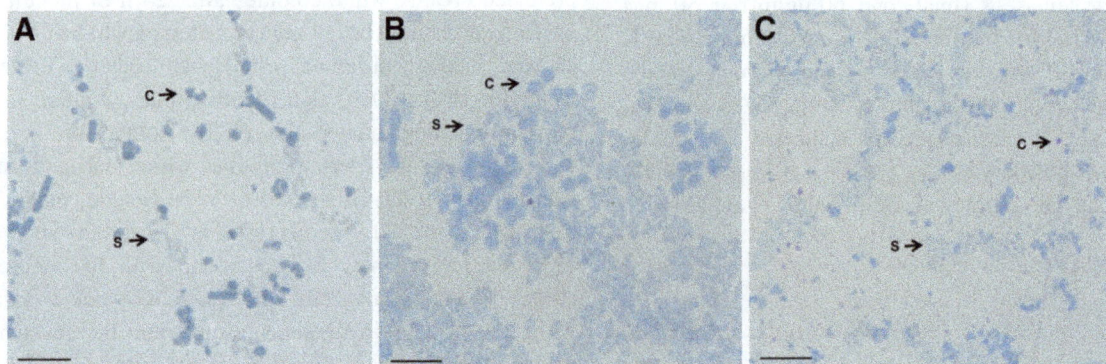

Fig. 1 Bright field microscopic observation of crystal morphology from parasporin producing Bt isolates. **a** Parasporin reference strain A1470. **b**, **c** Bt isolates T98 and T186, respectively. *c* crystal, *s* spore. *Scale bar* 20 μm

Fig. 2 SDS-PAGE analysis of spore–crystal mixture of Bt strains. *M* Genei Protein marker (Higher Range #105977). *Lanes 1–2* reference strain HD1 and 4Q7. *Lanes 3–6* reference strains of parasporin (PS4, PS3, PS2 and PS1). *Lanes 7–13* Bt isolates, T6, T37, T68, T461, T165, T186 and T98

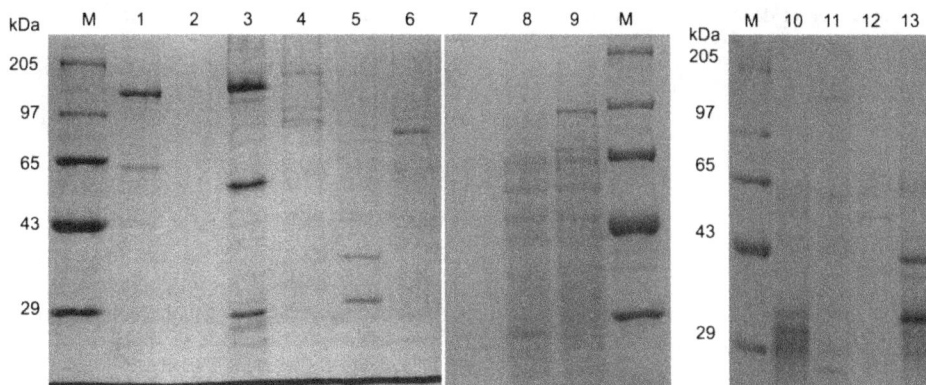

Table 3 PCR screening of new isolates of Bt for *cry* and *ps* genes

Isolate	*cry1*	*cry2*	*ps1*	*ps2*	*ps3*	*ps4*
T6	–	–	*	*	–	–
T37	–	–	–	*	–	–
T68	–	–	–	–	–	–
T98	–	–	*	–	–	+
T165	–	–	–	–	–	–
T186	–	–	–	*	*	+
T461	–	–	–	–	–	–

+ Present, − absent

* Unexpected size

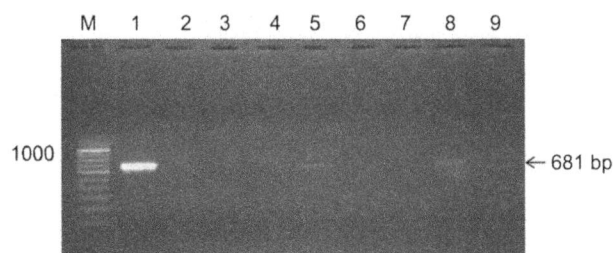

Fig. 3 Amplification of *ps4* gene from the test isolates of Bt. *M* 100 bp ladder. *Lane 1* Reference strain of ps4 A1470. *Lanes 2–8* Bt isolates, T6, T37, T68, T461, T165, T186 and T98. *Lanes 9* water control

amplification of expected size in the reference strain HD1 only. None of the seven new isolates gave amplification to both these gene families, indicating the absence of *cry1* and *cry2* family genes. Primers specific for *ps1, ps2,* and *ps3* genes gave amplification of expected sizes in the respective reference strains of Bt only. Primers specific to *ps4* gene gave amplification of expected size in the reference strain (A1470) and two new isolates, T98 and T186 (Table 3; Fig. 3). This result suggested the presence of *ps4* gene(s) in the two new isolates.

The partial *ps4* gene (681 bp) fragment amplified by gene-specific primers from the new isolates T98 and T186 were cloned into pTZ57R/T (T/A) cloning vector. The

transformants of *E. coli* were screened by PCR. The nucleotide sequence from the positive clone was generated from Eurofins Genomics India Pvt. Ltd., Bangalore. Sequence similarity analysis of nucleotide sequences of the partial *ps4* gene (681 bp) cloned from the new isolates T98 (KC832499) and T186 (KC832500) with that of *ps4Aa1* showed 100 and 99 % homology, respectively. Comparison of deduced amino acid sequence of T186 with that of *ps4Aa1* showed variation in two positions. At position 84, leucine is replaced by histidine, and at position 87 serine by threonine (Fig. 4). Thus, the sequence of partial *ps4Aa* gene cloned from the two new isolates showed high homology with its holotype *ps4Aa1*. It confirms the presence of *ps4Aa* type gene in the two isolates, T98 and T186.

A study on proteolytic processing of crystal proteins from the *ps4* harboring isolates T98, T186, and reference strain of *ps4* (A1470) by SDS-PAGE, showed a major polypeptide of 40-kDa; two prominent bands: one at >29 kDa and another at <29 kDa; and a faint band at 27-kDa in the solubilized protein of reference strain of *ps4* (Fig. 5) as reported earlier (Saitoh et al. 2006). Similar to that of the *ps4* reference strain, the proteinase K-treated protein of new isolates (T98, T186) also showed a faint band at 27-kDa and a prominent band of >29 kDa (31-kDa). The protein of PS4 and the proteinase K are of same molecular weight ~31 kDa (Saitoh et al. 2006). Hence the prominent band at 31-kDa in proteolytic processed samples corresponds to proteinase K. As reported earlier, 31-kDa protoxin of PS4 will be digested to a 27-kDa toxin upon proteolytic processing. Therefore, it can be suggested that the faint band of 27-kDa polypeptide in the isolates (T98 and T186) may be proteolytic product of the 31 kDa PS4 protein. This gives the evidence that the test isolates may be parasporin producers. In addition, a prominent band of 43-kDa is also seen in the isolate T98 which discriminates the isolate from T186.

Generally, the parasporin protein producing strains of Bt do not produce any insecticidal protein (Kitada et al. 2006; Mizuki et al. 1999a). The two new isolates of the present study, T98 and T186 (which showed presence of *ps4* gene) did not show toxicity on *S. litura* and *H. armigera*. Growth

Fig. 4 Comparison of deduced amino acid sequence of T186 and Ps4Aa1

```
Parasporin 1470D [Bacillus thuringiensis serovar shandongiensis] Length=275

Score = 457 bits (1177), Expect = 3e-161, Method: Compositional matrix
adjust. Identities = 224/226 (99%), Positives = 225/226 (99%), Gaps = 0/226
(0%)

T186    1    SGLQAHTGNYGRIYNYNMSVPDPIVTDNPTNAAMARGTTPNPTSQPIIRTISFNETHTDT   60
             SGLQAHTGNYGRIYNYNMSVPDPIVTDNPTNAAMARGTTPNPTSQPIIRTISFNET TD+
A1470   28   SGLQAHTGNYGRIYNYNMSVPDPIVTDNPTNAAMARGTTPNPTSQPIIRTISFNETLTDS   87

Query   61   QSTATEHGITAGAEVTVKSEAGLIFAKVGFEVKVSFQYNYTTTNTYTTETSRSWTDSLQI   120
             QSTATEHGITAGAEVTVKSEAGLIFAKVGFEVKVSFQYNYTTTNTYTTETSRSWTDSLQI
Sbjct   88   QSTATEHGITAGAEVTVKSEAGLIFAKVGFEVKVSFQYNYTTTNTYTTETSRSWTDSLQI   147

Query   121  TVPPGYVTEHTFIVQTGPYSKNVVLEADIAGHGWFNYSAPGYTGTGIVNITQVLYDNKVP   180
             TVPPGYVTEHTFIVQTGPYSKNVVLEADIAGHGWFNYSAPGYTGTGIVNITQVLYDNKVP
Sbjct   148  TVPPGYVTEHTFIVQTGPYSKNVVLEADIAGHGWFNYSAPGYTGTGIVNITQVLYDNKVP   207

Query   181  GVTPYPDNFYARFRGSGKLEGKMGLQSFVNLVERPLLGRAGQVREY   226
             GVTPYPDNFYARFRGSGKLEGKMGLQSFVNLVERPLLGRAGQVREY
Sbjct   208  GVTPYPDNFYARFRGSGKLEGKMGLQSFVNLVERPLLGRAGQVREY   253
```

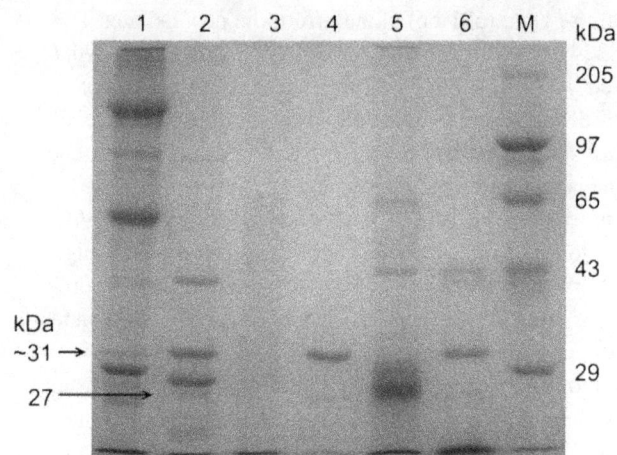

Fig. 5 Proteolytic processing of inclusion proteins of Bt strains. *M* Genei Protein marker. *Lanes 1, 3, 5* solublized inclusion protein. *Lanes 2, 4, 6* proteinase K-treated solublized protein. *Lanes 1, 2* reference strain of ps4 A1470. *Lanes 3 and 4, 5 and 6* Bt isolates T186 and T98, respectively

inhibition of insect larvae was also not observed in both the new isolates. The reference strains A1470 and 4Q7 also recorded the same results; whereas, the reference strain of *cry1* and *cry2* genes (HD1) showed 100 % mortality on both *S. litura* and *H. armigera*. Mizuki et al. (1999a) also reported that PS4 producers do not have insecticidal activity on lepidopteran (*Plutella xylostella* and *Bombyx mori*) and dipteran pests (*Aedes aegypti, Culex pipiens molestus, Anopheles stephensi, Telmatoscopus albipunctatus,* and *Musca domestica*).

Conclusion

Based on protein profile, PCR screening, nucleotide sequencing and insect bioassay, it is evident that the parasporin producing strains are members in *B. thuringiensis* populations occurring in natural environments of India. Cloning and characterization of complete gene (*ps4*) and evaluation of these isolates for their anticancer properties are required for identifying potential use of parasporin proteins in anticancer medical research.

Acknowledgments We thank Dr. Natsuko Kurata, Biotechnology and Food Research Institute, Fukuoka Industrial Technology Centre, Japan, for providing the reference strains of Bt for parasporin genes.

Conflict of interest The authors declare that they have no conflict of interest in the publication.

References

Adang MJ, Staver MJ, Rocheleau TA, Leighton J, Barker RF, Thompson DV (1985) Characterized full-length and truncated plasmid clones of the crystal protein of *Bacillus thuringiensis* subsp. *kurstaki* HD-73 and their toxicity to *Manduca sexta*. Gene 36:289–300

Aronson AI, Fritz-James P (1976) Structures and morphogenesis of the bacterial spore coat. Bacteriol Rev 40:360–402

Beegle CC, Yamamoto T (1992) History of *Bacillus thuringiensis* Berliner research and development. Can Entomol 124:587–616

Ben-Dov E, Zaritsky A, Dahan E, Barak Z, Sinai R, Manasherob R, Khamraev A, Troitskaya E, Dubitsky A, Berezina N, Margalith Y (1997) Extended screening by PCR for seven *cry* group genes from field-collected strains of *Bacillus thuringiensis*. Appl Environ Microbiol 63:4883–4890

Bernhard K, Jarrett P, Meadows M, Butt J, Ellis DJ, Roberts GM, Pauli S, Rodger P, Burges HD (1997) Natural isolates of *Bacillus thuringiensis*: worldwide distribution, characterization and activity against insect pests. J Invertebr Pathol 70:59–68

Carrozi NB, Kramer VC, Warren GW, Evola S, Koziel MG (1991) Prediction of insecticidal activity of *Bacillus thuringiensis* strains by polymerase chain reaction product profiles. Appl Environ Microbiol 57:3057–3061

Chambers JA, Jelen MP, Gilbert T, Johnson B, Gawron CB (1991) Isolation and characterization of a novel insecticidal crystal protein gene from *Bacillus thuringiensis* subsp. *aizawai*. J Bacteriol 173:3966–3976

Chaterjee SN, Bhattacharya T, Dangar TK, Chandra G (2006) Ecology and diversity of *Bacillus thuringiensis* in soil environment. Afr J Biotechnol 6:1587–1591

de Maagd RA, Bravo A, Crickmore N (2001) How *Bacillus thuringiensis* has evolved specific toxins to colonize the insect world. Trends Genet 17:193–199

Federici BA, Lthy P, Ibarra JE (1990) The parasporal body of *Bacillus thuringiensis* subsp. *israelensis*: structure, protein composition and toxicity. In: de Barjac H, Sutherland DJ (eds) Bacterial control of mosquitos and blackflies: biochemistry, genetics and applications of *Bacillus thuringiensis* and *Bacillus sphaericus*. Rutgers University Press, New Brunswick, pp 16–44

Gonzalez E, Granados JC, Short JD, Ammons DR, Rampersad J (2011) Parasporin from a Caribbean Island: evidence for a globally dispersed *Bacillus thuringiensis*. Curr Microb 164:3–8

Herrnstand C, Soares CG, Wilcox ER, Edwards DI (1986) A new strain of *Bacillus thuringiensis* with activity against coleopteran insects. Biotechnology 4:305–308

Hofte H, Whiteley HR (1989) Insecticidal crystal proteins of *Bacillus thuringiensis*. Microbiol Rev 53(2):242

Ibarra JE, Rincon MC, Orduz S, Noriega D, Benintende G, Monnerat R, Regis L, Claudia MF, de Oliveria M, Lanz H, Rodriguez MH, Sanchez J, Pena G, Bravo A (2003) Diversity of *Bacillus thuringiensis* strains from Latin America with insecticidal activity against different mosquito species. Appl Environ Microbiol 69:5269–5274

Ishiwata S (1901) On a kind of severe flacherie (sotto disease). Dainihon Sanshi Kaiho 114:1–5

Kim HS, Yamashita S, Akao T, Saitoh H, Higuchi K, Park YS, Mizuki E, Ohba M (2000) In vitro cytotoxicity of non-Cyt inclusion proteins of a *Bacillus thuringiensis* isolate against human cells, including cancer cells. J Appl Microbiol 89:16–23

Kitada S, Abe Y, Shimada H, Kusaka Y, Matsuo Y, Katayama H, Okumura S, Akao T, Mizuki E, Kuge O, Sasaguri Y, Ohba M, Ito A (2006) Cytocidal actions of parasporin-2, an antitumor crystal toxin from *Bacillus thuringiensis*. J Biol Chem 281:26350–26360

Kronstad J, Whiteley HR (1986) Three classes of homologous *Bacillus thuringiensis* crystal-protein genes. Gene 43:29–40

Lenin K, Mariam MA, Udayasuriyan V (2001) Expression of *cry2Aa* gene in an acrystalliferous *Bacillus thuringiensis* strain and toxicity of *Cry2Aa* against *H. armigera*. World J Microbiol Biotechnol 1:273–278

Logan NA (2005) *Bacillus anthracis, Bacillus cereus*, and other aerobic endospore-forming bacteria. In: Borriello SP, Murray PR, Funke G (eds) Topley & Wilson' Microbiology & Microbial Infections. Bacteriology, 10th edn. Hodder Arnold, London, pp 922–952

Lopez-Meza JE, Ibarra JE (1996) Characterization of a novel strain of *Bacillus thuringiensis*. Appl Environ Microbiol 62:1306–1310

Martin PAW, Travers RS (1989) Worldwide abundance and distribution of *Bacillus thuringiensis* isolates. Appl Environ Microbiol 55:2437–2442

Mizuki E, Ohba M, Akao T, Yamashita S, Saitoh H, Park YS (1999a) Unique activity associated with non-insecticidal *Bacillus thuringiensis* parasporal inclusions: in vitro cell killing action on human cancer cells. J Appl Microbiol 86:477–486

Mizuki E, Ichimatsu T, Hwang SH, Park YS, Saitoh H, Higuchi K, Ohba M (1999b) Ubiquity of *Bacillus thuringiensis* on phylloplanes of arboreous and herbaceous plants in Japan. J Appl Microbiol 86:979–984

Mizuki E, Park YS, Saitoh H, Yamashita S, Akao T, Higuchi K, Ohba M (2000) Parasporin, human leukemic cell-recognizing parasporal protein of *Bacillus thuringiensis*. Clin Diagn Lab Immunol 7:625–634

Ohba M (1996) *Bacillus thuringiensis* populations naturally occurring on mulberry leaves: a possible source of the populations associated with silkworm-rearing insectaries. J Appl Microbiol 80:56–64

Ohba M, Aizawa K (1986) Distribution of *Bacillus thuringiensis* in soils of Japan. J Invertebrate Pathol 47:277–282

Ohba M, Yu YM, Aizawa K (1988) Occurrence of non-insecticidal *Bacillus thuringiensis* flagellar serotype 14 in the soil of Japan. Syst Appl Microbiol 11:85–89

Ohba M, Vasano N, Mizuki E (2001) *Bacillus thuringiensis* soil populations naturally occurring in the Ryukyus, a subtropic region of Japan. Appl Environ Microbiol 72(2):412–415

Ohba M, Tsuchiyama A, Shisa N, Nakashima K, Lee DH, Ohgushi A, Wasano N (2002) Naturally occurring *Bacillus thuringiensis* in oceanic islands of Japan, Daito-shoto and Ogasawara-shoto. Appl Entomol Zool 37:477–480

Ohba M, Mizuki E, Uemori A (2009) Parasporin, a new anticancer protein group from *Bacillus thuringiensis*. Anticancer Res 29:427–434

Okumura S, Akao T, Higuchi K, Saitoh H, Mizuki E, Ohba M, Inouye K (2004) *Bacillus thuringiensis* serovar *shandongiensis* strain 89-T-34-22 produces multiple cytotoxic proteins with similar molecular masses against human cancer cells. Lett Appl Microbiol 39:89–92

Okumura S, Saitoh H, Wasano N, Katayama H, Higuchi K, Mizuki E, Inouye K (2006) Efficient solubilization, activation, and purification of recombinant Cry45Aa of *Bacillus thuringiensis* expressed as inclusion bodies in *Escherichia coli*. Protein Expr Purif 47:144–151

Patel RC, Patel JK, Patel PB, Singh R (1968) Mass breeding of *Heliothis armigera* (H.). Indian J Entomol 30:272–280

Ramalakshmi A, Udayasuriyan V (2010) Diversity of *Bacillus thuringiensis* isolated from Western ghats of Tamil Nadu State, India. Curr Microbiol 61:13–18

Rampersad J, Ammons D (2005) A *Bacillus thuringiensis* isolation method utilizing a novel stain, low selection and high throughput produced typical results. BMC Microbiol 5:52–63

Saitoh H, Okumura S, Ishikawa T, Akao T, Mizuki E, Ohba M (2006) Investigation of a novel *Bacillus thuringiensis* gene encoding a parasporal protein, parasporin-4, that preferentially kills human leukemic T cells. Biosci Biotechnol Biochem 12:2935–2941

Schnepf HE, Wong HC, Whiteley HR (1985) The amino acid sequence of a crystal protein from *Bacillus thuringiensis* deduced from the DNA base sequence. J Biol Chem 260:6264–6272

Schnepf E, Crickmore N, Van rie J, Lereclus D, Baum J, Feitelson J, Zeigler DR, Dean DH (1998) *Bacillus thuringiensis* and its pesticidal crystal proteins. Microbiol Mol Biol Rev 62:775–806

Widner WR, Whiteley HR (1989) Two highly related insecticidal crystal proteins of *Bacillus thuringiensis* subsp. *kurstaki* possess different host range specificities. J Bacteriol 171:965–974

Yamashita S, Katayama H, Saitoh H, Akao T, Park YS, Mizuki E, Ohba M, Ito A (2005) Typical three-domain Cry proteins of *Bacillus thuringiensis* strain A1462 exhibit cytocidal activity on limited human cancer cells. J Biochem 138:663–672

Yasutake K, Uemori A, Kagoshima K, Ohba M (2007) Serological identification and insect toxicity of *Bacillus thuringiensis* isolated from the island Okinoerabu-jima, Japan. Appl Entomol Zool 42:285–290

Permissions

List of Contributors

Suresh P. Kamble
Center for Biotechnology, Pravara Institute of Medical Sciences, Loni, Ahmednagar, Maharashtra, India

Madhukar M. Fawade
Department of Biochemistry, Dr. Babasaheb Ambedkar Marathwada University, Aurangabad 411004, Maharashtra, India

James J. Davis
Institute for Genomic Biology, MC-195, University of Illinois at Urbana-Champaign, 1206 W. Gregory Dr., Urbana, IL 61801, USA

Gary J. Olsen
Institute for Genomic Biology, MC-195, University of Illinois at Urbana-Champaign, 1206 W. Gregory Dr., Urbana, IL 61801, USA

Ross Overbeek
Fellowship for Interpretation of Genomes, 15W155 81st St., Burr Ridge, IL 60527, USA
Mathematics and Computer Science, Argonne National Laboratory, 9700 S. Cass Ave., Argonne, IL 60439, USA

Veronika Vonstein
Fellowship for Interpretation of Genomes, 15W155 81st St., Burr Ridge, IL 60527, USA

Fangfang Xia
Mathematics and Computer Science, Argonne National Laboratory, 9700 S. Cass Ave., Argonne, IL 60439, USA

Reda E. A. Moghaieb
Department of Genetics and Genetic Engineering Research Center, Faculty of Agriculture, Cairo University, Giza, Egypt

Etr H. K. Mohammed
Department of Genetics and Genetic Engineering Research Center, Faculty of Agriculture, Cairo University, Giza, Egypt
Plant Protection Research Institute, Agriculture Research Center, Ministry of Agriculture, Dokki, Giza, Egypt

Sawsan S. Youssief
Department of Genetics and Genetic Engineering Research Center, Faculty of Agriculture, Cairo University, Giza, Egypt

Kandasamy Kathiresan
Faculty of Marine Sciences, Centre of Advanced Study in Marine Biology, Annamalai University, Parangipettai 608 502, Tamil Nadu, India

Kandasamy Saravanakumar
Faculty of Marine Sciences, Centre of Advanced Study in Marine Biology, Annamalai University, Parangipettai 608 502, Tamil Nadu, India

Sunil Kumar Sahu
Faculty of Marine Sciences, Centre of Advanced Study in Marine Biology, Annamalai University, Parangipettai 608 502, Tamil Nadu, India

Muthu Sivasankaran
Faculty of Marine Sciences, Centre of Advanced Study in Marine Biology, Annamalai University, Parangipettai 608 502, Tamil Nadu, India

José P. Faria
Mathematics and Computer Science Division, Argonne National Laboratory, Argonne, IL, USA
IBB-Institute for Biotechnology and Bioengineering, Centre of Biological Engineering, University of Minho, Campus de Gualtar, 4710-057 Braga, Portugal

Janaka N. Edirisinghe
Mathematics and Computer Science Division, Argonne National Laboratory, Argonne, IL, USA
Computation Institute, University of Chicago, Chicago, IL, USA

James J. Davis
Mathematics and Computer Science Division, Argonne National Laboratory, Argonne, IL, USA
Computation Institute, University of Chicago, Chicago, IL, USA

Terrence Disz
Mathematics and Computer Science Division, Argonne National Laboratory, Argonne, IL, USA

Anna Hausmann
Fellowship for Interpretation of Genomes, Burr Ridge, IL, USA

Christopher S. Henry
Mathematics and Computer Science Division, Argonne National Laboratory, Argonne, IL, USA
Computation Institute, University of Chicago, Chicago, IL, USA

Robert Olson
Mathematics and Computer Science Division, Argonne National Laboratory, Argonne, IL, USA

Ross A. Overbeek
Mathematics and Computer Science Division, Argonne National Laboratory, Argonne, IL, USA
Fellowship for Interpretation of Genomes, Burr Ridge, IL, USA

Gordon D. Pusch
Fellowship for Interpretation of Genomes, Burr Ridge, IL, USA

Maulik Shukla
Virginia Bioinformatics Institute, Virginia Tech, Blacksburg, VA, USA

Veronika Vonstein
Mathematics and Computer Science Division, Argonne National Laboratory, Argonne, IL, USA
Fellowship for Interpretation of Genomes, Burr Ridge, IL, USA

Alice R. Wattam
Virginia Bioinformatics Institute, Virginia Tech, Blacksburg, VA, USA

Toshy Agrawal
Department of Plant Molecular Biology & Biotechnology, Indira Gandhi Krishi Vishwavidyalaya, Krishak Nagar, Raipur 492006, Chattisgarh, India

Anil S. Kotasthane
Department of Plant Molecular Biology & Biotechnology, Indira Gandhi Krishi Vishwavidyalaya, Krishak Nagar, Raipur 492006, Chattisgarh, India
Department of Plant Pathology, Indira Gandhi KrishiVishwavidyalaya, Krishak Nagar, Raipur 492006, Chattisgarh, India

Renu Kushwah
Department of Plant Molecular Biology & Biotechnology, Indira Gandhi Krishi Vishwavidyalaya, Krishak Nagar, Raipur 492006, Chattisgarh, India

S. Ashe
Division of Fish Health Management, Central Institute of Freshwater Aquaculture, Kaushalyaganga, Bhubaneswar 751002, Orissa, India

U. J. Maji
Division of Fish Health Management, Central Institute of Freshwater Aquaculture, Kaushalyaganga, Bhubaneswar 751002, Orissa, India

R. Sen
Division of Fish Health Management, Central Institute of Freshwater Aquaculture, Kaushalyaganga, Bhubaneswar 751002, Orissa, India

S. Mohanty
Division of Fish Health Management, Central Institute of Freshwater Aquaculture, Kaushalyaganga, Bhubaneswar 751002, Orissa, India

N. K. Maiti
Division of Fish Health Management, Central Institute of Freshwater Aquaculture, Kaushalyaganga, Bhubaneswar 751002, Orissa, India

Sujata Mohanty
Centre of Biotechnology, Siksha 'O' Anusandhan University, Bhubaneswar 751003, Odisha, India

Manoj Kumar Panda
Centre of Biotechnology, Siksha 'O' Anusandhan University, Bhubaneswar 751003, Odisha, India

Laxmikanta Acharya
Centre of Biotechnology, Siksha 'O' Anusandhan University, Bhubaneswar 751003, Odisha, India

Sanghamitra Nayak
Centre of Biotechnology, Siksha 'O' Anusandhan University, Bhubaneswar 751003, Odisha, India

H. C. Yashavantha Rao
Department of Studies in Microbiology, University of Mysore, Manasagangotri, Mysore 570 006, Karnataka, India

Parthasarathy Santosh
Plant Biotechnology Division, Unit of Central Coffee Research Institute, Coffee Board, Manasagangotri, Mysore 570 006,
Karnataka, India

Devaraju Rakshith
Department of Studies in Microbiology, University of Mysore, Manasagangotri, Mysore 570 006, Karnataka, India

Sreedharamurthy Satish
Department of Studies in Microbiology, University of Mysore, Manasagangotri, Mysore 570 006, Karnataka, India

Lucina Yeasmin
Department of Agricultural Biotechnology, Faculty Centre for Integrated Rural Development and Management, School of Agriculture and Rural Development, Ramakrishna Mission Vivekananda University, Ramakrishna Mission Ashrama, Narendrapur, Kolkata 700103, India

Md. Nasim Ali
Department of Agricultural Biotechnology, Faculty Centre for Integrated Rural Development and Management, School of Agriculture and Rural Development, Ramakrishna Mission Vivekananda University, Ramakrishna Mission Ashrama, Narendrapur, Kolkata 700103, India

Saikat Gantait
Department of Crop Science, Faculty of Agriculture, Universiti Putra Malaysia, 43400 Serdang, Selangor, Malaysia
Department of Biotechnology, Instrumentation and Environmental Science, Bidhan Chandra Krishi Viswavidyalaya, Mohanpur, WB 741252, India

Somsubhra Chakraborty
Department of Agricultural Biotechnology, Faculty Centre for Integrated Rural Development and Management, School of Agriculture and Rural Development, Ramakrishna Mission Vivekananda University, Ramakrishna Mission Ashrama, Narendrapur, Kolkata 700103, India

Bejoysekhar Datta
Department of Botany, University of Kalyani, Nadia, Kalyani, West Bengal 741 235, India

Pran K. Chakrabartty
Acharya J.C. Bose Biotechnology Innovation Centre, Madhyamgram Experimental Farm, Madhyamgram, Kolkata, West Bengal 700 129, India

Showkat Hussain Ganie
Department of Botany, Jamia Hamdard, Hamdard Nagar, New Delhi 110062, India

Zahid Ali
Department of Biotechnology, Jamia Hamdard, Hamdard Nagar, New Delhi 110062, India

Sandip Das
Department of Botany, University of Delhi, New Delhi 110007, India

Prem Shankar Srivastava
Department of Biotechnology, Jamia Hamdard, Hamdard Nagar, New Delhi 110062, India

Maheshwar Prasad Sharma
Department of Botany, Jamia Hamdard, Hamdard Nagar, New Delhi 110062, India

Ashwani Kumar
Molecular Biology Unit, Dairy Microbiology Division, National Dairy Research Institute, Karnal 132001, Haryana, India
Department of Biotechnology, Seth Jai Parkash Mukand Lal Institute of Engineering and Technology, Radaur, Yamuna Nagar 135133, Haryana, India

Sunita Grover
Molecular Biology Unit, Dairy Microbiology Division, National Dairy Research Institute, Karnal 132001, Haryana, India

Virender Kumar Batish
Molecular Biology Unit, Dairy Microbiology Division, National Dairy Research Institute, Karnal 132001, Haryana, India

S. Umesha
Department of Studies in Biotechnology, University of Mysore, Manasagangotri, Mysore 570006, Karnataka, India

P. Avinash
Department of Studies in Biotechnology, University of Mysore, Manasagangotri, Mysore 570006, Karnataka, India

Hadis Kord
Department of Tissue culture and Genetic Engineering, Agricultural Biotechnology Research Institute of Iran (ABRII), Karaj, Iran
Ramin University of Agricultural and Natural Resources, Ahvaz, Iran

Ali Mohammad Shakib
Department of Tissue culture and Genetic Engineering, Agricultural Biotechnology Research Institute of Iran (ABRII), Karaj, Iran

Mohammad Hossein Daneshvar
Ramin University of Agricultural and Natural Resources, Ahvaz, Iran

Pejman Azadi
Department of Tissue culture and Genetic Engineering, Agricultural Biotechnology Research Institute of Iran (ABRII), Karaj, Iran

Vahid Bayat
Department of Tissue culture and Genetic Engineering, Agricultural Biotechnology Research Institute of Iran (ABRII), Karaj, Iran

Mohsen Mashayekhi
Department of Tissue culture and Genetic Engineering, Agricultural Biotechnology Research Institute of Iran (ABRII), Karaj, Iran

Mahboobeh Zarea
Department of Tissue culture and Genetic Engineering, Agricultural Biotechnology Research Institute of Iran (ABRII), Karaj, Iran

Alireza Seifi
Department of Tissue culture and Genetic Engineering, Agricultural Biotechnology Research Institute of Iran (ABRII), Karaj, Iran

Mana Ahmad-Raji
Department of Tissue culture and Genetic Engineering, Agricultural Biotechnology Research Institute of Iran (ABRII), Karaj, Iran

Pushpender Kumar Sharma
Department of Biotechnology, Panjab University, Sector 14, Chandigarh 160014, India
Department of Biotechnology, Sri Guru Granth Sahib World University, Fatehgarah Sahib, India

Rajender Kumar
Department of Pharmacoinformatics, National Institute of Pharmaceutical Education and Research (NIPER), S.A.S. Nagar, Mohali 160062, Punjab, India
Computer Centre, National Institute of Pharmaceutical Education and Research (NIPER), S.A.S. Nagar, Mohali 160062, Punjab, India

Prabha Garg
Computer Centre, National Institute of Pharmaceutical Education and Research (NIPER), S.A.S. Nagar, Mohali 160062, Punjab, India

Jagdeep Kaur
Department of Biotechnology, Panjab University, Sector 14, Chandigarh 160014, India

Rama Raju Baadhe
Department of Biotechnology, National Institute of Technology, Warangal 506004, India

Naveen Kumar Mekala
Department of Biotechnology, National Institute of Technology, Warangal 506004, India

Sreenivasa Rao Parcha
Department of Biotechnology, National Institute of Technology, Warangal 506004, India

Y. Prameela Devi
Department of Zoology, Kakatiya University, Warangal 506009, India

Magda A. Pacheco-Sánchez
Plant Molecular Biology Lab, Centro de Investigación en Alimentación y Desarrollo, A.C., Carretera a la Victoria Km 0.6, Apartado Postal 1735, 83304 Hermosillo, Sonora, Mexico

Carmen A. Contreras-Vergara
Plant Molecular Biology Lab, Centro de Investigación en Alimentación y Desarrollo, A.C., Carretera a la Victoria Km 0.6, Apartado Postal 1735, 83304 Hermosillo, Sonora, Mexico

Eduardo Hernandez-Navarro
Plant Molecular Biology Lab, Centro de Investigación en Alimentación y Desarrollo, A.C., Carretera a la Victoria Km 0.6, Apartado Postal 1735, 83304 Hermosillo, Sonora, Mexico

Gloria Yepiz-Plascencia
Plant Molecular Biology Lab, Centro de Investigación en Alimentación y Desarrollo, A.C., Carretera a la Victoria Km 0.6, Apartado Postal 1735, 83304 Hermosillo, Sonora, Mexico

Miguel A. Martıínez-Téllez
Plant Molecular Biology Lab, Centro de Investigación en Alimentación y Desarrollo, A.C., Carretera a la Victoria Km 0.6, Apartado Postal 1735, 83304 Hermosillo, Sonora, Mexico

Sergio Casas-Flores
División de Biología Molecular, IPICYT, Camino a la Presa San José No. 2055, Lomas 4a sección, 78216 San Luis Potosí, Mexico

Aldo A. Arvizu-Flores
Departamento de Ciencias Químico Biológicas, Universidad de Sonora, Blvd. Luis Encinas y Blvd. Rosales S/N, 83000 Hermosillo, Sonora, Mexico

Maria A. Islas-Osuna
Plant Molecular Biology Lab, Centro de Investigación en Alimentación y Desarrollo, A.C., Carretera a la Victoria Km 0.6, Apartado Postal 1735, 83304 Hermosillo, Sonora, Mexico

Deepu Vijayan
Botanical Survey of India, Eastern Regional Centre, Shillong 793003, India

Archana Cheethaparambil
Crop Improvement and Biotechnology Division, Centre for Medicinal Plants Research, Arya Vaidya Sala, Kottakkal, Malappuram 676503, Kerala, India

Geetha Sivadasan Pillai
Crop Improvement and Biotechnology Division, Centre for Medicinal Plants Research, Arya Vaidya Sala, Kottakkal, Malappuram 676503, Kerala, India

Indira Balachandran
Crop Improvement and Biotechnology Division, Centre for Medicinal Plants Research, Arya Vaidya Sala, Kottakkal, Malappuram 676503, Kerala, India

V. Subashini
Division of Plant Biotechnology, Institute of Forest Genetics and Tree Breeding, Coimbatore 641002, India

A. Shanmugapriya
Division of Plant Biotechnology, Institute of Forest Genetics and Tree Breeding, Coimbatore 641002, India

R. Yasodha
Division of Plant Biotechnology, Institute of Forest Genetics and Tree Breeding, Coimbatore 641002, India

Bharat Singh
AIB, Amity University Rajasthan, NH-11C, Kant Kalwar, Jaipur 303 002, India

Ram A. Sharma
Department of Botany, University of Rajasthan, Jaipur 302 055, India

Vimal S. Prajapati
BRD School of Biosciences, Sardar Patel University, Sardar Patel Maidan, Vadtal Road, Satellite Campus, Post Box No. 39, Vallabh Vidyanagar 388-120, Gujarat, IndiaKamlesh C. Patel

Ujjval B. Trivedi
BRD School of Biosciences, Sardar Patel University, Sardar Patel Maidan, Vadtal Road, Satellite Campus, Post Box No. 39, Vallabh Vidyanagar 388-120, Gujarat, IndiaKamlesh C. Patel

Aravind Kumar Jukanti
Central Arid Zone Research Institute, Jodhpur 342003, Rajasthan, India

Ramakrishna Bhatt
Central Arid Zone Research Institute, Jodhpur 342003, Rajasthan, India

N. K. Lenina
Department of Plant Biotechnology, Centre for Plant Molecular Biology and Biotechnology, Tamil Nadu Agricultural University, Coimbatore 641003, India

A. Naveenkumar
Department of Plant Biotechnology, Centre for Plant Molecular Biology and Biotechnology, Tamil Nadu Agricultural University, Coimbatore 641003, India

A. E. Sozhavendan
Department of Plant Biotechnology, Centre for Plant Molecular Biology and Biotechnology, Tamil Nadu Agricultural University, Coimbatore 641003, India

N. Balakrishnan
Department of Plant Biotechnology, Centre for Plant Molecular Biology and Biotechnology, Tamil Nadu Agricultural University, Coimbatore 641003, India

V. Balasubramani
Department of Plant Biotechnology, Centre for Plant Molecular Biology and Biotechnology, Tamil Nadu Agricultural University, Coimbatore 641003, India

V. Udayasuriyan
Department of Plant Biotechnology, Centre for Plant Molecular Biology and Biotechnology, Tamil Nadu Agricultural University, Coimbatore 641003, India